John Croumbie Brown

Reboisement in France

Records of the Replanting of the Alps, the Cevennes, and the Pyrenees with Trees,

Herbage and Bush

John Croumbie Brown

Reboisement in France

Records of the Replanting of the Alps, the Cevennes, and the Pyrenees with Trees, Herbage and Bush

ISBN/EAN: 9783337407971

Printed in Europe, USA, Canada, Australia, Japan

Cover: Foto ©berggeist007 / pixelio.de

More available books at **www.hansebooks.com**

REBOISEMENT IN FRANCE:

OR,

RECORDS OF THE REPLANTING OF THE ALPS, THE CEVENNES, AND THE PYRENEES WITH TREES, HERBAGE AND BUSH,

WITH A VIEW TO ARRESTING AND PREVENTING THE DESTRUCTIVE CONSEQUENCES AND EFFECTS OF TORRENTS.

COMPILED BY

JOHN CROUMBIE BROWN, LL.D.,

Formerly Government Botanist at the Cape of Good Hope, and Professor of Botany in the South African College, Capetown, Honorary Vice-President of the African Institute of Paris, Fellow of the Royal Geographical Society, Fellow of the Linnean Society, &c.

SECOND ISSUE.

LONDON:
C. KEGAN PAUL & CO., 1, PATERNOSTER SQUARE.
1880.

PREFACE.

The following treatise owed its origin, and the first issue of it to a desire which I felt to show that it is quite practicable to prevent, or to moderate inundations at the Cape of Good Hope, such as occasionally occur there, destroying property of great value. For some years I held the appointment of Government Botanist in that Colony, and there saw something of the appearance of these inundations, and the serious consequences following. Of both I have given details in a volume, entitled "Hydrology of South Africa; or, Details of the former hydrographic condition of the Cape of Good Hope, and of causes of its present aridity, with suggestions of appropriate remedies for this aridity."* And in the preface to the former issue of this treatise, I had occasion to state—"I have before me details of destructive effects of torrents which have occurred since I left the Colony in the beginning of 1867. Towards the close of that year there occurred one, the damage occasioned by which to roads and to house property at Port Elizabeth alone was estimated at from £25,000 to £30,000. Within a year thereafter a similar destructive torrent occurred at Natal, in regard to which it was stated that the damage done to public works alone was estimated at £50,000, and the loss to private persons was estimated variously from £50,000 to £100,000. In the following year, 1869, a torrent in the Western Province occasioned the fall of a railway bridge, which issued in loss of life and loss of property, and personal injuries, for one case alone of which the railway proprietors were prosecuted for damages amounting to £5000. In Beaufort West a deluge of rain washed down the dam, and the next year the town was flooded by the waters of the Gamka; and the next year, 1871, Victoria West was visited with a similar disaster. Such are the sums and the damages with which we have to deal in connection with this question, as it affects the case; and these are only the most remarkable torrents of the several years referred to.

"Towards the close of last year, 1874, still more disastrous effects were produced by torrential floods. According to the report given by one of the Colonial newspapers, the damages done could not be estimated at much less than £300,000. According to the report given by another, the damage done to public works alone was estimated at £350,000,—*eight millions, seven hundred and fifty thousand francs*. And my attention was called anew to the subject.

* London: C. Kegan Paul & Co.

"Torrents have proved destructive on the continent of Europe by washing away fertile soil, by undermining houses and fields, and whole villages and towns, and causing their fall, by burying fields and vineyards and towns in the *debris* thus produced, and swept away, and by producing extensive inundations of lower lying level lands, drowning man and beast, and burying, washing away, or otherwise destroying the labour of years, and I would briefly advert to the remedial measures which have been adopted.

"One of the means employed to avert destruction when it was threatened, was the erection on the river-bed of protecting walls, and of advanced structures, to determine the current, and of continuous slopes to regulate its rapidity and force, and of combined and modified forms of all of these appliances, which manifested great art and skill, ingenuity, and power. It would be exaggeration to say they proved in every case an utter failure, but this would only be an exaggeration of what was the fact, which was, that in very many cases they failed to avert the evil, and in not a few cases they were carried away before the torrent like chaff before the wind, while the torrent seemed to laugh a loud and hollow laugh at the silliness of man's device.

"To prevent the destruction of land by inundations, the more promising measure of raising embankments based or founded on the dry land was adopted, and the river was thus chained within its bed, with only liberty of action within a limited space beyond. But what did the river do? It silted up its bed, and thus raised itself, and attempted to overflow the embankment. The danger was perceived in time, and the embankments were raised to a higher elevation. The river quietly repeated the silting up of its bed, which was met by a repeated addition to the embankment. This was done again and again. It was a continuous struggle between dead matter and living mind, carried on for years—for generations,—both refusing to give in. Meanwhile, as in the case of the River Po, not only the embankments, but the silted-up bed of the river was elevated considerably above the level of the country lying on one side and on the other, an aqueduct of earth overtopping and threatening with destruction houses and trees, and man and beast alike. Then it was a desperate and a deadly struggle, which many saw it would have been well it had never been entered on, while others looked on and said, It is evident that that is not the way in which the evil is to be averted. Meanwhile the struggle was continued, until a breach was at length effected in the embankment, and the river poured forth its torrent, inundating the country far and wide.

"While this contest was going on, the study of torrents in the Alps revealed the form of the bed of these to be a large somewhat semi-circular funnel-shaped basin, from the rainfall in which the waters were collected,— a channel more or less elongated, along which the waters flowed,—and a fan-shaped bed of deposit corresponding to the delta of a river, the whole being like to a river-bed reduced or contracted in length; it showed, further, that these torrents were to be met with in all stages of progress, from incipient information, throughout various stages of activity, to final extinction; it showed that in forest-covered mountain regions there were none; that in denuded mountain ranges they were numerous, and sometimes very destructive; that, where they were extinct, the forest had extended itself till it covered the basin and lined the banks of the channel; that, where they were in a state of progressive extinction, the forests were progressively extending themselves; and that this extension of the forest was apparently the cause or occasion of the extinction of the torrent.

"From what had thus been observed the inference was drawn that by artificial plantation the gradual extinction or the subjection of the torrent to control might be effected,—and numerous facts which had been long known were recalled to give their testimony in confirmation of the correctness of the inference drawn. Rain falling on a metallic roof rushes off, while the same rain falling on a thatched roof trickles down in drops; from the bared ground the rain runs off in streamlets long before it runs off in a similar way from the grass-field or the thicket; and the more the phenomena of percolation and drainage was studied, the more manifest did it become that vegetation retarded the flow and prevented the rush of water, retained it to moisten the soil, and extinguished the torrent, requiring the river to take days and weeks to carry away what the torrent carried away in hours, and thus scouring something like a permanent flow in what had become a dry channel, filled occasionally from bank to bank with a destructive torrent, converting the lion into a lamb. And now millions of francs are being spent on the work of planting trees, and herbage, and bush, with a view to preventing torrents and inundations destroying the land."

When the first issue of this volume took place, the inundation which had proved so destructive to Toulouse was engaging the attention of the General Directory of Forests in France, who were satisfied that they had the means of preventing the recurrence of such a catastrophe if they only had the money necessary for carrying out the necessary reboisement and gazounement of the mountains; and the works have been carried on with more or less energy ever since.

Amongst other important and interesting models exhibited by the Forest Administration at the *Exposition Universelle* of 1878, were models, and charts, and drawings of works of *reboisement* of mountains; and in the Budget for 1880, provision has been made for the work being carried out with still increasing energy. An application was made to the Chamber for a credit of about four millions of francs, well nigh £164,000, a million of francs or £111,667 above what had been asked for 1879, for the execution of such works. In making this application the Administration stated that after the disasters occasioned in 1875, by the overflowing of the Garonne and the Hernault, and their affluents, the Minister of Finance and the Minister of Public Works gave assurance that measures would be concerted between the departments over which they respectively presided, to be taken with a view to prevent the recurrence of such calamities. They stated that many surveys which were subsequently undertaken had been completed, but in the absence of funds the works of *reboisement* had not been begun. That, subsequently the Minister of Public Works had solicited their co-operation to enable him to give a specification of works actually called for in Savoie. That information supplied by the engineers of roads and bridges showed that the four torrents of Saint Martin, the Grillaz, the Pousset, and Saint Julien, all of them affluents of the Arc, were causing every year great destruction, which it was of importance should be arrested without delay. That according to information in possession of the Administration, the execution of the works in Savoie alone would absorb more than a million of francs.

The Budget Committee of the Chamber in reporting on the application, submitted a detailed statement of what had been done, and the results, giving tabulated statements as well as details, showing how effectually *reboisement* had arrested torrents, and showing further, that for the completion of the works in the Alps, in the Cevennes, and the plateau of Central France, and in the Pyrenees, there would be required 148 millions of francs, upwards of six millions sterling, and 72 millions more, upwards of three millions sterling, for the purchase of land. And they unanimously recommended that a credit for the whole amount applied for should be granted. "We are all," say they in the concluding sentence of their report, "deeply impressed with the thought—better far spend a million in *reboisement* than have to give such a sum to sufferers from inundations."

The credit applied for was unanimously granted by the Chamber, together with a grant made *sua sponte* of 5000 francs to be employed in developing roads, to facilitate the exploitation of communal forests, the effect of which it was anticipated might be to raise the average value of 360,000 hectares, or 900,000 acres of forests, from five francs to fifty francs per hectare. And in view of the importance of employing forest engineers of superior attainments in the works of *reboisement* in Savoie and in the basin of the Garonne, 50,600 francs, about £2,110, in addition to the credit for the material work was granted.

If such sums tell of the great expense at which these works are being executed, they tell of the importance attached to the execution of them, and of the perfect confidence which is felt in their immediate efficiency, and in their ultimately proving remunerative of the outlay.

I have retained the terms *reboisement* and *gazonnement*, because I know of no equivalent English terms by which they can be replaced. Both in India and in America the former term at least has been adopted; and I believe it will soon be naturalised among the English speaking population on both sides of the Atlantic.

JOHN C. BROWN.

HADDINGTON, *10th December, 1879.*

REBOISEMENT IN FRANCE.

INTRODUCTION.

ONE of the striking features of the scenery of extensive districts in the High Alps is that presented by numerous ravines, of greater or less depth and extent, furrowing the mountains, created by mountain floods. These are the Torrents of the High Alps. In the creation of these much valuable land, and in some cases houses and fields, have been undermined, precipitated into the water-course, and washed away; and land not less valuable has been devastated by being covered with the detritus. The most efficacious means of preventing the formation of torrential floods have been found to be what are designated *reboisement* and *gazonnement*,—the former being the replanting with woods lands in the districts formerly covered with forests which have been denuded of these, the latter the creating of a dense turf of herbage and bush upon adjacent ground.

Evils similar in kind but differing in degree are not unknown in several newly-settled lands. The success with which these remedial operations have been carried out in France may commend them as appropriate appliances to remedy these evils; and the magnitude of the evil which is being combated and remedied in France may be considered as calculated to speak encouragement to those who are called to meet only lesser forms of the evil. Under this impression I would here cite details which have been given of the form and magnitude which the evil had assumed, and in which it has been attacked with success.

The first I shall cite relates to the Dévoluy. Of this valley Surell writes,—" The Dévoluy forms to the west of the department of the High Alps an elongated valley, divided into two parts by a little *col* and circumscribed by elevated mountain chains. It is entered by five passages, which are gorges or cols which the horrors of the locality make impracticable for passage during a part of the winter. The mountains are bare,—eaten up by the flocks and by the sun; they are without shade and without verdure. The bases of the mountains are almost deserts, having been ruined by the deposit of material dejected from ravines. The aspect of this miserable country is oppressive to the soul: one would say of it, It is smitten with death. The pale and uniform colour of the soil, the silence which weighs on the fields, the hideous spectacle of these mountains flayed by the waters and falling into disintegration, and everything about them, announces a miserably ruined, decrepitated land, which does not appear even to struggle against, or resist, or resent its destruction. The unchanging serenity of the sky, which anywhere else would be a trait of beauty, adds here to the melancholy sadness of the country. I shall go over step by step the errors of man which have brought about this state of things.

"Everything concurs to show that in ancient times this country was wooded. There are dug up from its peat bogs buried trunks of trees—monuments of ancient vegetation. In the frame-work of old houses are seen pieces of enormous timber such as is not now to be found in the district. Many localites completely bare still bear, even to-day, the name of wood. One of these valleys (that of Agnères) is called, in old deeds, Comba-nigra, on account of its thick forests. By these evidences, and many others, are confirmed the traditions of the district, which are, on this point, unanimous.

"There, as in all the High Alps, the destruction of the forests began on the flanks of the mountains, and thence descended little by little towards the depths of the valleys, and ascended to the highest accessible peaks. Then came the late Revolution which caused to fall the remainder of the woods which had escaped the first devastation. This last destruction was accomplished under the eyes of some of the present population, and all the old men remember what the forest was in a former day."

He adds in a note:—"And many have told me that they have lost flocks of sheep straying in the forests of Mount Auroux, which covered the flanks of the mountain from La Cluse to Agnères. These flanks are to-day as bare as my hand."

"And," he resumes, "there, after the destruction of the forests, have come also the grubbing up of roots and the pasturing of flocks. They grubbed up the grounds nearest to the dwelling-places. They let the flocks go freely everywhere, wherever it was inconvenient or impossible to transport the ploughs. This proceeding, begun centuries ago, accelerated by the Revolution, has produced its inevitable fruits, and the inhabitants suffer sorely to-day from the improvidence of their fathers.

"The first evil to be noticed is the extreme rarity of woods. The communes are burdened with the purchase, at great expense, of the possession of distant forests. It requires in certain localities, as for instance at Saint Etienne, thirteen hours of fatiguing work to convey, on the back of a mule, a load of wood across the fearful precipices, and this without reckoning the time occupied in felling and cutting. Other communes, for example La Cluse and Saint Disdier, have preserved woods which, with the greatest economy, might suffice to meet their wants, but they are not more happy; and this fact makes it apparent that the forests have a function to fulfil here other than simply that of satisfying the daily wants of the inhabitants. For, first the clearances, then the plough and the flocks, have so dissipated the vegetable soil that there now remains no more of it than a thin bed formed by the disintegration of the rock which underlies it, and which now protrudes through it on all hands. Such is the mobility of this ground that it is washed away by the slightest showers and leaves an arid bottom in the place of cultivated fields. Every storm gives rise to a new torrent."

In confirmation of this it is stated by Marsh in his treatise on *The Earth as Modified by Human Action*,—"No attentive observer can frequent the southern flank of the Piedmontese Alps or the French province of Dauphiny for half-a-dozen of years without witnessing with his own eyes the formation and increase of torrents. I can bear personal testimony to the conversion of more than one grassy slope into the bed of a furious torrent by the baring of the hills above of their woods."

And Surell goes on to say,—" There can be shown here torrents, which have not been in existence for three years, which have destroyed the finest parts of the valleys. Whole villages have been almost carried away by ravines formed in a few hours; and the greater part of the torrents have not as yet received a name. Often the wild waters, flowing in broad sheets over the surface of the ground, without bed, without ravine, without torrent, have sufficed to soak and ruin whole districts which have been abandoned for ever.

" One may see also dispersed here and there on the brows of many hills (*revers*), traces of old fields and of old estates, the bounds of which are still marked out by thick dry stone-walls, but which no man has been near for a long time. Such are to be seen on the rising grounds of Agnères, and on the col of the Noyer. One can with difficulty imagine anything more distressing and more significant than the sight of these ruins; they have written on the brows of hills *(revers)* of the Dévoluy the future destiny of all the French Alps. And here again come into view proofs which do not admit of any doubt in regard to the destructive influence of flocks.

Some communes, dreading the future, have enclosed some quarters, as the mountain of Chaumette, quartier de Maniboux, quartier de Lierravesse, quartier de Auroux, near Saint Etienne. Immediately vegetation had again gained possession of the soil, the herbage, bushes, and shrubs have spread with wonderful rapidity, and formed what are called *blanches* in the country. Whole forests have sprung up on the soil of the forests which were destroyed at the Revolution, but which the inhabitants, now inspired with a better feeling, have subjected to a regular course of forest management. Finally, on the same mountain brows *(revers)* enclosed portions assume, by the end of two years, appearances different from that of those given up to the sheep. The latter are bare and cut into ravines; the former are covered with vegetation, the soil is consolidated, and the ravines, carpeted with tufted plants, look like cicatrices occasioned by wounds, which are under the benignant influence of a topical application. In the two quarters—the exposure, the slopes, the soil are the same; the mere fact of putting them *en reserve* has determined the difference. What can be objected to such facts? Are they not conclusive? Do they not give the clue to the system to be followed to put at last a stop to these calamities always increasing?

" To resume, we see here always the same effects resulting from the same causes. Let us follow them a little further and we find them become still more saddening.

" The country is being depopulated day by day. Ruined in their cultivation of the ground the inhabitants emigrate to a great distance from this desolated land, and, contrary to the general custom of mountaineers, many never return. There may be seen on all hands cabins deserted or in ruins, and already in some localities there are more fields than labourers.

" The precarious state of these fields discourages the population. They abandon the plough and invest all their resources in flocks. But these flocks expedite the ruin of the country, which would be destroyed by them alone. Every year their number diminishes in consequence of want of pasture-grounds. The number of sheep which was 53,000 twenty years ago are now only 36,000. One commune, Saint Etienne, which supported 25,000 sheep fifteen years ago, supports no more than 11,000 now. Thus the inhabitants, who sacrifice all their soil for the flocks, will not even leave this last inheritance to their descendants.

"Thus may one see clearly whither tends this fatal chain of causes and effects, which commences with the destruction of the forests and ends in suffering and misery for the population, condemning man also to share the ruin of the soil which he devastated.

"All these facts have been lately recounted by M. Morgue, the present Prefect of the High Alps, in a memoir which treats specially of this unhappy valley. 'The history of Dévoluy,' says he, in closing his memoir, 'will be that of the High Alps before five centuries have passed if the indifference of the Legislature go on, if the recklessness of the Administration continue, and if nothing occur to arrest the cupidity of the communes.' We may place side by side with these words those of a former Prefect of the Low Alps, M. Dugied, in a memoir on the subject. 'Such,' says he, 'are the causes of the sad condition of the department. *One may affirm with certainty that, if a remedy be not speedily applied, ere long the population in the upper portion will go on diminishing*, and that with a rapidity which can only be accounted for by that which went on before. I do not know if I deceive myself, but I believe it is possible to remedy the evil; and I believe, moreover, that it is high time to set about this. Wait a quarter of a century and perhaps it will be too late, because the best grounds which exist on the mountains furrowed by the storms may then have been carried away by the floods.'"

In accordance with the forebodings of Surell were the following forebodings of M. Jonsse de Fontanière, Inspector of Forests, embodied in a memoir, *Sur la degradation des forêts dans les arrondisements d'Embrun et de Briançon,* "From all that has been said the conclusion may be drawn that the department of the High Alps is the one, in all France, in which the cultivators of the land are most menaced in their fortunes, and that they will be compelled, and that sooner than they dream of, to abandon the places which were inhabited by their forefathers; and this solely in consequence of the destruction of the soil, which, after having supported so many generations, is giving place, little by little, to sterile rocks.

"It is the destruction of forests which will be the principal cause of the calamity. The torrents, becoming more and more devastators of the country, in consequence of the destruction of these, will bury under their deposits extensive grounds which will be lost for ever to agriculture. The hills, denuded of their vegetable soils, will no longer admit of the infiltration of water. Then sources of streams and rivulets will be exhausted, and the drought of the summers not being moderated by their irrigation, all vegetation will be destroyed.

"The destructive elements thus give birth one to another, and it is only necessary to notice what is going on to-day to foretell what will infallibly come to pass some ages hence—when the forests shall at last have entirely disappeared—fuel and water, the two first necessaries of life, will then fail from these desolated countries.

"The cupidity of the inhabitants, and the tenacity with which they hold to old usages, admit of no hope that any moral conviction in regard to their future will so impress them as to lead them to submit willingly to a temporary sacrifice. It is for the Administration, more enlightened than they on the state of things and on the consequences which are coming, to meet the evil by legislation appropriate to the requirements of the country."

Varied is the tone in which like forewarning was given by different far-seeing men, who gave their attention to the subject, about the time in which these forebodings were published.

To one unacquainted with the facts of the case such forebodings of evil may appear extravagant. To one knowing something of these facts they appear legitimate and true; and to one who has seen the region in some of its aspects they seem to be not unreasonable.

But the truth is not always truth-like, and to remove any lingering incredulity I may state that the torrents of the High Alps are equalled and even exceeded by torrents seen elsewhere. The traveller, Antoine d'Abadie, who was almost frozen to death in climbing the Wosho,—a mountain of Abyssinia, 5060 métres, upwards of 16,000 feet, above the level of the sea, gives the following picture of what he witnessed :—" Sometimes we would be going on in all security under a serene sky, when a native, hearing a strange noise at a distance, which quickly increased, would cry out with all his might, The torrent! and with all haste clamber up upon the nearest height. Thirty seconds would not have elapsed when the bottom of the valley totally disappeared under a sheet of water, which swept away with it trees, blocks of rock, and even wild beasts. These torrents, formed in a moment, exhaust themselves in the course of the same day, and leave no trace of their passage but debris of all sorts and pools of muddy water retained here and there in the clefts and hollows."

M. d'Abadie relates that one day he arrived at a spot just a little too late to see in all its grandeur one of these sudden inundations. He found only a native, looking with a dumfoundered air on the wet ground. " Good morning," said the traveller. " What has happened to you? Where are your arms? Can a man like you stand there without lance or buckler?" " Good morning," answered the African, " and health be yours! The torrent has carried off my lance, my buckler, my camel, and all my possession; my wife, and my children. Wretched me! Wretched me!"

Such are the torrents of Abyssinia.

The brothers Schlangenweit, writing of the energy of the torrents of the Himalayas, state it as their belief that they will cut gorges through that lofty chain wide enough to admit the passage of currents of warm wind from the south, and thereby modify the climate of the countries lying to the north of the mountains.

Morell, in his *Scientific Guide to Switzerland*, mentions that about an hour from Thusis, on the Spluegen road, " opens the awful chasm of the Nolla, which a hundred years ago poured its peaceful waters through smiling meadows protected by the wooded slopes of the mountains. But the woods were cut down, and with them departed the rich pastures—the pride of that valley—now covered with piles of rock and rubbish swept down from the mountains." And he goes on to say,—" The result is the more to be lamented as it was entirely compassed by the improvidence of man in thinning the forest."

Marsh, citing a **pamphlet** published at Brescia in 1851, entitled *Della Inondazioni del Mella nella notte del 14 al 15 Agosto 1850*, says,—" The recent changes in the character of the Mella—a river anciently so remarkable for the gentleness of its currents that it was specially noticed by Catullus as flowing *molle flumene*—deserves more than a passing remark. This river rises in the mountain chain east of Lake Iseo, and traversing the district of Brescia, empties into the Oglio after a course of about seventy miles. The iron-works in the upper valley of the Mella had long created a considerable demand for wood, but their operations were not so extensive as to occasion any very sudden or general destruction of the forests, and the only evil experienced from the clearings was the gradual diminution of the volume of the river. Within the last thirty years the superior qualities of the arms manufactured at Brescia has greatly enlarged the sale of them, and very naturally stimulated the activity of both the forges and of the colliers who supply them, and the hill-sides have been rapidly striped of their timber. Up to 1850 no destructive inundation of the Mella had been recorded. Buildings in great numbers had been erected upon its margin, and its valley was conspicuous for its rural beauty and for its fertility. But when the denudation of the mountains had reached a certain point, avenging nature began the work of retribution. In the spring and summer of 1850 several new torrents were suddenly formed in the upper tributary valleys, and on the 14th and 15th of August in that year a fall of rain, not heavier than had been often experienced, produced a flood which not only inundated much ground never before overflowed, but destroyed a great number of bridges, dams, factories, and other valuable structures, and what was a far more serious evil, swept off from the rocks an incredible extent of soil, and converted one of the most beautiful valleys of the Italian Alps into a ravine almost as bare and barren as the savagest gorge of Southern France. The pecuniary damage was estimated at many millions of francs; and the violence of the catastrophe was deemed so extraordinary, even in a country subject to similar visitations, that the sympathy excited for the sufferers produced in five months voluntary contributions for their relief to the amount of nearly 200,000 dollars, or £40,000."

The rendering of Job xiv. 18–19 in the Vulgate is,—

"*Mons cadens definit, et saxum transfertur de loco suo; lapides excavant aquae et alluvione paullatim terra consumitur.*"

"The mountain crumbling down comes to an end; and the rock is removed from its place; the waters undermine the stones; and by inundation little by little the land is laid waste."

This is accurately descriptive of the action of the torrent, and this the author of the pamphlet has prefixed as a motto to his narrative. By Mr Marsh it is stated,—"The recent date of the change in the character of the Mella is contested, and it is possible that though the extent of the revolution is not exaggerated, the rapidity with which it has taken place may have been."

From such independent testimony in regard to similar phenomena presenting themselves elsewhere, it may be seen that there is nothing incredible in the published reports of the state to which the High Alps had been brought before the operation of *reboisement* was commenced with a view to arrest the evil.

It is with a view of promoting the adoption of a similar remedy for corresponding evils manifesting themselves in other lands that this compilation has been made. Anticipating that the aridity and limited average rainfall on some lands, on which the remedy would not be inappropriate, may be considered a satisfactory reason for delay, I may state that I admit without hesitation that to produce such torrential flows as has been seen in the Alps the quantity of rain falling there must be very great; but I must add that the effect of the rainfall on water-courses depends more on its distribution over time and space than on its average annual amount, and that *orages*, or storms of rain, constitute one of the peculiar meteorological phenomena of the High Alps. M. L. Marchand, *Garde General des Forêts*, says on this subject,—" When the torrential rains of the Alps are made a subject of study it is soon seen that they are all of them occasioned by a particular wind called the *fœhn*. These winds are generally violent, and present almost always the character of *orages*, or storms of rains; it follows that great quantities of rain are poured down upon the soil; and to this may be attributed disasters sometimes coming upon spots which seemed to be placed in the best possible situation and circumstances to bear the most persistent rains.

" The *fœhn* is a wind which blows from the south, often with extraordinary force; it is peculiar to the Alps, and is felt throughout their whole extent. Having climbed over Italy where it is no other than the *siroco*, the following are its chief characteristics:—It comes from the south, but its direction is modified at every step, either by mountain chains or by valleys. Its origin is still a subject of discussion: according to some it originates in the Sahara, according to others it originates in the Gulf of Mexico. It gives to the sky a strangely-marked, peculiar, heavy, whitish aspect; and the rain falls on the second or third day following its appearance.

" The wind arrives on the Mediterranean coast loaded with vapour; it there encounters that immense calcareous simi-circular wall of the Maritime Alps, and it scales their higher slopes; but in consequence of their covering of forests, and the great heat concentrated by them, in doing so it only attains a higher temperature. It is rarely the case that the moisture is condensed or precipitated on these countries which it rapidly traverses; but it cools by degrees as it mounts the Maritime Alps, and on reaching the upper basin of the Var and its affluents it deposits an enormous quantity of water; then it continues to advance northwards to French Comté, before reaching which latitude it has lost much of its force.

" If a glance be cast over a map of the Southern Alps, it may be observed that from Mount Viso there part off great chains running perceptibly from east to west; the *fœhn* comes by the valleys of the basin of the Var, or of the upper sources of the Durance, it strikes upon the first chain parting from the col of the Pas-de-la-Cavale, or of the Grandes-Communes, taking a deviation to the north of Digne. It is against this chain that the first great storms of rain dash themselves. The clouds in passing over these mountains seek the cols or lower parts, and they arrive in the valley of the Ubaye by the openings of Grange-Commune, of Enchastrayes, of the Col d'Allos, of the Lawerq, of the Bas, and in fine, by the great passage of the mountains of the Seyne.

" The *fœhn* forces a passage for itself into the valley of the Durance; goes up this throughout its whole length; it makes its way also by some cols of the chains which separate this valley from that of the Ubaye, and more especially by those which are opposite Embrun.

" If now the forest chart of the country spoken of and the chart of the *fœhn* be compared, it will be seen that the mountains of Seyne have been cleared of woods, and that the whole southern upper slope of the valley of the Ubaye is devoid of forests; in a word, that all the parts which bear the direct attacks of the *fœhn*—those which arrest it—force it to ascend them, and to pour upon them masses of water, are all of them almost entirely cleared of woods. Here we have no longer, as is the case above Menton, a tropical sun to warm the soil; the wind has cooled down as it rose higher from the sea, and is obliged with fatal effect to precipitate in the form of rain the moisture it has borne thither; and at that place where the forests are an absolute necessity, and where the most considerable quantities of water fall, there it is that they have completely disappeared.

"This summary is incomplete, but it may suffice to render intelligible the general course of the *orages*, or storms of rain in the Alps, and the intensity of those on certain parts, which are generally those at which the *fœhn* is compelled to rise considerably or to change its direction. The celebrated torrent of Riou-Bordoux, near Barcelonette, in face of the opening at Allos, is exactly so situated. The portion of the Alps situated below the department of the Isère almost completely relieves the *fœhn* of its humidity, and this is the classic region of the *orages*.

"The *fœhn* does not confine itself to the production of torrential rains; it is not less terrible in its action on the snow, and on the glaciers. As has been stated it blows sluggishly and warm for one, two, or three days before the rain appears; if at this time the ground be covered with snow this is not slow to melt rapidly, and absorbing a great quantity of water it becomes like a sponge; then supervenes the rain which expedites the process and brings on a kind of *débâcle*, or breaking up, and the water arrives in great quantities in the valleys. If the rain do not supervene the action of the *fœhn* may suffice to cause all the snow to melt and to produce great consequent disasters. In 1856 the inundations of the valley of Barcelonette had no other cause of production; the maximum of the flood was attained under a magnificent sky, and all the water came from the melting of the snow which covered the mountain. In Switzerland the terrible inundations of 1868 had in general a double origin—with warm continuous rains were combined the melting of the glaciers. It is always in the spring, or with the first snows of October, that the latter torrents are to be dreaded if the mountains be not covered with glaciers; where this is the case the danger is constant.

"The *fœhn* sometimes produces general rains over the whole of the country over which it blows, but sometimes only local *orages*, or storms of rain. This can easily be accounted for when it is considered that the contour of the Alps admits of one current of air passing up a valley to be in its cause and in its effects quite independent of a current passing up a neighbouring valley, though they have had a common origin,—and that a difference in the cooling of the currents of air may occasion a precipitation of rain in one valley, while the neighbouring valleys, being warmer, are enjoying a cloudless sky."

Thus can the immense quantities of water poured down by these torrents be traced to their source, and thus can the immensity of the quantity of water producing these devastations be accounted for. The inquiry brings into view the fact that it is the temporary deluges of rain, and not the mean average annual rainfall, which occasion the torrential floods of the Alps. And there are countries in which the mean average annual rainfall may be very small, when an *orage*, equalling or exceeding any in the Alps, occurring once in a *decade*, may prove not less destructive than any torrent in that torrent-ravaged region.

PART I.

RESUMÉ OF SURELL'S STUDY OF THE TORRENTS OF THE HIGH ALPS.

Of numerous treatises on subjects connected with the natural history, and the arrest or control of torrents in France, that by M. Surell appears to have been that which has done most to give the direction to remedial operations which has been pursued thus far with the happiest results. There were writers before him who anticipated him in some of his suggestions, and there are writers of the present day who have suggested more advanced operations; but that the work of Surell to which I have referred had the effect I have indicated seems to be proclaimed by all. This work, entitled *Etude sur les Torrents des Hautes-Alpes*, was printed by order of the Minister of Public Works, and published in Paris in 1841. The author had been engaged in engineering work on the High Alps, and his first intention was to prepare a few notices of matters connected with engineering for insertion in the *Annales des Ponts et Chaussées;* but becoming interested in the subject, and being encouraged by the Prefect of the district, he was led to make a study of water-courses and every thing connected with them.

In the sequel I adhere not closely the order in which the several subjects noticed are discussed by him; but to some extent I follow that order, while the division adopted is my own.

SECTION I.—*The Phenomena of Torrents in the High Alps.*

M. Surell, to give precision to his treatise which relates to torrents alone, classifies the water-courses of the High Alps as—*ruisseaux*, or mountain streams; torrents; *rivieres torrentials*, or torrential rivers; and rivers: and states what he reckons the distinguishing characteristics of these. He refers also to glacier streams, and to what are known as *torrents blancs*, to point out wherein they differ from what are known as torrents.

In what are called *torrents blancs* the agency of water is scarcely perceived; it is in operation, but it occupies a very subordinate position; in torrents it is the one commanding power, acting with apparently resistless force.

From the glaciers there proceed currents of water, and by them are formed deposits of stones and rubbish, known as *moraines*, which might be mistaken for beds of deposit formed by torrents; but these have characteristics all their own by which they may be easily destinguished from those.

The *ruisseaux*, or mountain streams, are formed of a body of water, small in comparison with the torrents of which he treats, and may form cascades but not torrents, though they may become feeders of these.

He describes the rivers of the High Alps, of which he enumerates four, as flowing in wide valleys enclosed by elevated ranges of mountains or of hills, and as forming larger bodies of water which, when swollen, continue so for a time more or less protracted; the slope of their fall is constant throughout long stretches, and does not exceed 15 millimeters per metre, or a fall of 15 in a thousand. They are in many places characterised by a watercourse in a level bed of very great breadth, a small portion of which only is taken up by such a water-course, and this is liable to be forsaken and left dry, while the waters flow in another channel which they have formed for themselves, to be again changed for another, and that again after a time for another; by which constant changes there is frequently occasioned a great waste of land, and this, if cultivated, must be cultivated at the risk of the whole being swept away—crop and soil together.

Elsewhere he mentions that traces of the former existence of ancient lakes are frequent in these mountains, and that it is the constant rule for a water-course, whatever may be the class to which it belongs, when it enters one of these basins, to change its bed when traversing it; but while this happens once and again, perchance, with others of the different kinds of Alpine water-courses which he has enumerated, it occurs so constantly as a general feature of all the rivers, repeating itself unceasingly throughout the whole of their course, while in the other forms of water-course its occurrence is only occasional and as it were accidental, that he considers this one of the permanent and specific characteristics of the rivers.

Torrents, on the contrary, is a name given to what may be called a dry water-course, along which a tiny stream may be generally seen to flow, but which from time to time is filled with a rushing, roaring, resistless flood. They generally traverse very short valleys, which cut up the mountains into buttress-like projections. Their fall throughout the greater part of their course exceeds six centimètres per mètre, and it is never less than two centimètres per mètre, or two in the hundred. Changes in the slope of their fall succeed one another very closely; and there is given as a characteristic of them that they constantly, if they have not previously done so to a great extent, undermine the sides of their course at one place, and sweep away the debris and deposit it at another, and subsequently change their course above the place at which the deposit has been made,—giving occasion for the same process being again repeated at some other spot. By the rapid fall, the rapid succession of changes in the degrees of this, and their destructive effects, they are distinguished from rivers, and also from torrential rivers, in the technical classification of water-courses adopted.

Of torrential rivers, *rivieres torrentiales*, in the High Alps, Surell enumerates five, but he intimates that there are many more. They are affluents to the principal rivers. The valleys in which they flow are less extensive and more compressed, and they cut up the mountain range into spurs and lesser chains. Variations in the slope of their fall succeed each other more closely than do those of the rivers. They do not change their courses as do these, or they do so but little. Their fall is greater, but it does not exceed six centimètres per mètre, or six in the hundred. They have not the characteristics or specific characters assigned to rivers; neither do they present the characteristics or specific characters assigned to torrents; they present characteristics of both with characteristics peculiar to themselves; and they are classed apart that the field may be clear for the study of what are known specifically as *torrents*.

While the distinctions thus drawn between torrents and other watercourses is maintained in the treatise, it is stated that the different forms may be considered as passing, by intermediate gradations, into one another, and that the same body of water may in one part of its course appear in one of these forms, and in another part of its course it may appear in another.

The torrents thus specified he classifies under three heads, those of each category presenting characteristics by which they may be distinguished from the others. Torrents of the first class take their departure from a col in the mountains and flow through a valley. Those of the second class flow from the mountain-top and follow the line of greatest declivity. Those of the third class take their origin from the flank of the mountain at some distance below the summit.

Of these also there are intermediate varieties, and varieties assimilating them to some of the other forms of water-courses. The first class approximate in some of their features to those of torrential rivers; in the second class all the characteristics of the torrent are prominent, and to this type most of the torrents in the High Alps are conformed; and the third class often show ravines, with all the secondary characteristics of these.

The washing away of earth, and stones, and blocks of rock being one of the constant effects of torrents in the High Alps, and the deposits of the detritus presenting certain constant features whereby they may be distinguished at a glance, not only from the *moraines* of a glacier, but from the shot-heap of a land-slip, and from all other earthen mounds whatever, Surell has fixed upon the bed of deposit as the most characteristic indication of the previous action of a torrent, and makes the study of these beds of deposit, or *lits de déjection*, the point of departure in his study of torrents.

Of these torrents, he says, in the introduction to his work, "The department of the High Alps presents us with water-courses of a singular form. There is given to them in the locality the name of *torrents*, but with the term, as thus used there, there are associated peculiar characteristics which do not manifest themselves in the torrents of other countries.

"The sources of the torrents are hid in the recesses of the mountains, thence they descend to the valleys, on arriving in these they spread themselves out over an immensely extended convex bed, the convexity of which establishes a marked distinction between these torrents and most other water-courses.

"In these the waters always flow in a hollow which encloses them in such a way that a section of the ground in a direction perpendicular to their course would give a curve concave towards heaven, the lower portion of which was occupied by the waters. In the torrents, on the contrary, when they reach the plain, a similar section would show a curve convex towards heaven, and the waters confining themselves in their course on the summit of this. With the water flowing in a slight depression on the summit of a convex torrent bed, it may be imagined that there can be but little stability in the current; and such is the case. The most trifling rise or swelling of the torrent throws the water out of the depression, and it is scattered right and left, flowing away in streams which, however, still follow the line of the course of the bed.

"This instability renders the torrents very damaging, for they are ever breaking bounds at new points, and subjecting to their ravages immense areas of ground. Beds of torrents are to be seen exceeding 3000 mètres,

B

or about two miles, in breadth. It never happens, indeed, that a torrent covers at any one time the entire surface of this; but in going now here, now there, it threatens continually every part of it, and after some floods every part may be found to bear traces of its passage. Such are the torrents when they debouch into the valleys.

"When they are traced up into the mountain passes they are seen to bury themselves in between steep cleft banks, which rise to the greatest heights, and thus form deep gorges. These banks, constantly undermined at the base, give way, and in their fall drag with them cultivated fields and adjoining dwellings. When this water-course is traced up to the sources of the torrents, the ground there is seen to be spread out like an amphitheatre. It forms a sort of funnel, open to the sky, which receives waters from the rains, from the snows, and from the thunder-storms, and precipitates them rapidly into the gorge." By this gorge, as by the neck of a funnel, the water is drawn off and precipitated into the water-course opening upon the *lit de déjection*, or bed of deposit.

In giving additional details of the principal peculiarities or characteristics of torrents, he says elsewhere, "When one casts an eye over a map of the High Alps he sees a country furrowed with innumerable water-courses, which are spread over the ground in a kind of confusion. It is an aspect presented by all mountainous countries. Perhaps here the confusion is more manifest because of the little regularity in the arrangement of the mountain chains. These run in many different directions. They constantly cross each other's lines, break into each other, and disturb the straight line of the valleys. From these frequent intersections results a certain disorder which has for a long time engaged the attention of geologists, but no satisfactory explanation of the production of which has been produced. All the larger water-courses flow into the Durance, the Buëch, and the Drac, whereby are formed three distinct basins marked out by these rivers." In a note, it is mentioned that by one author, to whom I shall afterwards have occasion to refer—M. de Ladoucette, author of a work entitled *Historie, Topographie, Antiquités, Usages, Dialects, des Hautes Alpes*—there are reckoned five distinct basins; and by another, M. Hericart de Thury, there are reckoned eight; but the number might be increased indefinitely by considering every valley a basin. The three basins spoken of receive, he says, all the water-courses of the department with the exception of some insignificant streams which flow to the west. And he goes on to say, "When the three rivers named are followed beyond the boundary of the department they all three are seen to discharge their waters into the Rhone, the first retaining its name to the confluence, the other two previously losing theirs. And thus it appears that all the water-courses of the department of the High Alps belong to the great basin of the Rhone, one of the five great basins of France. Each of the three basins is traversed by a great valley, which rises by insensible degrees to the col, or neck, in the mountain, where it originates. It receives secondary valleys, into which descend other valleys smaller still, which may again be seen subdivided in a similar manner. These last being, like ramifications, indefinitely subdivided, of which the secondary valleys are the branches, while the principal valley forms the trunk.

"All of these valleys, whatever be their comparative magnitude, their relative rank, or their position, are watered or drained by a stream which indicates the *thalweg* or direction of the inclination of the valley; and if we look horizontally across the sweep of this *thalweg* we see in most cases a curve, evidently continuous, the inclination of which rises,—or, if the expression be preferred, a curve the tangent of which, by degrees, approximates the vertical as we approach the neck.

"The curve is convex towards the centre of the earth, and it may be remarked that the changes in the tangent are more rapid towards the neck than towards the base. In other words, the radii of the curve diminish in approaching the neck.

"This configuration," says he, "is remarkable. Why should the bed of the water-courses be disposed in the form of a continuous curve? Why is this curve convex? Why does the curvature vary more rapidly above? The answer is—All these peculiarities are combined in the exact curve which best suits the flow of a liquid the volume of the current of which increases with the length of the distance gone over. And he asks,—Does it not seem that the forms which are so perfectly adapted to the laws regulating the movement of water can be themselves but consequences of these laws? If it be supposed that the *thalwegs* have been brought into the state in which they are now seen by the same general cause, whatever it may have been, which created the mountains, why have they such regular forms, while the outlines of the summits, which, according to the hypothesis, would have been formed at the same time as they, show only capricious lines? By what chance, in an infinitude of possible forms, have they taken exactly such an one as the waters would have themselves created had they not found it already made? It is in these circumstances reasonable to conclude that a regulated cause has operated in the formation of the *thalwegs*, whilst the summits have been left to themselves; and it is equally reasonable to attribute this to the action of the waters as the cause.

"It is true that this supposition attributes to the waters a prodigious power, very different from the effects which they produce daily before our eyes, and therefore it is necessary fully to understand the manner in which they have been able to act in the formation of the curve of their bed, or in other words that of the *thalweg*.

"When we trace attentively the course of the Durance it is seen that the valley successively expands and contracts in such a way as to produce a succession of basins separated by connecting straits. These basins are elongated in the line of the river's course. The bottom of them is very level, and exhibits a clear and well-defined junction with the base of the enclosing mountain, giving to it an appearance suggestive of its having been in some measure reduced to level by water."

According to a generally received opinion, such elliptical basins are the basins, now filled up, of ancient lakes, and it may be that for a time the place of the river was occupied by a succession of such lakes or sheets of water appearing at different successive levels, communicating with each other by waterfalls or rapids, through which the waters then poured from the lakes, successively passing, as it were, from mill-race to mill-race. Little by little the beds or basins have been silted up, the rocks by which they were separated have been hollowed down, and the waters have at length come to flow in a united bed, and over continuous slopes. We have in our own day an example of such action in the consecutive lakes in the north of the United States, which seem destined to be lost one day in the River St Lawrence, and numerous illustrations of the same thing may be seen in Finland in all directions throughout the country.

There may be reckoned up on the Durance very distinct forms of five of these ancient lakes, extending from the neck of Mont-Genève, where its source is, to the boundary of the department. Vestiges of the same phenomenon are to be seen in the valleys of the Grand Buëch, and of the Petit Buëch. They are to be seen, again, in the valley of the Drac, and in that of the Romanche. In general, all the great valleys of the department present similar traces. Some of these lakes existed within historical times, and we may remark, in fine, that the same appearances have been observed in a great many other places, and on all sorts of rivers.

In this, then, we find a general mode of action, of which traces are constantly reproduced in a certain kind of valleys, to which may be attributed not only the formation of the valleys but also the formation of their *thalwegs*, which two things are, he states, distinct and different.

There are, he remarks, valleys which seem to have been created solely by the erosion effected by waters flowing at first in a simple depression in the soil; other valleys seem to have originated in dislocations of the soil opening clefts into which the waters have afterwards precipitated themselves. But in valleys of both formations the action of the waters has invariably been the same, and it has produced the same results. Thrown upon an irregular surface of soil, they have followed at first the line of the greatest inclination; then they have modified this. Whilst this was going on there has been thus formed the most stable curve of the bed; under the double influence of the friction of the waters tending to reduction to a minimum, and the resistance offered by the soil tending to a maximum: this curve, thus formed, is the *thalweg*.

Thus are brought together and harmonized a great many facts, the explanation of all of which are embodied in one formula—vague it may be—but unique, general, and of universal applicability.

If the valleys be studied in their topographical aspect several laws may be discovered, covered by this regulated appearance, which seem to be entirely the result of chance. Amongst these are two beautiful laws evolved by Brisson, which may be verified here in most of the necks of the mountains. I adduce only one illustration of each. The first is supplied by the col of the *Lauteret*, situated between two water-courses, parallel and flowing in opposite directions—*La Romanche* and *La Guisanne*. The other is supplied by the col of the *Bayard*, situated in the district where the Drac and the Durance, after they have both flowed from east to west, separate,—the one directing its course towards the north, the other towards the south. A highway which passes from the second basin into the first shows distinctly the *thalweg* passing by the col from the one into the other.

By this notice of the action of water in flood we are prepared for entering upon the more special study of torrents.

In the torrent, or what, in accordance with the English application of that term, may be called the torrent-bed, there are noticeable these three distinct parts,—the basin drained by the torrent or funnel-shaped hollow from which the waters are collected, called the *bassin de réception;* the gorge and channel by which the waters are carried off from this funnel-shaped basin, called the *canal d'écoulement;* and the deposit of detritus at the lower extremity of this, called the *lit de déjection*.

To this last great importance is attached, as by detritus borne down by torrents many fruitful fields have been buried under a layer of debris under which they have been lost for ever; in view of this next in importance is reckoned the ravages committed by the flood in undermining enclosing banks, and thus bringing down fields and houses to be washed away and added to the deposit of debris; and M. Surell, after having traced the evil to its source, returns to treat of the several parts of the torrent in what would probably be considered by some of my readers an inverted order,—treating first of this bed of deposit, next of the channel, next of the basin drained, and next of the flood creating the torrent. I find it more convenient for my purpose to reverse somewhat the order in which I bring forward his views, following that which I have adopted in enumerating these different parts of the torrent.

Looking at a bed of deposit, or *lit de déjection*, such as is often seen in the Alps, the question suggests itself,—Whence has come this detritus? Deep as may be the channel of the torrent, the *canal d'écoulement*, this alone could not have supplied such a mass of material as is generally found constituting a *lit de déjection*.

A study of the outline and soil of the *bassin de réception*, or basin drained by the torrent, with the information previously obtained, supplies the information desired. This is generally more or less of a funnel-shaped basin; the angle of inclination formed by its sides may be acute, very acute, or it may be obtuse, very obtuse,—but the resemblance to the sides of a funnel is marked; the curve may be more or less irregular, and the arc may be more or less nearly complete, but there it is, more or less distinctly perceptible. Here we have discovered what may have been both cause and effect of what we have seen,—an effect of the rapid rush of water, a cause of the increased fall, and of the increased flow, and increased velocity of flow, and thus of the increased ravages and increased deposit and devastation occasioned by the torrent; and here we have found what may have been the quarry whence most of the material deposited at the outlet of the gorge may have been obtained.

It is optional with any one to prosecute the enquiry thus suggested by himself alone, or to do so with the help of others who have gone over the ground before him. It is a matter to which Surell has given careful consideration. He has given as the result of his observations and thoughts that in order to the formation of these deposits there must have been in operation a great erosive force, acting on ground susceptible of erosion; and seeing these meet in the flow of the torrent of water, and in the character of the soil over which it flows, he attributes all the phenomena to the meeting of a copious rainfall and a friable soil, so situated that a rapid flow of the water and a consequent erosion of soil must follow; and I have cited in detail his exposition of the whole contour of the region being attributable to some such aqueous operation.

To follow him in his application to *bassins de réception* of the law thus evolved, it may be desirable to bear in mind that he speaks of three distinct forms of torrents, designated respectively torrents of the first, second, and third classes. The distinction is based entirely upon the position which their *bassins de réception* occupy in the mountains,—the first proceeding from a col or neck in the mountain range, the second from the mountain brow, the third from the mountain flank,—this difference of position to a great extent determining the differences seen in the aspect they present.

In torrents of the first kind, in which everything appears on the largest scale, the basin embraces vast ridges of mountains, and the outline may be traced on an ordinary map. The gullet is prolonged towards the lower part of the channel, forming a valley or rather a narrow gorge deeply embanked by the flanks of the mountains, and the length of which is often more than two leagues. It supplies, says Surell, the very best example I could give of valleys opened up or created by the action of the waters alone. In this gorge the hills are very abrupt, and *minceo par les pieds*, cut away at the base, and cut up by a great many ravines. They rise frequently more than 100 mètres, or 335 feet, above the bed. At different distances they are cut into by secondary torrents, which are lost above in the ramifications of the contour of the mountains, and they each bring into the gorge the waters collected from a part of the basin. These mountains furnish to the torrents a large portion of the matter carried away and deposited in the bed of dejection, and from their sides come the large blocks which fall here and there into the bed of the torrent.

He mentions that in the *bassins de réception*, or basins drained by torrents, of the first order, there are often seen on the sides of the mountains enormous blocks of stone, which sometimes fall into the beds of the torrents and are then carried far by the rush of waters. In some cases there may be seen standing in a vertical position, in the middle of a slope, what looks like an artificial obelisk ; such are almost always capped by some such large block, which one would almost say had been placed there by the hand of man. It is to this block, says Surell, that the obelisk owes its formation. Originally the block lay on the surface of the slope. In this position, when there came a sudden heavy fall of rain, and the water was rushing away in little streamlets on the face of the mountains, this stone presented a solid and indestructible obstacle which divided a current turning it off to the right and to the left. It may easily be conceived that in this manner it would protect the portion of the slope immediately beneath it, on which it rested ; this then would remain untouched and undisturbed, while the ground around it was being dug into and carried away. At last it would come to pass that the portion of the soil which had thus managed to keep itself above the level of the parts washed away, forming at first a ridge or a block of earth with a sharp angle, which became thinner and thinner by the action of time and atmospheric disintegrating influences, took the figure of a well-defined obelisk, standing out clearly from the slope.

These obelisks are known by the inhabitants of the country under the designations *demoiselles*, or young ladies, and *nonnes*, or nuns. They may be seen on the mountains of the torrent of the Graves, of that of Crevoux, of Rabioux, of Grenoble, of that near Briançon, etc., etc.

The throat or gullet widens upward at the spot where it joins the funnel, and this sometimes takes the figure of a col denuded of its covering of earth, which assumes the form of an amphitheatre before the embouchure of the gullet. At other times the col forms what is called a pastoral mountain—a name given to mountains appropriated to the flocks—furrowed by innumerable currents, which there spread themselves out in the form of the foot of a goose. The torrents of Rabioux and of Mauriand may be taken as types of such, and so may the torrent of Bachelard, abutting on the col d'Allos, in the Lower Alps. These vast depressions being situated in the higher parts of the mountains, the water supply during the greater portion of the year can only fall in the shape of snow. In this state it is not dispersed, or is but little dispersed ; it is retained, it accumulates, and if the warmth of spring supervene without a gradual preparation there is poured forth in the course of a few days the mass of water accumulated during months. This may be considered one of the principal causes of the violence of certain floods.

He cites the torrents which proceed from the Col Izoard towards Arvieux, to which reference has already been made, as presenting the most complete and perfect type of the gullet of a *bassin de réception*. There, as has been stated, more than sixty torrents, within less than 3000 mètres, or two miles, precipitate into the depth of the gorge the debris torn from the two flanks of the mountain.

In the torrents of the second kind the basin, instead of being cut out on the cols of the mountains, is formed by an indentation of their summits, and is hollowed out in their *revers*. It is in this kind of basin that it is easiest to trace the disposition to assume the funnel shape so characteristic of these basins, as the eye can take into one glance the entire course of the torrent, all the points of which are depicted before it. The torrent of Merdinal, at Saint-Crépin, may be cited as a type.

Lastly, in the third kind the basin is reduced to a kind of large bog, hollowed out by some ravine, and which in the country often bears the name of *combe*, as for instance the Combes of Puy-Saniere, the torrent of Combe-Barre, the torrent of Comboye.

It receives no affluents or feeders, and it collects little more, if any, than the waters which fall in the same enclosure as the depression. It is always dug out in the flanks of the mountains and below their summit; but it tends to grow, and it creeps up little by little towards the summit, which it reaches at last. This process goes on with greater rapidity in grounds subject to rapid disintegration, and thus is formed in the long run many of the torrents of the second kind. And one can, in many cases, follow the progress and the different phases of the formation of these, from their nascent condition on to their complete development.

Below the basin of reception, and in continuation of the gullet, is a region in which there is neither any more downfall of earth nor is there as yet any deposit. This is designated the *canal d'écoulement*. Of the three parts of the torrent this is the least marked by characteristics, and almost always the least extended. It is the longer the more gentle are the changes of inclination in its bed. And this is the reason why it is generally pretty lengthened in torrents of the first kind; it becomes shorter in those of the second; and lastly, in those of the third, it reduces itself almost to a vanishing point.

The *canal d'écoulement* is always contained between mountains well defined. In fact, when there are no mountains the slope does not suffice to prevent the torrent from spreading itself out; and in doing this it would lose velocity and it would cease to be.

The *canal d'écoulement* is the only part of the course in which the torrents do little damage. Unhappily it is the least extensive. It is here bridges should be located.

If we could artificially prolong this channel to its confluence with the river, maintaining throughout its slope, its section, and its course, we would stop the ravages. And this is the problem in the embankment of torrents.

The *lits de déjections*, or beds of deposit, at the mouth of the torrent next demand attention. The aspect of many of these is suggestive at first sight of a vast ruin, and several torrents have obtained their names from a perception of this resemblance. Thus is it with the torrent *de la Ruine*, at Lantaret, the torrent *de la Ruinasse*, at Monestier, and the torrent *de Ruinance*, on the Lower Alps.

The deposit is a heap of pebbles and of blocks of stones, scattered over a vast extent of ground—an arid region devoid of culture, of vegetation, and even of vegetable soil—and it suggests to the mind the idea of some great catastrophe having occurred. In sight of this enormous mass of debris, one finds it difficult to perceive or admit that it can be the work of the paltry thread of water—a mere streamlet—which is seen oozing through among the rocks. Examined more carefully, it is seen that these heaps, which seem scattered there in so much disorder, are disposed in accordance with mathematical laws.

The general outline of elevation is that of a very much flattened hillock; the outline of shape is that of a half-expanded fan extending from the mouth of the gorge and leaning on the mountain like a buttress. Projecting lines, which mark on the surface of this cone the lines of greatest declination, are arranged very regularly, following the gentle slopes, which bend inwards a little towards the bottom, but maintain withal a perfect continuity,—all taking their departure from the mouth of the gorge forming the apex of the cone. Further on they diverge somewhat further horizontally, with an outline so distinct that if made with a ruler it could scarcely have been more so, and thus is completed the resemblance first suggested—that of an expanded fan, the joint of which is represented by the mouth of the gorge, and the scales of the fan by these rays, somewhat raised towards the middle, as is the back of an ass, and presenting an appearance such as may be supposed to have been produced by the natural slope of a semi-fluid or viscous body flowing out of the mountain and escaping by the gorge.

The whole aspect of the mound is so peculiar that it reveals from a great distance the existence of a torrent before any other indication has been seen to awaken a suspicion that such may be there. It stretches often more than three-quarters of a league in breadth, and its height above the level of the valley may exceed 70 mètres, or 230 feet. Nothing can better prove the force of these torrents in action than those immense deposits formed entirely of what has been ejected by them.

When one looks, says M. Surell, at the slope presented by these beds of deposits at the water level, following with the eye the central ridge of the cone-shaped group of these, he may perceive them to manifest the following three laws, which may be seen regulating the deposit beds of all torrents reproducing the same or similar effects everywhere with the greatest constancy:—(1) The longitudinal profile forms a continuous curve convex towards the centre of the earth,—that is to say, to express the fact in other terms, that the slope becomes less, diminishing in proportion as it goes down towards the river; (2) The changes in the declivity of the fall are more rapid towards the top than towards the bottom; (3) The declivity of the fall, or slope, varies with the nature of the deposits. It is never under 2 centimètres per mètre, nor above 8 centimètres—2 and 8 in the 100; and it is constant for all the torrents of the same locality, and which have their origin in the same mountain range.

It is then shown by the author that that curve is the natural result of the action of the flood; and he proceeds to discuss the causes and the consequences of the formation of these beds of debris deposited by them.

Two distinct causes concur in the formation of these deposits. First, the torrents proceeding from a confined channel in the mountain come into a valley, in which, being all at once deprived of the side support of sustaining banks, they diffuse themselves, losing velocity and depth. And then the passing from the steep declivity of the mountain to the gentle declivity of the plain proves a second and an additional cause of loss of velocity and of depth. The two causes are distinct and altogether independent of each other, and importance is attached to this circumstance. The tendency is to form a continuous curve from the *canal d'écoulement*, corresponding to the angle of stability. Where this has been done the first cause alone will make additions to the bed. Where this limit of slope has not been created deposits will be continued in virtue of the operation of the second cause. From which it follows that some torrents may be confined by artificial structures, but not others; and that in the former case, other things being equal, the effects will be probable in proportion as the diminished slope may be continuous with that from the gorge, as this continuity is a presumptive proof that the curve of the bed has been definitely taken to such an extent that the dejected matters have reached the limit of their slope, which is to them in the circumstances the angle of stability.

Detailed information is given in regard to the effect of the current in giving to the bed of *déjection* its peculiar form, with such variations as have been noted, and in regard to the effect of this upon the current.

There are next described the materials brought down by torrents—clay, gravel, shingle, and blocks of stone. The laws regulating the deposit of these are noticed; and the injuries which are thus done are detailed.

Every thing connected with the phenomena of the *bassin de réception*, the *canal d'écoulement*, and the *lit de déjection*, having been discussed, attention is given to the phenomena of the flood of water by which the damage and devastation are occasioned. This he traces to two sources—first, the melting of snow towards the beginning of June, and second, storms of rain occurring towards the end of summer. Those occasioned by the latter are by far the most awful, and by far the most injurious

In general, says he, the rain of such a storm gives rise to a much more terrible swelling of the torrents than does the melting of the snow. Rains are rare in these mountains; but when they do fall they fall in tremendous showers, like waterspouts. Their action is instantaneous and cannot be foreseen. The snows never melt so suddenly and quickly as come the deluges of rain, and they produce more prolonged but less sudden swellings of the torrents. Besides this, they may be foreseen and anticipated, for they come at known times. The *torrent de l'Ascension* owes its name to the regularity with which it flows about the time of Ascension day. And the melting of the snows produces a general swelling of the torrents and rivers, which causes all to overflow at the same time. The swellings caused by storms of rain are local; one torrent becomes furious, while another quite near to it remains dry. The time of the melting of the snow is that for the highest floods in all the water-courses in all the department; and for all, without exception, the time for low-water is towards the end of autumn.

The phenomena which accompany the swelling of torrents are very varied. It may be said that each torrent in its manner of flooding has something which is peculiar to itself, and which is not found in any of the others. It must be so, for all the torrents have not the same distribution of slopes;

and the same thing may be observed in all rivers, each of which has a character of its own.

Sometimes the swelling occurs gradually; the waters rise; clear at first, they become more and more turbid, and then throwing their strength into their velocity, rolling along stones which strike each other with a dull sound, they end at last by overflowing their banks, and then begin the ravages and additions to the deposit in the bed *de déjection*.

At other times they come suddenly, and all at once is seen instead of water the black lava-like flow of stones, the slow progression of which has nothing like to the flow of liquid.

At other times, again, we find the torrent falls like thunder. It is announced by a rumbling roar in the interior of the mountain range, and at the same time a furious wind escapes from the gorge. These are the precursory signs. In a few instants the torrent appears in the form of an avalanche of water, rolling before it a heaped-up mass of blocks of stones. This enormous mass forms a moving barrier, and such is the violence of the impulse that the stones may be seen leaping before the waters become visible. The hurricane which precedes the torrent is accompanied by effects still more surprising It makes stones fly in the midst of a whirlpool of dust; and there have been seen sometimes on the surface of a dry bed blocks moving as if propelled by some supernatural force.

All these statements, incredible as they may appear, are attested by a host of cases. I quote some of these, but I shall afterwards have occasion again to call attention to the subject.

"In 1837 several carriers, and at the same time a *Conductuer des Ponts et Chaussées*, were stopped during a storm at the place where the torrent *La Couche* crosses the highway, No. 94. The torrent was then dry. All at once a whirlpool of dust descended along the river-bed, and before their eyes some lumps of stone cleared the road at a bound.

"In 1821 the roadway of the bridge at Boscodon was swept away by a blast of wind coming with fury from the gorge of the torrent. Immediately the waters arrived, tearing along between the abutments of the dismantled bridge. This event occurred within ten minutes after the Prefect of the Department had passed, and under the eyes of a great number of country people engaged in harvest work in the field above. The Prefect, questioning the fact, caused several of these people to appear before him, when he questioned them, and held a kind of formal inquiry, which established all the details which have been reported.

"At Guillestre, in 1836, there was a frightful overflow of the stream Rif-Bel, which flows through the middle of the market-town. Several persons were standing near a bridge, listening to the noise made in the mountain, when an enormous stone was, without apparent cause, thrown to their feet, more than 4 mètres, 13 feet, above the bed of the stream.

"The torrent of the Moulettes, which threatens the market-town of Chorges, overflows every year, and it gives every time an opportunity of verifying facts of the kind stated. In July 1838, a little rain having fallen on the summits of the mountain, this drew some of the inhabitants on to the embankment to see the torrent. Soon the blast of wind—the *avant-courier* of what was coming—made the stones roll with such violence that all these people, drawn thither by curiosity, drew back in haste. In a moment the embankment which they had just quitted fell down as it were, so to speak, under their heels. It was a massive wall built of stone and lime, 2 mètres or nearly 7 feet thick, and 5 mètres or 17 feet in height. The breach, extending 25 mètres or 83 feet in length, fell with a crash which was heard more than 3000 mètres or 2 miles off. It raised a cloud of dust through which was seen the lava-like stream making straight for the town."

Another case, which shows how sudden these irruptions are, was this:—" In 1837 the village *des Crottes* was encroached upon by a small torrent of the third kind, which no one had ever feared. In an instant the cellars and the tortuous streets of the village were inundated with mud and blocks of stone. A great many cattle were smothered. With difficulty many people escaped with life, and a child perished in a stable."

The following additional facts, relative to the avalanche form taken by the torrents, are given :—" At the bridge over the little torrent-stream of Chaumateron, in June 1838, the road-labourer heard the precursory sound. Aware of the danger he moved away. He had gone but a step or two when he saw coming the torrent tumbling over itself. It threw itself in one mass over the bridge and broke it. The elevation of the roadway of the bridge above the *radier* plate was 5 mètres or 17 feet.

"The village of Saint-Chaffrey is traversed by a small torrent. The *bassin de réception* is hollowed out of a bed of gypsum. It flows over a steep declivity at the foot of solid banks, but not very high. At every rise or swelling of the torrent it comes tumbling over itself like a ball, 8 mètres or 25 feet in height, and a portion of the hemisphere appears above the banks. It is formed of liquid thickened with gypsum, and brings in its train a great current of water, which tears along with violence, but following ordinary laws. With these examples (says he) I stop. They might be multiplied indefinitely, for they are renewed every year."

My purpose in citing these details is, first, to make my readers acquainted with the facts stated ; next, to give confidence in the man who could bravely grapple with the question,—How shall such torrents be bridled and tamed ? and beyond this, to give confidence in the application, to what may be considered as the torrents of a mill-lead in comparison with these, of measures deemed, and proved by recorded results, to be sufficient to prevent so much as the formation of torrents so irresistible in their might as these. To this I have referred in the introduction, and I refer to it again. My fear, as stated then, is that to many the statements will appear incredible, and that thus the end and object I have in view will fail to be accomplished. Statements of fact, far surpassing what may have come under the experience or observation of a reader, may arouse suspicion in regard to much besides what may be stated in connection with what thus startles, and may call forth resistance to the truth advanced. The rise of such incredulity may perhaps be prevented, if I shew that these statements are in accordance with what has been stated by others of what has come under their observation elsewhere. To those who are conversant with the literature of the subject there is nothing startling in such statements. Theories may be questioned, but the facts are accepted.

I shall afterwards have occasion to cite at some length the statements made by M. de Mardigny in a *Mémoire sur les Inondations des Rivieres de l'Ardèche ;* here I cite only one. Of the tributaries of the Ardèche he tells that they often hurl into the bed of that river "enormous blocks of rock, which this river in its turn bears onwards and grinds down at high-water, so that its current rolls only gravel at its confluence with the Rhone."

The expression "enormous blocks of rock" may seem vague; I can be more explicit. Coaz reports that at Renkenberg, on the right bank of the Vorder Rhein, in the flood of 1868, a block of stone, computed to weigh nearly 9000 cwt., was carried bodily forwards—not rolled—by a torrent a distance of three quarters of a mile. Coaz, *Die Hochwasser im* 1868, p. 54, cited by Marsh, by whom also is cited the following statement from *Die Oesterreichischen Alpenländer und ihre Forste*, by Joseph Wessley, a work published in Vienna in 1853 :—" The terrific roar, the thunder of the raging torrents, proceeds principally from the stones which are rolled along in the bed of the stream. This movement is attended with such powerful attrition that, in the Southern Alps, the atmosphere of valleys where the limestone contains bitumen has, at the time of floods, the marked bituminous smell produced by rubbing pieces of such limestone together."

Occasionally it happens that after a temporary suspension of the flow, the torrent of water, and mud, and stones, burst forth afresh. These explosive gushes of mud and rock appear to be occasioned by the caving-in of large masses of earth from the banks of the torrents, which dam up the stream, and check its flow until it has acquired volume enough to burst the barrier, and carry all before it. In 1827, such a sudden irruption of a torrent, after the current had appeared to have ceased, swept off forty-two houses, and drowned twenty-eight persons in the village of Goncelin, near Grenoble, and buried with rubbish a great part of the remainder of the village.

From these statements it will be seen that similar phenomena have occurred elsewhere; and we may thus be prepared to follow Surell in his study of the phenomena reported by him.

"There are," says he, "in these irruptions an action like to that of the avalanches. The inhabitants of the district designate them by this term; it is not a mere figure of speech; there is in reality an identity of cause, as there is a similitude in the effects. When a great mass of water suddenly pours into the gullet of a *bassin de réception*, resting on a very steep slope, and confined in a deep gorge, this mass no longer flows in accordance with the peaceful rules of hydrostatics. It rises behind to a great height, rolls over on itself, and thus descends the gorge with tremendous rapidity—far beyond that of the regular current of water which is flowing before it towards the bottom. It must then overtake in succession all the points of that current; it absorbs all its waters, which it hurries along with itself, and which it assimilates to its own mass. In this course its volume swells in proportion to the distance traversed, and when it debouches in the valley it arrives charged with the whole mass of water which was contained in the bed of the torrent from its birth to its exit from the gorge. It is in reality the whole mass of the torrent heaped up and concentrated simultaneously in a single wave. This phenomenon is identically that of the avalanche, with only this difference, that the water, fluid in the first case, is in the state of snow in the second. By this explanation may be understood the short duration of certain floods,—for instance, an hour after the catastrophe at the bridge of Chaumateron, mentioned above, the bed was dry as it was before.

"Another fact, not less singular, is that of the hurricane which precedes the torrent. Let us try also to explain this. All the examples of a hurricane which I have been able to collect relate to those floods following storms of rain during the close heats of summer. Let us suppose that in one of those sultry days, so common at this season in this part of the Alps, a thunder-shower, storm of rain, or water-spout falls on the *bassin de réception*;

there is immediately poured a great mass of cold air over the whole extent of this region. This, specifically heavier than the rest of the atmosphere, can neither rise nor spread out, because it is imprisoned in a kind of funnel, which constitutes always the form of the basin. It escapes then by the gorge, following the line of greatest declivity, as every fluid must, and is precipitated to the bottom of the medium of lesser density. The phenomena of this efflux becomes in every respect similar to that of water.

"But there are causes which must prodigiously accelerate the velocity. The column of water carries with it a great volume of air incorporated with it, which it pours with violence into the gullet. At the same time it does not cease to press with all its weight on the volume of air, which has been engulfed in the gorge as in a closed channel. There is there, then, a double action, the force of which is extreme; one may form some idea of it by comparing it to that exercised by the *trombes d'eau*, which serve as blast-engines to the works established in the mountains. It is necessary to imagine the air escaping by the gorge of the mountains as by the nozzle of the bellows of a gigantic forge, and then there will be no wonder that it produces the effects I have described, which are all the consequences of excessive rapidity."

This may require some explanation or illustration.

Marsh, citing *Wanderungen durch Silicien und die Levant*, by G. Parthey, a work published in Berlin in 1834, gives the following singular instance of unforeseen mischief, following from an interference with natural arrangements, which may be considered a natural illustration of the application of force referred to by Surell in his allusion to the application to blast-furnaces of what is called a *trombe d'eau* :—"A land-owner at Malta possessed a rocky plateau sloping gradually towards the sea, and terminating in a precipice forty or fifty feet high, through natural openings in which the sea water flowed into a large cave under the rock. The proprietor attempted to establish salt-works on the surface, and cut shallow pools in the rock for the evaporation of the water. In order to fill the salt-pans more readily he sank a well down to the ocean beneath, through which he drew up water by a windlass and buckets. The speculation proved a failure, because the water filtered through the porous bottoms of the pans leaving little salt behind. But this was a small evil compared with other destructive consequences which followed. When the sea was driven into the cave by violent west or north-west winds it shot a *jet d'eau* through the well to the height of sixty feet, the spray of which was scattered far and wide over the neighbouring gardens, and blasted the crops. The well was now closed with stones, but the next winter's storm hurled them out again, and spread the salt spray over the grounds in the vicinity as before. Repeated attempts were made to stop the orifice, but at the time of Parthey's visit the sea had thrice burst through, and it was feared the evil was without remedy."

Something similar to this is the action referred to by Surell. The analogy holds only in the compression of air by the pressure of water following upon it quicker than it can escape, and the force developed by its elasticity where space is found for its subsequent expansion.

M. Surell enters into several computations to determine the rapidity of the flow of torrents, from which it appears that while the flow of the most rapid rivers does not exceed 4 mètres, or 13 feet, per second, both calculations and observations shew the flow of these torrents to be sometimes about 14·21 mètres per second—nearly 15 mètres, or 50 feet,—which is the velocity of a strong wind. Applying this to a torrent through a canal 8 mètres, or 27 feet in breadth, and 2 mètres, nearly 7 feet in depth, he shews that it gives a flow of 228·48 cubic mètres per second, while the Garonne gives only a flow, in the ordinary state of the river, of 150 cubic mètres, and the Seine of 130 cubic mètres, per second; and thus is the brief duration of the flow of a torrent accounted for. The calculation is founded on a formula given in D'Aubuisson *Hydraul*—(p. 133), in which, representing the fall per mètre by P, the section of the body of water by s, the *perimetre mouille*, or circumference of the wheel, by c, the velocity = 51 square P S ÷ c.

It is founded on the observation that in such rapid currents the resistance to the flow is proportional to the square of the velocity; and extending the computations to determine the size of blocks of stones which may be carried down by such torrents, he shews that such a torrent as is supposed is capable of moving a stone of the heaviest kind equivalent to a cube of 5·15 mètres. But referring to the circumstance that a torrent 2 mètres, or 7 feet in depth, could not act on such a block over the whole of its side, he shews that this will give only an equivalent of 2·74 cubic mètres; and then he states that, in accordance with this, it is not rare to find blocks of 20 cubic mètres near slopes of 6 centimètres per mètre; and that in the last preceding irruption of the torrent of Chorges the waters left on the bed *de déjection* a hundred blocks of 30 cubic mètres, and some even which measured upwards of 60 cubic mètres.

SECTION II.—*Natural History of Torrents in the High Alps.*

The most striking and characteristic feature of torrents—understanding by that term what in English would be called the bed of the torrent—is, according to M. Surell, the deposit known technically as the *lit de déjection*, though this can only be considered a product of the flow of water by which that bed of the torrents is produced, for, if the waters had not carried off the material deposited, then there could have been no deposit; and by this are supplied indications of the comparative age or antiquity of many torrents now extinct.

Often, says M. Surell, are we struck, in passing through the department, with the appearance of a flattened mound, situated at the opening of a gorge, presenting a fan-shaped surface with very regular slopes,—it is the bed *de déjection* of an ancient torrent.

"Sometimes careful continued observation is requisite to the discernment of the original form, concealed as this is by massive trees, by cultivated fields, and often even by houses and towns. But when it is examined with care, and looked at under different aspects, the outline so characteristic of beds *de déjection* comes out at last most clearly, and it becomes impossible to mistake it. Along this mound flows a little streamlet which proceeds from the gorge, and peacefully traverses the fields. It is this which has ormed the ancient torrent, and in the depth of the mountain may be discovered the old basin *de réception*, recognisable also by its form.

"These extinct torrents, if such a phrase may be used, are more numerous than one at first thought would expect. When once the key to be employed in the search has been obtained, and attention is directed to them, great numbers are discovered.

"The site of the market town of Savines may be adduced, amongst others, as a very remarkable example of this kind of formation. The whole town, along with a part of its fields, stands on a bed of ejected deposit, the breadth of which exceeds 1500 mètres, upwards of a mile, covering fields once of great fertility. The nature of this ground is no more doubtful than is its origin. It has been excavated to its greatest depth in digging foundations and in sinking several of the wells in the town; and the drains of a highway lately put in order have disembowelled it in all directions. Below that town the Durance has cut out a channel and bed on some banks more than 70 feet in height, which forms a sort of natural cutting across the bed. It surmounts and overlooks the whole place, and towards the west, at the extremity of the town, there flows the stream by which all the deposits have been produced; this is confined between high banks adorned with meadows, and flows deep down in its own earlier alluvial deposits.

"It is thus open to the day on all sides, and may be studied with the greatest ease. Everywhere it is composed of rolled stones, agglutinated by a lime-like mud. This pudding-like matter is spread in regular beds parallel to the curvature of the surface. It becomes harder and coarser as we get further down, and ends in forming a very compact mass. As to the characteristic form, it may be distinguished from a distance, especially on the east side. The town is built on the highest portion, and the fields lie scattered around it. In the background rises the mountain, Le Morgon, in which the basin of reception is covered or buried now under black forests of firs.

"It may be remarked that the extinction of this torrent, although of a very old date—dating as it does from a time beyond the memory of man —must nevertheless have occurred after the first establishment of human habitations in this mountain range, for hearth-stones and lumps of charcoal have been disinterred from great depths in the pudding-like mass. These fragments show that men had been then in the locality while, anterior to historical times, the torrent in full action was making this bed of deposit; and the name of the stream seems to indicate that the stream must have retained its violent character till times less remote from our own." In a note it is stated it is called *Branafet*, which seems to be a corruption of *Bramafam*, Howling Hunger, a name already mentioned as common to many torrents; and it seems as if in losing its violence it had lost also the name which spoke of it.

"The details mentioned leave no doubt in regard either to the fact or to the interpretation put upon it; and they are applicable, not to a single isolated case, but to an order of things which is quite general, the examples of which are widespread, and would each of them furnish materials for observations precisely similar. Names are given in a note of several, with references to more. It must therefore be admitted as an established fact, that the violence of torrents is not of interminable duration, but that it may be arrested—be it by the accomplishment of a definite effect, or be it that the torrent has been brought under some influence by which it has been stifled.

"The torrents which present these features are probably the most ancient. To render this conjecture more probable, I proceed by a bound to the opposite end of the scale. We find villages standing in the place where torrents in full action *debouch* from the mountains. Thus is it with *Les Crottes*, and with the market-town of Chorges. It is most probable that these towns were built where they stand before the torrents by which they are now threatened made their appearance; for, on the one hand, these towns are very ancient—Chorges, for instance, dates certainly from before the commencement of the Christian era; on the other hand, the two torrents which now severally threaten these towns cannot have acted long with the energy which they at present manifest. Their slope is abruptly broken at the issue of the gorge; their bed of dejection is not yet regularly formed, and that of Chorges has risen 6 mètres, or 20 feet, in the course of the last fifteen years.

"If this process had been going on at the same rate for only a thousand years the market-town would have been buried long ago under a mountain of deposit. That of Crottes, again, is a large ravine, which has only within the last few years given occasion for disquietude. There are cases yet more conclusive in regard to the comparatively recent formation of some torrents which can be adduced. A church in the valley of Dévoluy is threatened by a torrent which flows directly towards the building, and is only kept in check by an embankment constructed about twenty years ago; and we cannot suppose such an edifice, the construction of which seems to have been attended to with all care, to have been erected in the very mouth of the torrent! The style of its ornamentation is that of the beginning of the thirteenth century. We know well with what precautions Christian architects have surrounded their edifices, and we infer that this torrent did not exist when this church was built in the thirteenth century, and if so there have been torrents formed in historic times. And, without quitting this same district of the Dévoluy, we can cite examples of formations of a still more recent time. In this district completely organized torrents have been developed under the eyes of the population of the present day. Several have not yet even received names, and they commit already fearful ravages.

"In travelling through other localities like observations may be made. Recent torrents are ploughing out for themselves their courses on all hands. Everywhere new cases are surging up, which prove the abundance and the rapidity of these formations; and one is soon brought to a stand in consternation before this accumulation of facts, which present a bad omen for the future of the country."

In a note it is added,—"Immediately in front of the esplanade of Embrun is seen a mountain cut by a number of torrents of the third kind. They grow, so to speak, under the very eyes of the town. One of them, called Piolet *(petit lit)*, which was only a little ravine about thirty years ago, when it received this name, has become a large and perfect torrent. The mountain, which extends from Orcières to the valley of Champoléon, on the right bank of the Drac, is being ravaged by such a number of torrents that it seems as if it must be swallowed up in a mass by the river. These torrents are for the most part recent, and the old men of the country have seen them born, and seen them develope themselves to their present magnitude."

"Thus does it appear that torrents may be formed in our own day; several are of an age quite recent, and besides these, as if not to leave a single link in the chain of ages awanting, there are torrents existing which, judging by their form, their appearance, and their effects, may be placed as intermediate in age between the extinct torrents, and the torrents still in full activity. These are not yet confined within a stable course in the middle of the deposits; but they overflow only a small part of their bed. The rest is covered with cultivated spots, woods, and houses, and seems to have been abandoned by the torrent from time immemorial. And torrents in all stages of the transition, which begins in the establishment of the extreme limit of the slope, and ends in complete extinction, are met with. Stability begins, generally, first to show itself towards the extremities of the bed, and vegetation establishes itself there, advances, and ends in invading the whole surface of the deposit." Names of several torrents, illustrating what is said, are given.

These observed facts are followed up by M. Surell with reflections on the age of the torrents themselves. Specifying and detailing the peculiar characteristics of three recognised forms of torrents, and generalizing the whole, he concludes,—" The action of torrents may thus be divided into three periods, corresponding to three different ages of growth and development and having each an end to accomplish, and distinct effects which they severally produce.

" The first period embraces the creation of the curve or general sweep of the bed of the torrent.

" In the second period the curve or sweep is determined, created, fixed, but the course or channel is not yet fixed; and it is changed from time to time as if by accident, but all in accordance with law.

" Finally, the third period is that of a stable *régime*. The course or channel as well as the curve is permanent, or as permanent as manhood is in comparison with childhood and youth.

" But many things remain yet to be explained.

" Why do extinct torrents, when they are confined within banks of their own deposits, plough up the very slopes over which they themselves immediately before flowed without having strength to scoop out of it a bed for themselves? The reason is a very simple one. In proportion as the torrent was becoming extinct the waters became more and more limpid. They took then on the same slopes a greater velocity than that which they had when they came charged with alluvial matter, and they then could scoop out where they had previously been depositing.

" By what cause, again, are new torrents produced? One cannot at all see why waters which have respected a district during long ages should begin to attack such district now, if all things continue as they were. Those causes which operate to produce a new torrent ought to have formed it from the first day of the creation of the mountains. How could the district of itself change its form or nature?

" It is evident that foreign influences must have interfered, which have modified the primitive conditions. We are thus brought into contact with a new order of facts which demand attention."

It is then stated that when we examine grounds, in the midst of which are torrents of recent origin, we find them always devoid of trees and of every kind of robust vegetation; and when, in some other localities, we look to *revers*, the sides of which have been recently deforested, we see them to be cut by a great many torrents of the third class, which aparently could only have been formed within a few years before; and extended observations bring under consideration a great many corresponding facts.

" There exist many *revers* formed by the detritus of the vertical rocks which generally crown the summits of the mountains. In these mobile soils vegetation takes root with power, and vigorous forests of larch and firs have clothed the sides of the mountains. But the axe, little by little, has decimated the trees; the fellings, made without plan, have opened across the forest large open spaces running with the slope of the *revers*, this arrangement being that which renders exploitation most easy. Now, wherever the woods have been cleared in this manner, at the place of each clearance the vegetable soil has been carried away by the waters; a furrow is formed there, of little depth at first, but which digs away more and more, extends itself upwards, enlarges itself, and soon constitutes a complete torrent. In the intermediate stripes, where the trees have been spared, it is seen to be altogether different. There—with the same soil, under the same exposure, under the same slope, and this often very steep,—the ground has been held firm, and the contour has been respected by the waters. In going over the forest we often traverse thus a succession of zones, the differences of which are striking. We may even catch sight of intermediate shades, which fill up the contrast. We see nascent ravines in parts where the stumps thickly standing bear testimony to a recent destruction of trees. We see completed torrents in other parts, where the indications of the ground, and the information given by the inhabitants, bear testimony to trees having been destroyed in times more remote. We are thus well assured that we are not taking the effect for the cause, when we affirm that it is the destruction of trees in the clearance which has formed the ravine, and not the ravine which has formed the clearance."

As is the case with the gorges, so does it appear to be the case with the *bassins de réception*. There is no question in regard to the fact that the effect of such a conformation of the basin drained by a torrent, as has been described, is to bring a large body of water, falling over a great extent of surface, to concentre in the orifice of the gorge; but the allegation of Surell is that the form of the basin is itself the product of the long-continued violent action of waters, collected first in a recess of the mountains, and flowing over a soil of little compactness and cohesion; and he accounts for the absence elsewhere of certain characteristics of the torrents of the High Alps by stating that, where the ground presents more resistance, and where the climate is less rigorous, there may be formed only brooklets and mountain streams. Similar torrents are not met with in the Vosges, in the Cévennes, or in the Auvergne. In the Lozère there are *vallats* which are not without characteristic features of torrents of the third class, such as are of frequent occurrence between Briançon and the Monestier, and along the Guissanne; but these, through their weakness, scarcely resemble true torrents, though, compared with *vallats*, they are torrents of great energy.

The torrents of the Pyrenees, generally called *Gaves* in the district, are very rapid water-courses in deep cuttings, often losing themselves in subterraneous canals, but they should be classed with mountain streams or torrential rivers. And no torrents are met with in the mountains of La Corse, or in those of the Jura. But torrents similar to those of the High Alps are found in a portion of the mountains of the department of the Isére, of the Drôme, and of the Lower Alps, which belong to the same formation.

A chapter is devoted to the consideration of climatal or atmospheric influence, and another to the effects attributable to the character of the geological formations of the locality.

In regard to climate, he shows that the elevation of the High Alps brings them into the region of snow. When this accumulates all winter over an extensive area, and under the powerful rays of spring melts in great quantities all at once, the process being often accelerated by the arrival of warm southerly winds, so much so that sometimes in two days' time the breaking up is finished and the whole of the snow has disappeared, this is one powerful cause of disintegration more energetic there than elsewhere; but it is trifling compared with others,—in illustration of which he refers to the clear blue sky of the High Alps, a district in which fogs, and mists, and long-continued drizzling rains are unknown, though these are throughout a great extent of France the normal characteristics of the atmosphere during six months of the year. "Nothing," says he, "can equal the purity of the air, the unchanging serenity of the heavens, there. But this dryness of the air and this cloudless sky are dearly purchased, for the rains, if less frequent, are the more tremendous."

M. Dugied, author of a *Memoire* entitled *Projet de boisement des Basses Alpes*, to which I shall afterwards have occasion to refer more in detail, says, in writing of this,—"It is thus that it comes to pass that the Alps are sometimes months, sometimes years, without rain. Then all at once the clouds gather as if from all points of the horizon, pile themselves up as if pressed by opposing winds, and empty themselves in torrents which sweep away everything in their course."

M. Surell says,—"It is an admitted fact that the quantity of water which falls annually in a mountainous country—other things being equal—is greater than in the country of the plains. It is also an admitted fact that the quantity is augmented as we approach the tropics. It follows that there ought to fall here a quantity of rain equal at least to what falls in the same time in Paris. But while the fall in Paris is distributed over a period of six months, here the whole quantity is used up in some few rain-storms." This makes all the difference; and thus, to some extent, is the soil made more mobile than it is elsewhere, and of this the following illustration is given:—

"There is a transition point very remarkable where the climate changes all at once from that of Provence to that of the north; it is the col *du Laterat*. In proportion as we rise towards this neck, in ascending the valley of the Durance, and then that of the Guisanne, its affluent, we see the serenity of the heaven disturbed, and rainy days become more and more frequent. When the neck is passed, and we penetrate into the gorge of Mallaval, dug out by the Romanche, in following this water-course into the country called the Oysan, which is a portion of the department of Isère, there the change of climate is complete. The rains are extremely frequent, and instead of falling in what seem like thunder-plumps they are prolonged, and fall continuously as drizzling rain. The air is almost constantly moist, and loaded with clouds. One sees the mists creeping over the sides of the mountains, to catch upon the projecting rocks, and often to envelope the valley completely. In a word, we have entered the climate of the north, the same as prevails at Grenoble, and which differs in a striking manner from that of Embrun, where fogs are a phenomenon almost unknown.

"From this difference in the climate follow corresponding differences in the action of torrents. The mountains which enclose the valley of the Romanche present in many parts the same kind of ground as do those of the basin of Embrun; it is a flaky, black, calcareous earth, remarkable for its excessive friability. But this same soil, which in the Embrunais is furrowed by a multitude of formidable torrents, shows in the Oysans only a few torrents, almost effaced, without energy, and in no repect to be compared with those. In the latter country the mountains are seen clothed on the steepest slopes and covered with vegetation over all their height; and although they may be stripped of trees, they are scarcely furrowed by a few thread-like streams. In the Embrunais, on the contrary, where the forests have disappeared from the sides of the mountain, these never fail to become the prey of the torrents.

"Such is the hygrometric action of the climate. There, where the soil is constantly bathed in a humid atmosphere, the summits carpet themselves with verdure, and the torrents have no more aliment. Here, where the air is always dry, vegetation proceeds with more difficulty, and the storms of rain sweep from the surface the soil to the extent to which vegetation has fixed it there.

"Thus the moisture of the climate impedes the action of torrents in two ways equally effective; first, it makes the rain storms more rare and less violent; secondly, it renders the soil more fixed by covering it with more vigorous vegetation; it diminishes thus as by one stroke two causes of erosion.

"If there still remain any doubt as to the active part played by the climate in the production of torrents, I would cite a general observation which has been made for a long time in these mountains:—When one traverses the valleys running east and west, or the reverse, he sees that the slopes on the north side are generally wooded, or carpeted with vegetation, whilst those which look towards the south are denuded and arid. He sees, at the same time, that the former are much less cut up with torrents than the latter; and the contrast is often such that he sees the one slope horribly disfigured by torrents over against another on which there exists not one, as, for instance, in the valley of Orcieres, in the Vallonise.

"Now it is evident that such a difference in the whole character of two slopes, which are almost always formed of the same banks of earth, cannot be explained but by the influence of the exposure. And how does the exposure act but by moderating in the slopes directed to the north the effects of the noon-tide sun? They protect for a longer time the snow, retain more humidity, are protected from the scorching winds of the south, enjoy all the advantages of shade and coolness, &c. All these effects combine and actually submit these slopes to climatal influences different from those which act on the opposite slopes, although they may both be situated under the same atmosphere."

Enumerating the geological formations of the High Alps, he shows that the most abundant are comparatively recent formations, many of them so friable that they crumble through exposure to the sun's rays, without the super-added action of either frost or moisture; that limestones presenting all the appearance of great hardness, and selected on this account for *enrochements*, were found to be reduced to earth in two years; that others were not only liable to be disintegrated, but, efflorescing with what seem crystals of alum, lose at once their coherence and their chemical constitution. And the torrents are found to abound in the mountain chains of unstable mineral composition; they are more rare and less formidable in mountains of more compact constituents; and in mountains of primitive rock they are altogether absent.

Nowhere are torrents more furious or more numerous than in the valley of Embrun, extending over the whole land from Gap and Tallard to the village of St Crepin. Throughout the whole of this basin the base of the mountain is of a slaty limestone, manifesting in a high degree the character given above. It is in this formation that innumerable ravines cut into the dry and bluish-tinted hills, which give to the mountains of Embrun their peculiar aspect. These hills are crumbled to such an extent that in trying to climb them one sinks often to the knees in the detritus. And this valley is situated in what may be called the point of intersection of the atmospheric and geological causes of the formation of torrents. To the north we travel over similar formations, but under a different atmosphere; to the south we travel under a similar atmosphere but come upon soil of a different character,—and in both directions the number of the torrents is diminished, as are also their effects. Other illustrations of the same fact are given.

Studying thus the natural history of torrents, he attributes their appearance to the simultaneous operation of several causes in combination. There appears to be (1) a geological cause—the nature of the soil; (2) a topographical cause—the superficial aspect assumed by the country; and (3) a meteorological cause—the rainfall in the locality. And the question next raised is—Is the second of these seen in the existence and form of the *bassin de réception*, or basin drained by the torrent, to be considered a primary, or only a secondary cause of the torrent?

Surell maintains it is a secondary cause—itself a consequence, effect, or product of that to which it ministers. Were it otherwise, he says, in order to this being the case, it would be necessary that the cause which created these mountains should have moulded and shaped at one stroke these basins, according to the characteristic figure which they present to-day; it would be necessary that this form, shape, and outline should have preceded all the action of the waters collected from them; that these, from the first, should have found all the ground so moulded and prepared for them; and that they should have produced, from the first day, all the phenomena which they continue till to-day to present before us.

But it is impossible, says he, to admit such a supposition. The *bassins de réception* are evidently the result of the violent and long-continued action of the water collected at first in a simple recess of the ground, and flowing over a soil deprived of coherence and consistency.

What proves this decidedly is the presence of the larger *lits de déjection*, which have been formed entirely and exclusively at the expense of the lower-lying lands, whence the torrents issue. Every day, moreover, we see the *bassins de réception* increasing in magnitude. These effects follow on with such rapidity that a limited number of years should have sufficed to have produced enormous modifications in the original outline of the land. We have then only to carry back, so to speak, into olden times the action going on to-day under our eyes, supposing that present phenomena are the continuation of an action begun some centuries ago, and the digging out of the basin finds a ready explanation. And he refers to the facts already cited, that there are torrents of quite recent formation; that new ones are being formed constantly; and that these then aid in the formation of basins in the midst of grounds in which there was nothing of the kind previously existing.

He goes on to say,—" I know well that I may seem to have exaggerated this action when there is considered the vast extent presented by the basins of certain torrents, and the profound depths of their declivities forming veritable valleys. But there should be taken into account, on the other hand, and at the same time, the enormous cubical contents of the deposits produced by them, which can have been obtained only from the erosion of such basins; and at the same time it should be recollected that the cubical measurement of these is still far from representing all the mass of material which the torrent has drawn away from the mountain, since a portion of this has been swept into the river, which has widely dispersed it far away. By an effort of thought let us transport the mountain formed by ¦this deposit to the upper part of the torrent; let us throw this into the hollow constituting the basin; let us add to this all that has been carried away by the river, and we shall not be far from having filled up those deep excavations which we hesitated just now to attribute to the digging away of the waters. And we may come in this way to comprehend that there is no exaggeration in alleging that the whole valley of the torrent, from its birth to its junction with the *thalweg*, is the work of the waters alone."

Of the correctness of this view there are numerous corroborative indications or proofs referred to; and as the result of the whole of these observations the natural history of many of the torrents in the High Alps appears to have been this: a deluge of rain such as is brought by the *fœhn*, falling on an exposed bare spot of greater or lesser extent on the col, or the summit, or the flank of a mountain, has washed away soil and formed thus a hollow basin with an outlet on its lower-edge, the water flowing off by this has made a little runnel carrying away, along with the earth washed out of the hollow, earth which impeded its progress; and as more and more fell into the runnel, through the undermining of its tiny banks, carrying this off also and depositing the detritus, whenever a reduced inclination of the ground reduced the velocity of the flow, and forming thus and there a tiny bed of deposit. But the operation—the process thus begun—goes on widening deepening, extending the basin drained, and the gorge or channel, and adding to the deposit, increasing both its depth and extent, till they have each of them attained the fearful aspect they now present.

But there have been similar torrents in the same region in former times—which are now as innocuous as the extinct volcano—they too, to borrow the term, have become extinct; and the brushwood and trees growing on the bed of deposit tell by their age that these torrents have been extinct long. And while the *lits de déjection* are now covered with vegetation, and in some cases with fields, and houses, and towns, the basin and the gorge have also been covered with forests. May not this have been the cause of the extinction? The more closely and the more extensively the subject is studied the more manifest does it appear that it must be so. Thus may it have been in the olden time. In more modern times the destruction of trees has preceded the formation of torrents; and the spread of the forests seems to have preceded the extinction of those of an older creation. This is in accordance with everything that is known in regard to the action of trees in promoting the infiltration, retention, and percolation of water through the soil, and subsoil, on which they grow.

With the light thus obtained, we are enabled to trace back the natural history of the existing torrents to the destruction of herbage and trees formerly growing on the bare and exposed spots, from which these torrents have originated,—a destruction of which in some cases historical records direct us to the time in which it occurred; while in other cases it has occurred within the memory of the present inhabitants of the district.

The student of physical phenomena may meet occasionally with what seems paradoxical facts, which do not appear to be in accordance with the law he thinks he has discovered. A modification of that law may, in some instances, be necessary to enable him to embrace by it all the facts of the case; but there may be other instances in which a more comprehensive view of the matter may show that the apparently paradoxical fact, so far from vitiating, establishes the law. Thus is it here. It is mentioned by M. Surell that there may be named a good many rivers which were navigable formerly, but are no longer so on account of the condition of their lower stream; this may seem to be inconsistent with the general law which has just been propounded, but the study of the phenomena presented by some torrents supplies a solution of the paradox.

To cite a case in point, the *revers* on the left bank of the Durance, from Savines to the river Ubage, is formed, it has been stated, by a succession of beds of dejection belonging to ancient torrents, which became extinct after a time. The whole district was covered with forests, but these have been cleared away in a great measure, and the torrents resumed their ravages.

Many rivers have attained to the state of stability, in the same way that many torrents have done so—by the spread of vegetation over the whole area of the grounds, through the midst of which their waters flow. If this vegetation were destroyed by any means, the soil being again left free, the stability would be interrupted, and *devagation* would be recommenced by the rivers, with effects similar to those connected with the devastations of the torrents. So that the undesirable change which has taken place in the permanent flow of some rivers may be attributed to the denudation of their basin.

This explanation, he says, has been frequently given, but without power to adduce direct proof of its correctness. But now the rekindling of extinct torrents by deforesting operations supplies the desiderated demonstration of an analogous fact. It may be considered, in some sorts, a special experiment on a small scale under exaggerated conditions, to render the effects more striking and more quickly produced. And thus may we obtain, from what has been termed the study of these torrents, information which may be turned to practical account in dealing with torrential floods in other lands, and in other circumstances.

The peculiar characteristics of the torrents of the High Alps, consequent on the combination of atmospherical influences on the mineral composition of the mountains, seems at first to place them apart from all other analogous water-courses. But the study of these has revealed the homology which subsists and seems to run through the whole of these, making it appear that in the torrents of the High Alps we have only one excessive development of what is common to all,—which, having arrested the attention of Surell, has enabled him by this excessive development to study it without difficulty in all its details, and to show in them what may be seen in a degree less manifest, and it may be less developed, but not the less really existent, in all mountain streams, and to show that rivers also are only homologues of these.

Comparing rivers with torrents, he finds and shows that the law of development of both is the same, marked by the same three stages, possessing the same characterestics, attained in the same way, the most stable in their course, having attained this stability after and by means of similar *devagations*, or changes of channel. And he goes on to say,—" When we consider the wide-stretching valleys in which flow the Rhine, the Nile, we Mississippi, and the greater part of the large rivers which diversify the surface of the globe; when we observe that the bottom of these valleys is flat, levelled by the waters, and entirely formed by their alluvial deposits; when, going back to the most ancient historic times, we see in Egypt, in China, in India, &c., the first societies of men, descending little by little from the heights, occupied in struggling against the inconstancy and the tremendous overflowings of these rivers,—may we not believs that all these courses have had, during a long course of centuries, changes in their channels such as those which the Durance exhibits now? But, gradually, the field of these *devagations* has been confined, as is seen so distinctly in the case of torrents, and like these they have ended in being confined within their present banks. The Durance, on the contrary, is still existing in its second stage—that of instability—which has succeeded to the first, characterized by a succession of lakes, and to which in course of time a period of stability will gradually succeed."

And inferring that the most stable rivers of to-day have passed through an epoch of change, of course corresponding to the second period of torrents, he goes on to say,—" In the study of these same rivers there have been collected a multitude of observations which show that they have had in a former age to open their *thalweg*, and to create their slopes the same as we have said has been done by the Durance, and the same as we see being done under our eyes by the torrents in the interior of the continents; they furrowed continents, they filled up basins, and the traces of these phenomena are still very apparent. In approaching these as they cast there immense deltas—ever enlarging deltas—on which sites for entire kingdoms have been found, which deltas constitute true beds of dejection. Thus have these rivers at a certain epoch of their existence acted as the torrents have done in the first period of their history."

And he goes on to say,—" Resuming this discussion, I will undertake to show in the action of torrents a faithful and miniature image of that which passed or will pass in all rivers in general.

" In all I see three consecutive stages, succeeding each other in the same order, and dividing their existence into three distinct periods—First, a period of corrosion and elevation, which prepares the bottom of the *thalweg* and puts throughout its course the slopes in equilibrium with the resistance of the soil and the friction of the waters. It has for its end to determine the longitudinal profile of the water-course.

" Second, a period of devagation, when the rivers seek that form and those bendings of the course which correspond to the greatest stability (for the rectilineal course is not the most stable, since it does not necessarily lead the current over those points where the bank is most solid and least likely to be changed). In this the action of the waters is confined to going hither and thither on the same level without perceptibly carrying away or elevating the bottom; it is the liquid mass which displaces itself rather than the soil. The result of this stage is to fix the laying out of the line of the course, or, if the expression be preferred, to determine the plan of it.

" Third, in fine, a period of permanence, when the waters may overflow their banks but ever return again to their place in an unchanging bed.

" The violence of torrents in the first period has been seen. There ought to be the same in the first period of rivers; and this analogy may serve to explain the formation of those alluvial deposits spread out in such a mass in the greater part of extensive valleys. If it be true that the mountains have been elevated successively in the midst of convulsions of which nothing can give us an idea, the waters have necessarily found in this chaos the matter of these enormous alluvial deposits. The rivers were acting at that time as our torrents do now—that is as these torrents do which have for their basins of reception entire chains of mountains, and which precipitate themselves across a soil newly disturbed and susceptible of being washed away, quite otherwise than that of our Alpine hills. Many hypotheses have been proposed to explain the origin of the Alpine pudding-like deposits. Along the Durance banks of these are met with which rise to upwards of 100 mètres, or 330 feet, above the actual level of the waters. But the dejections of extinct torrents are, relatively to the trifling streamlets which now furrow them, deposits still more surprising, and of an appearance more inexplicable; we are, nevertheless, well assured that they are the work of these streamlets in the first period of their action. Why then may it not be the same in regard to the puddings being the work of rivers in a period in every respect similar?

"I point out these things in passing, not daring to stop to develope and to follow out the views they suggest. This would take me too far away from my subject. Everyone can understand that a mass of water rolling over the soil must act in the same way and conform to the same laws, whether it form a torrent or constitute a great river. Now, as we see formed before our eyes the bed of torrents, we may infer that the bed of rivers has been created in the same manner. And this presumption is accordingly confirmed by the study of such rivers as show traces of their action in bye-gone times in the soil of the valley they have formed."

In more than one of the British Colonies, and in other newly settled lands—using that phrase as applicable to the immigration and settlement of more highly-civilized nations than the native tribes—and in lands which have not been so colonized, are rivers in some of the earlier forms of development referred to. Now, dry channels, or channels threaded by a tiny stream, and now filled from bank to bank—a mighty rushing flood—carrying all before it, undermining banks and washing away the *debris*, the analogues of the torrents studied by M. Surell, having like them their *bassin de réception*—one of immense extent—covering it may be thousands of square miles, and embracing numerous secondary basins drained by affluents, a thunder-shower falling in any one of which may produce a torrential flood,—having their *canal d'écoulement*, their water-course through which the waters roll their flood along towards the sea, and their *lit de déjection*, or bed of deposit, though this it may be is in the ocean-bed near to, or remote from, the shore, contributing in the former case to augment the bar which bars the river's mouth. And it may be inferred that the application of like remedies may produce like effects. What bridles the torrent like a young lion in its fury may bridle the torrential river subject only occasionally to fits of rage.

SECTION III.—*Remedial Appliances to prevent the Destructive Consequences of Torrents.*

The natural history of torrents is suggestive of a most efficient remedy, but it is only of late years that it has been applied, and for its adoption we are indebted greatly to the study of these torrents by Surell, though he was not the first to advocate its application. Until the natural history of these torrents was studied and made known special applications were in use, but a remedial cure seems not to have been attempted. What was tried was to prevent inundations, and the washing away of lands, and the deposit of detritus on fertile land. What is now being done is to extirpate the occasion of these.

In the low-lying plains, at some distance from the mountains, it was the destructive effects of inundations which commanded attention; in the Alps it was the ravages of torrents on the land which did this.

"The torrent which dashes a great body of water over very steep slopes (says Surell) undermines and eats away with fury the base of the banks. These fall in, and little by little pull down towards the bed the adjoining property, which is finally engulfed by the waters. As the banks are generally very deep their fall brings in its train effects the results of which extend far from the spot. All the surrounding land is disturbed. Some portions undermined subside, others slip, others break away, leaving deep crevices. Along the two banks of the torrent may be seen large chinks or rents running parallel to the bed. These subsidences, these rents, and this disturbance spread from place to place, repeat themselves to incredible distances, and end by including the whole sides of the mountain within the range of the effects. There are many quarters which can be named which the erosion of torrents have made so unstable that it has become impossible to build upon them. On the left bank of the torrent *Les Moulettes* there may be seen houses belonging to the village of *Les Andrieux*, which have been rent at a distance of more than 800 mètres from the bed. On the highway, No. 91, opposite *Les Ardoisières*, we have an example of a considerable *revers* of a mountain eaten away by the Romanche and tormented by continual movements of the soil. The instability of the soil has compelled many families to abandon cottages situated at a great distance from the river. One could scarcely comprehend that that could be the cause of movements so remote, if the analogy of facts and other evidences had not proved it to be so in a manner the most irresistible."

Numerous cases are referred to in a note followed up with the remark,— "I have thought it right to multiply citations, because the cause of these movements has been often misapprehended, and notably so in the case last mentioned. The inhabitants attribute it to some particular character of the ground. Having under their eyes only the case of their own locality, they are not aware that it is a phenomenon quite general and common to all torrents."

He specifies movements of the soil in the mountain of Saint Sauveur, over against Embrun, brought about by the torrent of Vachères, and by a great many other torrents of the third class, similar movements in the district of Vabries, mined by the torrent Crevoux on the left bank, and in the district of Villard Saint André, by the same torrent on its right bank; it is stated that this ground had become more mobile subsequently to the formation of a canal for irrigation; accounts are given of similar movements attributable to the torrent of Sainte-Marthe, near Caleyères, in connection with which it is stated that there was there a mill apparently on the point of being engulfed, and of movements attributable to the torrent Merdanel, above Chadenas; and it is stated that very violent movements have been **observed in the portions of the Diveset, of Labéoux, of the Rabioux, of**

Boscodon, of the Ruisseauioux (Lauterat), &c., &c. And he goes on to say,—" There are whole villages built in *bassins de réceptions* which are threatened to be engulfed in this manner by the torrents. Every year the torrent acquires more of the ground, and the village abandons to it several cottages. These facts demonstrate the encroaching march of these water-courses. Little threatening at first, they increase in size, they extend themselves, and soon they reach the habitations built without mistrust at a great distance from their banks. There was, before the thirteenth century, on the borders of the Ralioux, near to Châteauroux, a monastery inhabited by the Benedictines. At a later period the monks deserted it through fear of its being engulfed, and now one sees the ruins suspended in the middle of the river's bank.

" There are threatened with a similar fate the village of Lacluse, by the Labéoux (Dévoluy); that of the Hières, by the Mauriand; that of the Arvieux, by the Moulettes; the hamlet of the Marches, and the hamlet of the Maisonnasses, by the torrent Rousensasse, on the right bank of the Drac (Champsam)."

Having specified these as villages or hamlets exposed to a fate similar to that of the Benedictine Monastery, whose history is given, he goes on to say,—" Most frequently the undermining of the soil is done gradually, and this action is the more slow and the more regular in proportion to the extent of the region. The great mass of ground deadens the movements, and impresses them with a kind of continuity. But at other times also the soil detaches itself suddenly, as if through the effects of a blow. It is thus that in the valleys of the Dévoluy, some years ago, a fragment of the mountain Auroux, covered with cultivated fields, precipitated itself, in one block, into the gorge of the torrent Labéoux. The commotion occasioned by this frightful fall was felt at a considerable distance in the village Lacluse, and the inhabitants attributed it to an earthquake. The cause was no other than erosion by the torrent, which had sapped the base of the ground.

" This may demand some explanation.

" Many lands are formed of parallel banks, disposed in flat layers and raised up on great inclinations. Often an interposed bed, more soluble or less tenacious, is decomposed or disintegrated by infiltration. If it happens at the same time that the under banks be attacked by the current at the foot, an enormous weight of ground finds itself suspended over an abyss; the force of adhesion being weakened, it no longer suffices to keep together this mass and to attach it to the body of the mountain; it is then detached in a mass, and it slides over the surface of the decomposed bed as on an inclined plain. One may indeed see similar land-slips frequently occurring in the limestones of the lias formation, which decompose with the greatest facility, and which often present a schistose stratification; this kind of ground extensively prevails here. In other cases the grounds have been formed of the debris of the upper parts of the mountains; they compose a rough mass without stratification, and most frequently without consistency, covering the stratified nucleus of the mountain, and forming on its surface beds of great thickness. It rarely happens that a *bassin de réception* does not contain within its circuit a large strip of this quite recent formation, for it is into the scooped out parts that the debris have had to roll and rest. And one may easily see that the erosions which take place in such grounds, when they attack the foundation of very high banks, must force the soil to detach itself in great masses; and the fractures will take the form of immense prisms, in accordance with laws similar to those regulating land-shoots *(pousée des terres)*. So that it is in the abundance of certain kinds of grounds, and in the composition of the soil itself, that we find the secret of the principal power of these torrents.

"And this is the evil to be met."

With these destructive effects of the torrents are conjoined the devastating effects of the deposit of debris covering up fertile soil with barren sand, and gravel, and stones,—and, in some places, overwhelming not only cultivated ground, but houses and property not less necessary for the maintenance of the life of man, his wealth, and his comfort. M. Surell brings under consideration the several defensive appliances which had been employed in the bed of the torrent to prevent those destructive effects, and describes the respective merits of these.

The first of these brought under consideration is a wall built along the base of banks in danger of being undermined; and the impotency and inefficiency of such a defence is exposed. The second consists of stone erections or wears raised across the bed of the torrent, to create an artificial fall diagonal to the torrent's course, diverting it away from the ground which it is desired to protect; such erections, it is stated, operate beneficially, and do so in two different ways,—they retain the bed of the torrent, and they diminish the velocity of the torrent for some distance above them. The first action prevents the sweeping away of the ground, the second deadens the violence of the current, thus not only preventing erosion, but destroying the cause of erosion. And details of their structure, of the extent of some, and of beneficial results which have followed the erection of them, are given. References are also made to *fascinages*, structures of fascines, or bundles of bushes, and to *pallissades clayonnées*, or stockades of wicker work, which are successfully employed elsewhere—but not there.

In another chapter are discussed the defences employed in the valleys. Amongst the mountains, as has been intimated, the evil against which protection is sought is the erosion, and subsidence, and destruction of the ground; in the valleys the evils to be guarded against are those resulting from the deposit of the debris of the mountain in places where it does harm. Of the magnitude of these evils illustrations are given; and the defences employed are classified under two heads,—*epis*, blocks or piles, and longitudinal dams. The effects of a single *epi*, and of a line of these placed diagonally across a portion of the stream, are described, as are also the structure and effect of dams, and the structure and effect of a third defence consisting of a combination of the two. A chapter is then devoted to the more full discussion of *endiguements*, the designation given to embankments designed for the defence of one bank of a river, the designation *encaissement*, or enclosing banks, being applied to structures designed for the simultaneous defence or protection of both banks.

In regard to these effects, it is stated that, whenever in the bed of a water-course a resisting obstacle to the flow of the water presents itself—be it the projection of a rock, be it the bluff side of a mound or hillock, or be it an artificial obstruction—two effects manifest themselves. (1) The current is directed towards the obstacle and maintains this flow; (2) The current is thence reflected and directed against the opposite bank. The **hurtful consequence of this reaction is constant, and it is so serious that it** has called for special legislation; and to this legislation on the subject a chapter is devoted. The legislation referred to is embodied in the *Décret du 4 Thermidor an XIII. relatif aux torrents du department des Hautes Alpes.* It is given in full in an appendix to the work, with much additional information in regard to the subject; and in the text is given a succinct account of the working of the law, with illustrations in justification of the same. From this it appears that when a new bank of a certain extent is ravaged by a torrent, the proprietors meet together and constitute a *syndicat*, or court, a requisition is addressed to the prefect, he commissions a civil engineer, officially connected with the department, intrusted with the construction and conservation of roads and bridges—*ingenieur des ponts et chaussées,*—to examine the ground, and, if it be necessary, to report the works proper for the defence of the bank.

The work is executed in accordance with the adjudication; the engineer superintends the construction, and sanctions the delivery of it; and the expense is borne by those interested, shared according to a scheme of division prepared by the syndic.

A translation of the decreet will be given in the sequel.

Attention is next given to the different modes of constructing the defences—(1) *Levée en Perré;* (2) Walls built with lime; (3) Drystone walls; (4) *Chevalets;* (5) Coffres.

The first is employed by preference in longitudinal embankments; the three last mentioned are rarely employed but in the construction of épis; lime-built walls are employed in both forms of defence; the chevalet is a wooden erection of three pieces of timber stuck into the ground, apart below, meeting above, and sustained by a fourth piece stuck into the ground behind them, meeting them at the apex of the angle formed by them; coffres are quadrangular structures of timber, the interior of which is filled with stones; the *levée en perré* is an embankment of earth faced with stone.

A chapter is devoted to the consideration of a form of embankment called *Dique éperonné* or spurred embankments.

Another is devoted to the consideration of the *encaissement* or confining of torrents, the outline to be given in the *encaissement* in section, the direction to be given to the axis of the course, and the declination to be given to it. This is followed by a chapter devoted to the consideration of different systems of defence which have been proposed; and three chapters which follow are occupied with the condition of roads swept by these torrents, details of what measures are requisite to remedy existing evils, and of measures to be adopted in erecting bridges over the torrents. These constitute the third part or division of the work.

The ground being thus cleared, M. Surell proceeds, with a view to the adoption of less objectionable and more appropriate remedial applications, to bring under consideration the causes of the formation and of the violence of the torrents, and with this the fourth part of the work is occupied.

In discussing the foreign influences which have modified the primitive condition of the Alps, and produced definite effects on the formation or extinction of torrents, he gives prominence to the influence of forests. In successive chapters he discusses the influence of forests on the formation of torrents, and the influence of forests on the extinction of torrents, the decay of forests, and the influences of forest clearings and pasturage,—following the whole with a chapter devoted to illustrations and applications of the warning to be derived from the case of Dévoluy, which I have previously cited.

The whole tone and spirit of these chapters produces an impression that the exposition of the view given is not only the result of a prosecution of the study of the subject, but probably an exposition of what first gave to him a clue to the discovery of all he subsequently discovered in regard to the natural history of torrents, and the appropriate measures for extinguishing them and preventing their ravages.

I have often pictured him to myself as one day plodding along, gradually ascending a mountain valley, in the discharge of his professional duties, his thoughts being at other times full of the subject of torrents and their numerous phenomena, but on this occasion thinking on anything but these. When, standing for a moment to rest and wipe away the sweat from his brow, looking back he sees what he cannot but perceive is an old torrent deposit—a veritable *lit de déjection*—though overgrown now with shrubs and herbs, with here and there cottages, and cottage gardens, and cotter's fields. There it is! He feels he cannot be mistaken. Who would have thought to see it there and see it thus? But there is the cone-like formation, the fan-like expansion spreading from the outlet of the gorge! Here is food for thought, and he goes on his way rejoicing. He comes upon a lesser *lit de déjection* of recent formation; how like and yet how different! Here all is desolation; there all was clothed in living green, and the opening beyond showed a young and vigorous growth of trees. But stop! May not this have had something to do with the extinction of the torrent, and that more as cause than as effect? This is something to be thought about—I leave to others to follow out the train of thoughts thus begun. I find no difficulty in doing so till I picture to myself Surell master of the whole subject in all its details, and it is with these, his matured views, irrespective of the way in which they have been attained, that we have here to do.

Writing on the influence of forests, or of the absence of forests, on the formation of torrents, he says,—" When we examine the lands in the midst of which are scattered the torrents of recent origin, we see them to be in every case stripped of trees and of all kinds of arborescent vegetation. On the other hand, when we look at mountain slopes which have been recently stripped of woods, we see them to have been gnawed away by innumerable torrents of the third class, which evidently can only have been formed in later years.

" See then a very remarkable double fact: everywhere where there are recent torrents there there are no more forests; and wherever the soil has been stripped of wood recent torrents have been formed; so that the same eyes which have seen the forests felled on the slope of a mountain have there seen incontinently a multitude of torrents."

The names of numerous mountains and torrents, illustrative of both allegations, are given.

" The whole population of this country may be summoned to bear testimony to these remarks. There is not a commune where one may not hear from old men, that on such a hill-side, now naked and devoured by the waters, they have seen formerly fine forests standing, without a single torrent.

" Observations which are reproduced so often, and with characteristics so constant, can we explain as simply the result of chance? Do they not force us to admit that forests exercise a powerful influence on the production of torrents, whether it be by standing on the soil they defend it against their approach, or, obliterated by the hand of man they leave to them an open field which they are not slow to devastate?

"It is of importance to establish the fact of this influence by direct and positive proofs. Here we are almost embarassed by the very amount of evidence. It should be known that this influence manifests itself here in so many varied circumstances, in such a variety of forms, and with such a force of truth, that assuredly not one man throughout the whole country would dare to dispute it. It is only necessary to spend one day traversing these mountains to be struck with an infinity of facts fitted to produce conviction in opposition to the most rooted prejudice to the contrary. All of those who know the country can have, on this point, but one opinion. All the observations on this matter which have been published are of one accord, and the authors have had no other trouble than to verify the public opinion, nor other merit than to express by the pen that which has been for many ages in all mouths and in all minds."

In face of a belief so universal, so little disputed, and so indisputable, one finds himself at a loss when he tries to reduce it to a kind of demonstration; he knows not how to select one from so great a number of cases, which corroborate one another, and the force or power of which lies in their cumulation; and he thus writes on the influence of forests on the extinction of torrents:—"In examining the basins drained by great extinct torrents, there are almost always found there forests, and often dense forests. There may be observed also, along wooded *revers*, a number of small torrents of the third class, which appear as if stifled under the mass of vegetation, and are completely extinct. Now this second observation, which can be verified by a multitude of examples, supplies a demonstration of a fact of which the first only permitted us to entertain a suspicion in a vague way:—it is, that the forests are capable of bringing about the extinction of a torrent already formed. Indeed, it is impossible to admit that the small torrents, dug for the most part in mobile and friable ground, can have died of themselves, so to speak, in their very birth, and through the effect alone of that equilibrium to which reference has already been made.

"Stability cannot establish itself so speedily on beds which are scarcely formed, and in the midst of lands which offer still so much food for erosion by the waters; it is a work which demands time, and which is never entirely consummated until the mountain has been gnawed away to the quick, and to its last ridge.

"Amongst the great number of extinct torrents, the basins of which are wooded, there are some the forests of which have been subjected to the commune *régime*, and have fallen in part under the axe of the inhabitants. Very well, the result of this destruction of trees has been to rekindle the violence of the torrents, which only slumbered. There have been seen thus peaceful streams give place to furious torrents, which the fall of the wood had re-awakened from their long sleep, and which vomited forth new masses of *déjection* on beds of deposit, which had been cultivated without suspicion from time immemorial. This is what has been remarked more especially after the excessive destruction of woods which followed the first years of the Revolution; the devastations of many great torrents only date from this epoch. It is from this time that the torrent of Merdanel has advanced towards the village of Saint Crépin, the inhabitants of which are to-day within a little of being ruined. The same observation has also been made on the Lower Alps. We may cite as an example of what has been said the whole of the *revers* which are situated on the left bank of the Durance, from Sabines to the river Ubaye. It is formed exclusively by a succession of beds of dejection belonging to ancient torrents, which had been extinguished after having gnawed away a great portion of the mountain of Morgon. The whole of this district was covered with forests, which have been cut up with clearings, and which continue to be impoverished still further every day. The torrents also have commenced their devastations, and, if the destruction of woods be continued with the same recklessness, this *revers*, fertile to-day, will speedily be ruined, as so many others have been.

" This last fact completes all that need be said in regard to the influence of forests. In seeing these show themselves almost everywhere on the body of extinct torrents, one may suppose that these had first died, and that the woods had then seized upon them when the extinction had been completed, and when the soil of the neighbourhood, become stable, permitted vegetation to develope itself in safety: the forest would then only have been one of the effects of the extinction of these, instead of being the cause of it. But then the destruction of the woods would only have restored things to their primitive state, and the torrent ought to have been able to continue extinct after the taking away of the woods as it was before their appearance there —and this is exactly what does not happen. It has sufficed to clear away the woods to see the devastations immediately reappear. It must be then the forests which, by their permanent appearance on the soil, hindered the devastations, and it is the forests, in taking possession of the soil, which have again caused them to cease—and the extinction of the torrents is so completely their work that it begins, continues, and disappears with them, the effect ceasing immediately with the cause.

" One sees by this that the action of forests is not confined to preventing the creation of new torrents, but that it is sufficiently energetic to destroy torrents already formed. One sees also that the injurious result of the removal of woods is not only to open everywhere the soil to new torrents, but that it augments the violence of those which exist, and resuscitates those which appear completely extinct. We may then sum up the influence which forests exercise on torrents already formed in two facts, parallel to those which sum up their influence on lands where the torrents have not yet appeared. (1.) The presence of a forest on a soil prevents the formation of a torrent there. (2.) The destruction of forests leaves them subject to become the prey of torrents. Nor is there in this any thing for which we may find it difficult to account.

" When the trees fix themselves in the soil the roots consolidate this, interlacing it with a thousand fibres; their branches protect it, as would a buckler, against the shock of the heavy rains; and their trunks, and at the same time the suckers, brambles, and that multitude of shrubs of all kinds which grow at their base, oppose additional obstacles to the currents which would tend to wash it away. The effects of all this vegetation is thus to cover the soil, in its nature mobile, with an envelope more solid and less liable to be washed away. Besides, it divides the currents and disperses them over the whole surface of the ground, which keeps them from going off in a body in the lines of the *thalweg* and meeting there, which would be the case if they flowed freely over the smooth surface of a denuded ground. Finally, it absorbs a portion of the waters which are imbibed in the spongy humus, and so far it diminishes the sum of the washing away forces.

"It follows from this that a forest, in establishing itself on a mountain, actually modifies the surface of the ground, which alone is in contact with atmospheric agents, and all the conditions find themselves then modified as they would be if a primitive formation had been substituted for a formation totally different. Whence it is not more astonishing to see the same soil alternately cut up or free from torrents, according as it is despoiled or clothed with forests, than it is astonishing to see torrents cease when we come to primitive formations, or reappear suddenly on friable limestone.

"In accordance with this we find—first, the development of forests brings about the extinction of torrents; second, the destruction of forests redoubles the violence of torrents, and may even cause them to reappear. And nothing is more easy than to explain these new actions. It will be remembered what are the causes which call forth and maintain the violence of torrents: it is, on one hand, the friability of the soil; and, on the other, the sudden concentration of a great mass of water. Now, we know already that the forests render the soil less liable to be washed away; we know also that they absorb and retain a portion of the rainfall, and prevent instantaneous concentration of the portion which they do not absorb. Consequently they destroy both the one and the other cause. They prolong the duration of the flow, and they render the floods at once more prolonged, less sudden, and less destructive.

"It may be understood from this how forests, in invading the *bassins de réception*, may have contributed powerfully to stifle certain torrents. Whilst the waters were creating for themselves the most convenient slopes, the forests were retaining the soil which was ready to go, was rendering it more solid, was consequently diminishing the mass of earth washed away, and above all was opposing itself to the concentration of currents. They were augmenting all the resisting, all the existing, obstacles, and were diminishing all the motive powers; and they were coming thus to hasten by a double efficacy that epoch of stability in which the force of the waters would find itself in equilibrium with the resistance of the soil. There is one circumstance which ought to render their triumph still more speedy,—it is, that the torrent, in proportion as it is enfeebled, abandons to them a soil more and more stable and favourable to vegetation, in such a way that this augments every day their forces in proportion as the torrent loses force. In fact, if the expression may be allowed, it is reinforced by the effect.

"By this I do not mean to say that the torrents can never become extinct of themselves. That would be in contradiction to what I have said, and at the same time to facts observed, for there are examples of torrents being extinguished without the presence of forests, and solely through the erosion of the mountains—as, for instance, the torrent of Saint Joseph, near Monestier. But I say that the forests expedite the accomplishment of this effect, and that they can produce it where the other circumstances are not yet producing it.

"Thus nature, in summoning forests to the mountains, places the remedy side by side with the evil. She combats the active forces of the waters; to the invasions of the torrents she opposes the aggressive conquests of vegetation. On those mobile *revers* she spreads a solid layer which protects them against external attack, somewhat in the manner that a facing of stone protects an earthen embankment. It is worthy of remark, that the little cohesion of limestones, which is opposed to the fixing of grounds, which renders them so mobile, and draws torrents thither, is precisely the quality which renders them favourable to the development of vegetation. The same cause which multiplies the torrents ought then to multiply also the robust forests, and to cause productiveness to succeed in the long run to barrenness, and stability to disorder. Not that, strictly speaking, there can be in nature anything otherwise than orderly, for there is nothing which is not subject to the rule of immutable laws, but in popular phrase the term disorder has also its meaning.

"One is struck with the illustrations of the observation which has just been made in going over certain forests in these mountains. One sees the vegetation doubling its profusion and energy in grounds torn by ravines, and crumbling on all hands, as if it were mustering its last efforts to retain a soil escaping from it. To cite one example: in the forest of Boscodon may be seen the vigour and tenacity of the vegetation contending against a friable soil composed of schist, tufa, and gypsum. It is, in fact, the lands which are the most mobile which are at the same time the most fertile, and the hard rocks on which vegetation has no hold, brave also the effort put forth by all the causes of destruction. The mountains, if they were abandoned quite naked to external influences, would soon be levelled or cut up into bits, and they would offer to man nothing but a heap of cleft rocks, uncultivated and uninhabited.

"It is vegetation which prevents this ruin; and as there can be no vegetation without water, it is on the mountains that nature has poured out the water in the greatest profusion. We have already called attention to the remark, that there falls more rain on the mountains than on the plains. The mountains attract and retain the clouds [?]. Snows and glaciers crown their summits as immense reservoirs, whence trickles out a perpetual moisture, and whence flow innumerable streamlets which fertilize their sides, and distribute fertility, from brow to brow, down to the very depth of the valleys. Thus, the waters which are the most energetic means of destroying the soil are at the same time the most active in its conservation. In drawing on vegetation, they preserve the soil against their own attacks, and the more they have of power to destroy, the more vegetation they cause to spring up to preserve. It is in this way that nature imposes on all her forces moderators which counterbalance them and keep them from acting always in the same way; and this must end in bringing everything to a state of restored peace."

And dwelling on the thought of self-adjusting provision for the natural extinction of torrents, he thus, in something like a burst of enthusiasm, gives expression to his feelings in view of the thorough and efficient way in which torrents had naturally become extinct, and the contrast thus presented to the puny endeavours of man to restrain their ravages: the natural and the artificial; God's way of doing it, and man's way of doing; the work of God and the work of man; and the results: success, perfect and complete; and success, partial and imperfect!

"Let us go back for a moment," says he, "and compare these effects of vegetation with those exercised by the different systems of defence hitherto devised. The result of defences like that of vegetation is to arrest the ravages of torrents; and how powerless appear all embankments by the side of those great and powerful means which nature employs when man ceases to oppose her, and when she patiently prosecutes her work throughout a long series of ages! All our paltry works are nothing but defences, as their name indicates; they do not diminish the destructive action of the waters, they only keep it from spreading beyond a certain boundary. They are passive masses opposed to active forces; obstacles, inert and decaying, opposed to living powers, which always attack, and which never decay. Herein is seen all the superiority of nature, and the nothingness of the artifices devised by man.

"I make not here a barren comparison. I wish to let it be seen that it is better to bridle the torrents than to erect at great expense masonries and earthworks, which will always be, whatever may be done, expensive palliatives, better adapted to conceal the plague than to eradicate it. Why then does not man ask assistance of those new powers, the energy and efficacy of which are so clearly revealed to him? Why does he not command them to do yet again, and that under the directions of his own genius, that which they have already done in times long gone by on so many extinct torrents, and that under the prompting of nature alone?"

With the views thus expressed he proceeds to discuss more thoroughly the measures to be adopted for opposing, counteracting, subduing, and taming torrents. He argues that the continued application of such measures of defence as have been referred to must necessarily fail; and he alleges that prevention—not cure—must be attempted. This, says he, resolves itself into two distinct problems—(1) To prevent the formation of new torrents, and (2) To arrest the ravages of torrents already formed.

But the remedy proposed by him, as applicable to both, is the same—namely, the extension of vegetation. "All the facts which have been adduced," says he, "carry with them the conclusion to which they lead, and it would be superfluous to go back upon them. It is vegetation which is the best means of defence to oppose to torrents." And starting with this idea, the two problems resolve themselves into the discussion of the proceedings to be followed to throw the greatest possible mass of vegetation either on to the lands threatened with torrents in the future, or on to lands surrounding existing torrents.

"In doing this, art," says he, "should confine herself to imitating nature, to mastering its forces, and skilfully to opposing one of these to another. All that we are about to undertake nature has already done before us in time past, and she does it over again to this very day under our eyes whenever we leave her free to work. We are assured, then, beforehand of success, since all we have to do, to a certain extent, is to recommence experiments already made, and the success of which has been complete. Whence also it follows it is no longer a system of defence we have to seek, but a system of extinction."

As a preliminary measure, he argues the reservation, by legislative enactment, of certain portions of the soil; and a limitation or restriction of the number of the flocks and herds within what the reproductive vegetable power of the district can sustain. He recommends that the land to be defended against the ravages of the torrents should then be marked out by tracing, on each bank of the torrent, a continuous line, following all the windings of its course, from the highest point of its commencement to its issue from the gorge. "The strip of land comprised between each of these lines, and the summit of the mountains, would constititute (says he) what I would call a *zône de defense*, enclosed against flocks and herds. The zones of the two banks, following the outline of the basin, would meet in the heights, and would begird the torrent like a girdle. The breadth, varying with the slope and with the consistency of the soil, would be about 40 mètres, or 130 feet, below, but it would increase rapidly as the zone rose on the mountain side, and it would end in embracing a space of 400 or 500 mètres, or from a quarter to a third of a mile.

" This outline would require to follow, not only the principal branch of the torrent, but also the different secondary torrents which degorge into the first; following then the ravines which each of the secondary torrents receives, and going on thus, from branch to branch, it would go on to the birthplace of the last threadlet of water. In this way the torrent would find itself begirt thoughout the most minute of its ramifications. These zones of defence, in penetrating the *bassin de réception*, will be enlarged; while, on the other hand, as the ramifications are in this part more multiplied and more approximated, it will come to pass that neighbouring zones will join and even overlap each other, and their outlines will be lost in a common region, which will cover the whole of this part of the mountain, without leaving there a void space. The zones of enclosure being thus determined, the first part of the operation is finished. But this is in some respects only the outline of the periphery of the work which is to be done.

" We have next to do with what may be the most active and prompt means of drawing vegetation over the whole surface of this enclosure. For this purpose it should be sown and planted with trees; where it may be impossible to raise trees at once, the growth of shrubs, bushes, and thorns should be stimulated; but on the height, where the zones include the whole extent of the *bassin de réception*, it is a forest which must be created. The best adapted kind of trees must be selected; recourse must be had to all modes of procedure, indeed, even to modes of procedure which have yet to be discovered, and which go beyond experience. The work must be done any way and every way; and the end aimed at in these works ought to be to cover the *bassin de réception* by a forest which will every day become more dense, and which, extending itself step by step, will end in spreading even into the most hidden depths of the mountain.

" If the vegetation thus developed over the zones of defence be protected against flocks, if it be protected against the depredations of the inhabitants, if it be tended, maintained, stimulated by all means possible, it will ultimately envelope all the parts of the torrent by a very dense thicket, and thereby will be realized two effects at once, both of them equally salutary.

" First, this will arrest the waters which trickle down the surface of the soil, and will keep them from entering the torrent; or, if it do not prevent them doing this, it will at least retard them, and we know that this result is in every way a happy one. From the time this is done the torrent will only receive the waters which fall vertically from the sky into its bed; and this will diminish its volume in the same proportion as the proportion which exists between the extent of the general basin of the mountains and that of the stringently reduced opening presented by this bed. From a consideration of the great difference in extent of these two surfaces may be understood how great should be the reduction of the body of waters thus effected. And next, the ground of these zones can be no more washed away by the rains, and swept away by the torrent, and thereby will be diminished so far the mass of deposited matter. It is true, it may indeed be swallowed up little by little if the foot of the banks be undermined by the waters, but this constitutes another point to be attended to, and one to which I shall attend immediately, and on which, until I do so, I crave for a moment a suspension of judgment.

"To return, I give in one word the effect of these arrangements. I may say that the torrent will find itself placed in the same conditions as if it issued from the bosom of a deep forest, which will surround it in all its windings, and in which it will be as if it were drowned. Elsewhere I have described the results to which such a condition of things gives birth. It may be remembered as the forest struggling with the water ends in extinguishing the torrent, the same effects will reproduce themselves here, and it is unnecessary to repeat them.

"By the same analogy it may be understood that the vegetation advancing always, and gaining each day upon the ground, should descend on the banks and carpet them almost to the bottom of the bed, as has happened in extensive torrents; but the giving of permanence to the banks is a result of too great importance to be left thus to the caprices of the soil, and of the free will of nature. We come thus to a third department of the work. It is one in which it is especially necessary to redouble care and to multiply devices.

"To draw the vegetation over the banks they should be cut with small canals of irrigation derived from the torrent. These will impregnate with fertilizing humidity the land now rent and dry; they will break also the slope of the declivities, and serve to render them more stable, and soon they will disappear under the tufts of various plants brought to light by the water.

"The formation of these canals being extended ultimately to the summit of the bank, the water will thence penetrate the zones of enclosure and fertilize their soil. It is in the retention of the water, and in the possibility of opening everywhere and multiplying almost indefinitely provision for this, that rests in reality the whole future of the work.

"In fine, I pass to the fourth phase of the work, which is also the last. Whilst all these plantations retain the grounds through which the torrent flows, the undermining may be prevented by the construction of artificial *barrages*, or wears.

"We thus borrow from existing systems of defence that which is most efficacious in them; but in doing this how greatly have we ameliorated the circumstances in which we set to work!

"Indeed, we shall find in the plantations, everywhere where it is thought fit to establish these works, the best material for their construction. The young trees will supply stakes, prunings and bushes will supply facines. We can then construct the barricades of facines, or the wicker palisades recommended by Fabre. These works will cost little for manufacture, the materials will cost absolutely nothing. They will be cheap; and they do not present the dangers which accompany walls of masonry. One can then multiply them everywhere without any inconvenience, and almost without expense.

"These barricades will be like the completement of the works of extinction; they will serve to defend certain banks till the vegetation has reclothed them over all their extent, and till the torrent itself shall have lost the greater part of its violence. They can be employed also to stop up the secondary ravines, to intercept the little ramifications, to fill up small holes; in fine, to lead over the surface of the soil, and thus completely efface those innumerable streamlets divided like the hair-like fibres of a root, which are really and indeed the root of the evil.

" Behold the work completed !

"In recapitulating what has been said it will be seen that it resolves itself into four parts—first, the tracing of zones of enclosure ; second, the covering of these with trees ; third, the extension of vegetation over the banks ; and fourth, the construction of barricades of facines, of brushwood, or of wicker-work.

"One thing remains yet to be adverted to. I must speak for a moment of the order in which the work should be advanced. This order, far from being arbitrary, is an element of first importance, and a most essential element of success. I have already so often, in the course of this work, brought forward the necessity of attacking the torrents at their source that I believe it to be unnecessary to dwell upon it now. Thus, then, it is in the highest parts that the works should be first undertaken, thence to be extended to the parts of a lower level. Not only should a commencement be made by planting the *bassin de réception* before giving attention to the lower zones, but even in this basin the commencement should be made in its highest ramifications. One should go above the last traces of the bed, up to the abrupt slopes furrowed with ravines which the waters form and deform with each storm of rain,—it is there that the first works should be established ; one should afterwards—but only afterwards—carry them lower, but making sure first that the parts left are quite consolidated."

A chapter is devoted to the discussion of the practicability of carrying out such measures ; and another to the consideration of the legal difficulties in the way of this being done.

In a *résumé* of the work proposed, he concludes his recapitulation, saying, —" The definite result of the whole will be the creation of forests ; the whole work may be summed up in one sentence :—Reclothe with woods the more elevated parts of the mountains. If it be true, that forests exercise an influence on the climate, the effect of this extended mass of new woods will be to render the showers of rain less heavy, the rain-storms more rare, and the whole atmosphere more moist and more showery ; the climate will then, by insensible degrees, be changed at the same time as the surface of the soil ; and thus the two causes of torrents will be destroyed at one and the same time, and a general result will have been obtained while seeking at first only to remedy a particular evil."

But, he goes on to say, the work of reclothing the heights with wood will not of itself render unnecessary the construction of dams and wears ; and he proceeds to indicate the application of embankments, which would meet the requirements of the case with which he had to do—the prevention of ravages by torrents.

The question of expense is then discussed ; reasons are adduced to show that the expenses should be borne by the State. And, in a recapitulation and conclusion, the various measures proposed are reviewed and defended against such objections as it was thought possible might be brought against them.

Such is an analysis of the Study of the torrents of the High Alps, to which may be traced the commencement of the works of *reboisement* and *gazonnement* which are now being carried on, on a gigantic scale, in the Alps, the Cevennes, and the Pyrenees. But it is by no means the only work advocating such measures ; and I proceed to supply information in regard to other works, treating of the same subject, published before and after this work of Surell's.

PART II.

LITERATURE RELATIVE TO ALPINE TORRENTS, AND REMEDIAL MEASURES PROPOSED FOR ADOPTION TO PREVENT THE DISASTROUS CONSEQUENCES FOLLOWING FROM THEM.

The subject treated so exhaustively by Surell has commanded the attention of many besides him.

In 1797 was published an *Essai sur la theorie des torrents et des rivieres*, by M. Fabre, an engineer referred to by M. Surell, who had made these his study. The following are translations of some of his propositions relative to them, and to appropriate remedies for them.

"144. *The destruction of the woods which cover our mountains is the primary cause of the formation of torrents.*

"The reason is apparent. These woods, be they timber forests or be they high coppice, intercept by their foliage and by their branches a considerable portion of the water falling in rains and in thunderstorms. The remaining portion, which they could not retain, falls only drop by drop at intervals sufficiently long to let them have time to filtrate into the earth. On the other hand, the bed of vegetable earth, which goes on increasing annually, imbibes a considerable quantity of these waters. In fine, tufts of herbage and bush break and destroy at their origin the torrents which might have been formed notwithstanding all these obstructions. The woods being destroyed, the waters of a storm no longer meet with anything in their fall to intercept them. They cannot, by reason of their abundance, be absorbed by the ground as they fall. They flow over the surface, and meeting no more tufts which might have broken and divided their courses, they form torrents, as has been said.

"145. *The clearings on the mountains are the second cause of the formation of torrents.*

"We have shown that a torrent will be formed with so much the more facility in proportion as the matters which compose the mountain shall have less tenacity. Now the clearings, in rendering the earth friable and mobile, have diminished this tenacity; and thus they have favoured the formation of torrents.

"One may see from this how ill-advised and inconsiderate was the law given under the ancient régime, which authorised clearings, provided there were constructed at intervals walls of support to keep the earth on the slopes of mountains. It was not seen that in a great many countries the people confined themselves to raising two or three crops on a clearing, and that they then abandoned it. It was natural, this being the usage, that the sustaining walls, coming to cost more than the crops would repay, would not be constructed; and this is just what has happened. But there has already resulted from this, and there will result from it in the future, frightful disasters, as we shall now see.

"146. *The first disaster produced by the two causes of which we have just spoken is the ruin of our forests.*

"If there had existed wise laws, and these had been carefully executed, we should have had now building timber in such quantity as to permit of exportation. We should also have had in abundance wood for carpentry and fire-wood. It is felt that both of these things are essentially necessary in a well organized state. But they fail us to such a degree that in a great number of communes there is not even fire-wood. The evil has been long felt, and the necessity of remedying this is urgent.

"147. *The second disaster is the destruction in a great many places of the bed of vegetable soil with which our mountains were covered.*

"This bed would otherwise have produced abundant pasturage for the sheep, but, carried away by the storms and torrents, there remains at present on these mountains only a naked and dry rock. From this results necessarily a diminution of the small number of cattle which France might have been able to support if these pasturages had continued to exist.

"148. *The third disaster is the ruin of the domains which lie upon the rivers.*

"We have seen that the swellings of the torrents were stronger in proportion as the mountains were less wooded and more impoverished. These swellings are then greater now, through the operation of the two causes mentioned above, than otherwise they would have been. They ought, therefore, to cause, and they do really cause, much more havoc to the domains along their course than they otherwise would have caused.

"On the other hand, we have seen that it might happen, as it has in effect happened too often, that the torrents in issuing from their bed or channel would cover adjacent domains situated at the foot of the mountains with deposits, which absolutely alters their nature. Now, this never happened until that by the operation of the two causes mentioned above the torrents were formed.

"149. *The fourth disaster is the drainage, experienced in the navigation of the rivers, by the divisions in the water-courses, which are the consequents of great floods.*

"150. *The fifth disaster consists in the strifes and contentions, between the proprietors on opposite banks of the river, to which the divisions in these water-courses give rise.*

"151. *The sixth disaster results from the deposits which they make at the mouths of the streams, which often intercept the navigation.*"

Each of these three statements is illustrated in detail.

"152. *In fine, the seventh disaster consists in the diminution of the sources which feed the streams and the rivers in their ordinary state.*

"We have seen that springs, the sources of streams, are formed from the rains which filtrate through the earth and meet in the subterranean reservoirs, whence they escape by minute channels, and make their appearance at the surface of the ground. Now, if the mountains be despoiled of their bed of vegetable earth, and there remain only the bare rock, it is evident that the water of the rains will no longer filtrate through the soil, but will flow quite superficially; thence it follows, that as the fountains diminish so must the rivers which they feed; and a time will come when even the rivers which at present are navigable will cease to be so. True, indeed, that time is still distant, but sooner or later it will arrive if the cause which produces this effect be not destroyed."

With these views M. Fabre urged then the planting of trees, or the *reboisement* of the mountains, and the protection of these throughout their growth. He thus states his opinion:—"We have said that the destruction of the woods which were covering the mountains, was the primary cause of the formation of the torrents. To destroy the effect, the cause must be exterminated. Therefore, if there be still vegetable earth on the mountains, it will be well to leave these to become clothed again with wood, by leaving them in fallow, and with a view to the same end, it may be well to remove everything which might damage the young trees. For this reason, most rigidly should be carried into execution the laws relating to the prohibition of goats, for it is known that the tooth of this animal is murderous to young trees. It is not less essential to provide for the conservation of existing woods, since these woods, which have kept the torrents hitherto from being formed, are to us a sure guarantee that they will prevent the formation of them in the future.

"Clearings are the second cause of the formation of torrents. It is necessary, then, that after having been too extensively tolerated by the ancient laws, these should be restricted within prescribed limits. In consequence, we consider, that in this respect, they should be conformed to the following rules:—First, a clearing ought never, under any pretext whatever, to be permitted on the slope of a mountain, which has less than three of a base for one of vertical height, *i.e.* a slope of one in three.

"Second, the clearance might be permitted on one of less declivity, but only under the restrictions we are about to state.

"Third, the clearance ought never to be authorised, but on the verge, or in transverse horizontal strips, or on a level, or what is nearly such.

"Fourth, in this case the strips of fallow should be separated from one another by other strips, likewise horizontal or level, left uncultivated, on which the wood should be permitted to grow.

"Fifth, these uncultivated strips should be made to take the place of the sustaining walls, prescribed by the law previously spoken of. It appears that they should not be less than five *toises*, or thirty feet, in breadth, to enable them, in case of need, to destroy a torrent which might be formed on the strip of fallow above it.

"Sixth, the breadth of the strips of fallow should be only five *toises*, or thirty feet, where the slope of the mountains may be one in three; but it appears that it may be increased with the diminution of the slope, until a slope is arrived at, which leaves no cause of fear of the formation of torrents, in which case the breadth may be unlimited.

"Seventh and lastly, the clearings should in no case be permitted without the authorization of the respective municipal authorities, and after the specification and plan, which shall have been previously made by a public official of what is proposed in each commune.

"Every one must see that by some such regulation we may escape for the future all the disasters produced by arbitrary clearings, almost always ill-arranged, both as they affect the interests of the community and those of the individual. Nature is only the more active when aided by human industry, and so in cases in which it is wished to hasten on, on certain mountain slopes, the increase of woods, it would often not be bad to sow acorns and beech-nuts, or seed of any species of trees which may be presumed to be proper to the localities. There is more than one country where they are quite accustomed to do so.

"There are cases where there remains so little earth on the mountains as to lead one to conclude that wood will there make but little increase; such grounds may be laid with turf, and sown with seeds of plants which may be deemed most proper for the localities. The superficial tissue formed of turf will be an obstacle to the formation of torrents, and by this means besides will be created useful pasturage.

"These are the means of preventing the formation of torrents on the mountains. It remains for us to see those which must be employed to destroy, when the thing is possible, the torrents already formed."

The views advanced by M. Fabre have never, so far as I know, been subverted; and by subsequent studies of the phenomena many of them have been confirmed. But it has been objected that the subject was not one which admitted of being discussed in such precise propositions as those in which he invested his views—that some of his propositions were based on deduction rather than founded on an induction of fact—and that, in the absence of facts, adduced to establish or support his deductions, there was an element of uncertainty thus introduced into his conclusions, which prevented them being made the ground of extensive practical undertakings, involving great expenditure, until they had subsequently been verified by renewed observations of facts systematically conducted.

This circumstance makes the work more valuable to any one desirous of studying the subject in all its aspects. It is a work to which Surell often appeals, as a work the value of which was indisputable, and as the only work going to the root of the matter in discussing a subject not exactly the same but one nearly allied to that to which he was giving attention. And he mentions that Fabre had himself announced, that no work on the subject had previously been published, praying that the imperfections of his work might be borne with in view of the novelty of the matter.

Surell speaks of Fabre as an engineer who had occupied himself with this study, and he says of the work by him, that it contains a complete description of torrents, with just, and often ingenious, remarks on their action; but he states that he considered the form of aphorism in which his observations are couched a defect, exposing them to the objections I have cited. He states further that it is clear, from many passages, that the torrents seen by M. Fabre were not those of the High Alps, which were those which were the subjects of his own study, though they were similar to them in many respects; and that his theory, when applied to them, was not always borne out by the phenomena presented by them, or did not cover these: that it was evidently based on the observation of torrents, which devastated the South of Provence, and more especially the torrents of the Var, where he was *Ingénieur en chéf*.

But all of these considerations make his observations and conclusions the more valuable to any, who may be studying the subject, with a view to the discovery of remedial measures, appropriate to countries very differently situated from the ravaged and devastated regions of France. We find in Fabre and Surell, men of different casts of mind, belonging to different generations, following their professional pursuits in districts far apart and differently situated, propounding doctrine essentially and substantially the same. With regard to the deficiency of observations as a foundation of M. Fabre's counsels, such observations were greatly desiderated by him; he stated that no work on the subject had been published, and he craved that the defects of his work should be excused in view of the novelty of the subject of which it treated.

In 1804 there were published *Reserches sur la formation et l'existence des ruisseaux, des rivieres, et torrents*, by M. Lecreulx. The design of this publication was to refute the views advanced by Fabre; but it has been alleged that apparently the author did not know the kind of water-courses to which Fabre in his work had a reference.

On this point M. Surell writes,—"I scarcely know whether Lecreulx meant positively to dispute the position that woods have an influence on the production of torrents. In attacking Fabre on this point all that he does is to bring to light his complete ignorance of the kind of mountains and of the kind of water-courses which Fabre had specially before him. Lecreulx had always before his mind the case of the Vosges, which comes up in every page of his book. I know the Vosges well, and I can affirm that these mountains no more resemble the High Alps than the German patois, spread over several of the valleys, resembles the provinçal dialect which is here the general language of the country."

In 1806 appeared a *Ptomographie des cours d'eau du Dèpartement des Hautes Alpes*, by M. Héricart de Thury, in which are pencilled rapid sketches of the geological characters of the beds of the water-courses of the Department, and his work supplies data valued by students of the country, seeking to discover the cause or occasion of the ravages which these water-courses make, and a remedy for the evil. He reckons eight distinct basins or river valleys in the High Alps. Surell reckons three, but this affects not the facts recorded: it resolves itself into a mere question of judgment in regard to the best division to be made.

The views are in accordance alike with those advanced ten years before by Fabre, and thirty years later by Surell. Of the vicinity of Embrun he writes,—" In this magnificent basin Nature, has been quite prodigal of her blessings. The inhabitants have enjoyed her favours with their eyes shut; they have slept on in the midst of her beauties. Ungrateful for all, they have inconsiderately carried the axe and the fire into these forests which shade the steep mountains—the ignored source of their riches. Soon were these emaciated peaks ravaged by waters, torrents swelled and precipitated themselves with fury on the plains; they have cut down, torn away, and undermined the foundations of the mountains; grounds of great extent have been carried off; others have been entombed; these have been covered with rocks, those show nothing but stones and gravel. The ravages are still going on; no obstacle is opposed to their fury—soon in Crevoux, Boscodon, Savines, and all the country around, the torrents will have utterly destroyed all this fine basin, which but lately would have borne comparison with all possessed by the richest countries—with the most fertile and the best cultivated of them all."

The warning was sounded in vain. It was drowned in the roar of cannon carrying into other lands devastation, and death, and mourning, and woe; but after the men of that generation had mostly died away, and another generation had taken their place, the subject was again brought under consideration.

In the *Annales des Ponts et Chaussées*, for *1833, 2d Semestre*, is a paper by

M. Montleuisant, entitled *Note sur les Desséchements, les Endiguements, et les Irrigations*, which is not without its bearing on the subject in hand. And about the same time a Memoir by M. Delborgue Cormant, *Ingénieur en chêf des Ponts et Chaussées*, on embankments.

In 1834 appeared a second edition of a work previously published—*Histoire, Topographie, Antiquites, Usages, Dialects, des Hautes Alpes*—by J. C. F. Ladoucette, who had been prefect of the department, and who had been eulogised as the best prefect the High Alps ever had had. A statue of him erected in Gap speaks of the high estimation in which his labours for the good of the department were held. By him the number of basins, or river-valleys, in the High Alps is reckoned five, while by M. Héricart de Thury they had been reckoned eight, and by M. Surell they were afterwards, as we have seen, reckoned to be three; but, as has been stated, such enumerations are mere matters of judgment in regard to what are entitled to be considered separate basins, and to be entitled to this designation. This work did not contribute much information in addition to what was previously known on the particular aspect of the subject which connected it with forest science.

It was otherwise with another work by one who had also held the office of prefect, a memoir, entitled *Projet de boisement des Basses Alpes présenté a S. E., lê ministre secretaire d'etat de l'Interieur, par M. Dugied, ex-préfet de ce département*. The following is a translation of a statement of his views:—

" More than half of the department of the Low Alps is covered with arid and unproductive soils. These are increased by numerous torrents, which, descending there into the fertile valleys, complete the ruin of the country.

" Two causes have contributed more especially to bring about this sad state of things,—the destruction of forests on the one hand, and on the other the rage for clearing land by grubbing up the roots, and herbs, and bush. It is high time to apply remedies, for later to remedy the evil will have become impossible.

"To bring about a restoration of the department, three measures should be adopted—(1) To prevent additional grubbing, and to restore to the grubbed lands their primitive consistency; (2) To plant the summits and sides of mountains with trees; (3) To enclose the torrents. We shall remark on each of these three measures in succession.

" *First Measure.*—Grubbings may be prevented by enforcing the ordinance of 1667, which pronounced a penalty of 3000 francs against all those who should grub ground free of wood on declivities. And grubbed lands may have their primitive consistency to some extent restored by compelling the proprietors to convert them into artificial meadows, be it by the power of the tribunals, or be it by administrative action. (The author cites an experiment, from which it appears that sowing the grounds with sainfoin, *Hedysarum Onobrychis*, had completely consolidated a land previously subjected to extensive waste).

" *Second Measure.*—It follows, from statistical estimates, which have been prepared, that the area of the ground in the Low Alps, which we may hope to replant with trees with success, amounts to 150,000 hectares. It may be accomplished by each year taking of this surface from two to three thousand hectares, say 1200 acres, which it might be required of the proprietors of the soil to replant. But here there presents itself more of a difficulty. First, the great subdivision of the properties which will multiply the cases of resistance, and the little revenue which the proprietors of each will draw from the plantations during the earlier years of their growth. And, secondly, it is the case that the gross expense of the plantations, will not on all grounds be compensated by proportionate future products.

"These difficulties are very serious, and they cannot be overcome but by one expedient, *the intervention of the State.* This may consist, 1st, in premiums given to the planters; 2nd, in the gratuitous distribution of seeds; and 3rd, in a remission of taxes in favour of the planters.

"A premium should be granted to every proprietor whose sowings have been successful. The verification of this must be made by a commission, and the success stated in a minute addressed by this commission to the prefect. The value of the premium might be 20 francs per hectare, and it should be paid by the State conjointly with the department, the State paying three-fourths and the department one-fourth of the amount. Thus, on the supposition of two thousand hectares being sown annually, the department would disburse each year in prizes 10,000 francs, and the public treasury would disburse in the same way 30,000. . . . The grounds on which I propose that the department should not pay more than 10,000 francs a year are, (1) that the department is far from being rich; (2) that it will not recover payment of the sums it furnishes, whilst the Government will recover all its advances; *and* (3) *in a word, that without such advances on the part of the Government, there is no reason to hope that the operation will ever be carried out.* No doubt the department will derive very great advantages from the work; but the sacrifices which it will make to contribute to the success will not be the less real sacrifices." . . . M. Surell says,— "This, which was a weighty reason at the time when M. Dugied wrote these lines, has become, since the law of the 10th May, an absolute necessity." M. Dugied goes on to say, "The second mode of intervention, consisting in the gratuitous supply of seeds, should be wholly at the expense of the State. Let us suppose that there are sown 2000 hectares annually, and that they are divided, in regard to kinds of trees, in the following way: 600 hectares in acorns; 600 in beech; 800 in firs and pines—in all, 2000 hectares. The whole expense of the seeds, carriage included, should be about 23,400 francs. The expense would rise to 35,100 francs if there should be sown 3000 hectares per annum instead of 2000.

"The Administration, by delivering the seeds gratuitously, will have it in its power to determine that the different kinds of trees have been distributed with intelligence, and that each kind of soil has only received those for the growth of which it is best fitted. Declivities too steep should be sown with box trees and brooms.

"The sowing will also require to be protected against cattle and against plunder. It will be necessary to secure a very active and very strict surveillance on the part of the forest officials, who may remain charged with watching the future forests; their number should be augmented, their organization perfected, and at the same time their condition raised and their circumstances improved.

"In conclusion, passing on to the third means proposed—the remission of taxes. Each proprietor, after an examination and approval of his sowings, at the end of five years might have a remission of taxes for the period of ten years.

"Such are the sacrifices which impose themselves on the State to secure, by degrees, the *reboisement* of the mountains.

"*Third Measure.*—This relates to the enclosing of the torrents by embankments. This enclosing should not be commenced until the forests shall have produced their effects—that is to say, in fifteen or twenty years after the first plantings. The engineers of roads and bridges should prepare the plans of the works to be executed. The expense should be borne by the proprietors interested, and by the State, which should assume the responsibility for half the outlay. The effect of the dykes should be at once to protect the river lands and to acquire new lands." . .

The author calculates that the enclosing of the Durance between Sisteron and the Pertuis des Mirabeau would cost at most from 4 to 5,000,000 francs; and that the area of land acquired would be 10,000,000 square toises or fathoms, which would be worth, at the end of three years, at least 10,000,000 francs. The capital in this undertaking would thus be doubled at the end of three years.

In a second division of the work, M. Dugied endeavours to show benefits resulting to the State from such undertakings, which might induce them to enter into this expenditure, doing it in such a way that the first expenses could not be in excess of the sums to be repaid.

"The mortgage of the sums expended by the State," he says, "will resolve itself into an augmentation of the imposts to which should be subjected waste lands converted into forests. Strictly, and according to the rules adopted in the assessments of imposts, the augmentation should be for the advantage of the Department, and should lighten the manorial tax of the other proprietors. But it may be believed that the General Council will consent to the addition which may be made to the manorial contribution of the Department; and it is on this augmentation, on the assumption of this consent, that we can base our calculations.

"The contribution allotted to waste lands is upon an average twenty-two centimes per hectare; that on forests is seventy-two centimes. When, then, a hectare of waste lands shall have been converted into forest, it will produce an augmentation of contribution equivalent to fifty centimes. It is this difference of fifty centimes which will constitute the funds for repayment. It must be observed that the fifty centimes will not be touched until ten years after the sowing, if the State have granted to the sowers a remission of taxation during this period of time. It must also be taken into account in the calculations that all the sowings will not be successful, and that a portion of the seeds delivered gratuitously by the Administration, and paid for by it, will have perished. It is supposed that the loss of sowings may be about a fifth of the whole.

"From these data there can be formed tables which will give, year by year, a statement of the expenses, or of the returns, of the Government; and it may be seen in this way that for a sowing of 20,000 hectares, the expenses of the Government at the end of ten years will have amounted to 534,000 francs, but that at the end of eighty-six years it will have recovered all these advances. Moreover, it will have acquired an annual bonus of 8000 francs, seeing that the contributions will continue to run on.

"If one extends the calculations to 150,000 hectares (that is, to the whole of the area to be re-wooded), and if we suppose that the sowings will extend over fifty years, it will be found that the State will have recovered these advances at the end of one hundred and ten years, and that it will enjoy thenceforward an annual bonus of 60,000 francs. It follows from this that it is for the interest of the State to give to these operations the greatest extension possible.

"It is also necessary that the State should recover the advances which it will have made for the construction of dykes. And it will find the means of liquidating the amount sunk in the work, first, in the profit calculated above, as resulting from the 50 centime augmentation of impost on the land as wooded, and further, in the proprietorship of a certain portion of the lands acquired. As it will have furnished the half of the expense to which the acquisition owes its existence, it is just that it should obtain possession of half of the lands acquired."

M. Surell says,—"Such is the system developed by M. Dugied in his Memoir *Sur le Boisement des Basses Alpes.* This work produced no fruit. It did not for one moment stop the abuse. The Administration is not yet aroused from its indifference; and the devastation of the torrents, and the miseries which this brings in its train, and the daily progressive ruin of the country, go on still, as in the past, before unpitying eyes.

"The efforts of M. Dugied have been but little appreciated; and the country, in favour of which he was the first to raise his voice, has not been more just in regard to him than was the Administration of the Restoration which deposed him from the prefecture of the Basses Alpes, which he had not occupied more than a year, and where he would probably have rendered eminent services to the country. His work has called forth ridiculing criticisms. They have referred the execution of his project to the Princes of the Arabian Night Entertainments. I must confess that the extravagancies of the project of M. Dugied has entirely escaped me. I only see in it an operation, sufficiently simple at bottom, which could not fail to develope, on a vast scale, what is practised every day by private parties; an operation, the execution of which is evidently possible, and the expense of which has nothing surprising in it when I compare it with those which the Administration entrusts every year to the engineer of the smallest arrondissement. Certainly, it would read as a romance, much more extravagant than the alleged *palingenesique* romance of M. Dugied, if one would turn over the leaves, mastering the same, of the report of the 120,000,000 francs worth of works executed every year, on all the bridges of France, under the *Direction des pouts et chaussées!* This speaks of the sea imprisoned in harbours, roads tunnelled through rocks, rivers confined by embankments or by bridges, lighthouses erected on rocks in the midst of tempests, canals transporting boats across the summit of mountains. I see in these works, works more difficult, more costly, and more marvellous by far, than the *reboisement* of some nooks of mountains. And if any come to discuss in the Chamber seriously, and like people who are ready to put hand to the work, the enormous budget of a milliard and a half, which certain economists tell us to be necessary for the establishment of a complete network of railways, what will be thought of this other prodigy, which was held to be only fabulous not more than thirty years ago? When we shall have multiplied by ten, or by a hundred, the figures given by M. Dugied, we shall not yet have come to expenses like to those of a great number of our public works, which are ten times—or, for that matter, a hundred times —less useful, and which do not frighten us, accustomed as we are, for a long time, to open our purses for their execution.

"Will any one undertake seriously to deny the possibility of the *reboisement* proposed by M. Dugied? . . The proofs which have established this possibility are too numerous, too palpable, for this. Everybody admits that the Alps were wooded long ago; and this is itself a proof that woods may yet be made to reappear there. The first forests which nature cast on these mountains had to clothe a soil more naked, more sterile, more irregular, than the actual soil of the present. And if vegetation has already triumphed a first time in this struggle against destructive agents, why should she succumb to-day? It will be said that she was assisted by time! It is so. But to-day she will be assisted by man, and that assistance, in my opinion, avails more than that of some four centuries. There are here and there, in the bed of the Durance, conquests over the waters made by the effort of nature alone; but long ages have scarcely sufficed to ensure vegetation there, and some portions of it remain eternally sterile. When man undertakes like conquests he finishes them in three years; three years suffice for him to make fields to flourish on the very place where the waters rolled pebbles and barrens sands. This miracle is renewed every day, and under the eye of all. Is not this a more marvellous triumph than it would be that man should succeed in reforesting lands which, for the most part, have been covered with forests before.

"If I wished to criticise the work of M. Dugied," says he, "I would not bring against him such objections. But whilst entirely approving the basis and the end of the project, I would condemn some few details of execution. M. Dugied has comprised, under the designation of torrents, the Durance, the Verdon, the Cleone, which are rambling rivers, and on which the *reboisement* of the mountains could only have a *detournée*, and secondary influence in affecting the water-course. And in making the embankment of these water-courses a corollary of the plantation of forests, he has coupled together two distinct operations. From this it follows that his project is in some respects too ample and exaggerated, and at the same time in some measure defective. And this impression of vagueness is deepened when it is seen that M. Dugied does not attach to forests any action on the torrents other and beyond that effected by a climatal change. As this influence is rather uncertain, and very difficult to be cle*a*rly demonstrated, one cannot understand how the author came to build on it such great expectations, and that he should make of *reboisement* a preliminary operation, without which the embankment of rivers would not be undertaken with success.

"But there is a point in which his project seems to me defective in its very foundation—it is this, he makes the execution of it to rest entirely on the gooodwill of the proprietors. If the enterprise be really a thing of public utility, as the author says it is—if it truly have the degree of importance and necessity which he attributes to it—how does he come to leave it at the mercy of the first peasant—stupid or stubborn—who will refuse to take part in it? It showed little knowledge of the spirit of the inhabitants of the country, to believe that a premium will suffice in every case to overcome the natural apathy, and above all, the obstinacy of such, if once they stubbornly determine not to give in to the undertaking. Now, this will certainly occur oftener than once, if it do not become even generally the case. The twenty francs of premium per hectare, which M. Dugied tenders to them, would not always appear to them a sufficient indemnity to compensate the trouble which the sowings might entail, and the loss of their pastures, of which M. Dugied says nothing, and of the numerous interferences which will follow from the operation. These works, besides, will not succeed but through the expenditure of sustained and intelligent exertions, which the peasants will not make. They will soon have invented a thousand artifices to gain the premiums, without having done anything to deserve them.

"It is thus indispensable that the State undertake the charge not only of the expense, but also the execution of the works; and ex-appropriation or confiscation will furnish them with a legal means to bring down all possible resistances.

"It seems that M. Dugied has recoiled from urging this, most possibly because he was afraid of the expense; but I have shown that this will be somewhat reduced. Besides, does not the State acquire every day for roads, and by the same means, fields far more costly than the waste lands of these mountains? And in that case the possession of the soil brings to her nothing, or at least procures for her only a change of advantages. Here it buys the lands at a low price, it exploits them, it gives them value, and by that means she increases her domain if she retains them in her own hands, or the revenue from the taxes if she restores them to the inhabitants."

Of this work of M. Dugied, Surell says,—"It is the only memoir known to me which treats specially of the means to be employed to counteract and oppose the scourge of the torrents." And he adds,—"What is proposed by M. Dugied is conceived in a comprehensive spirit; but the characteristic peculiarities of the torrents are neither analysed nor described by him; the work is addressed to those to whom the torrents are already perfectly known."

In 1841 appeared his own work, *Étude sur les Torrents des Hautes Alpes*, of which a *résumé* has been given in Part I. of this compilation.

On my first perusal of this work, knowing as I did how much damage was done by torrential floods at the Cape of Good Hope, my feeling was a desire that I could make the substance of it my own, and give forth anew the observations, and the reasonings, and conclusions of the author, for the information of my former compatriots in that Colony, and of others in other lands exposed to such torrential floods as there alternate with severe and long-continued droughts. But this was impossible; and, moreover, I have often found excerpts from the work of an original thinker far more satisfactory, and often far more suggestive, than any digest of it given by friend or foe. Often, on reading some such digest, I have felt disposed to cry out, Give me his own words, for no words can better tell what he says than the words he has himself used in the collocation of them which he has given! but to do this was also impossible; and I have done what I consider most likely to be satisfactory at once to M. Surell and to students of the subject of which he treats, at the Cape or in other lands, in which the English language must be the medium of communication.

The work was published by order of the *Administration des Ponts et Chaussées*. Public opinion was not then so advanced on the subject as to prompt to action, and his services were put in requisition for the carrying out of the system of railways, which seemed to demand more immediate attention. While rejoicing in his honours and usefulness as *Ingénieur en chef des Ponts et Chaussées*, and *Directeur des Chemins de fer du Midi*, some regret may be felt by those who are alive to the importance of *reboisement* as a means of stifling torrents that scope was not found for his energies in originating and carrying out works such as he had advocated.

After the work was out of print, many solicitations were addressed to him to issue a new edition. But from this he shrunk. The state of things depicted by him had, to a great extent, ceased to be, his suggestions had been carried into effect, and a new state of things had come into being. But he was relieved of embarrassment by his comrade and friend, M. Cézanne, agreeing to prepare a supplemental volume, and the two were published conjointly,—the first volume, the *Étude* of M. Surell, in 1870, and the Supplement, by M. Cézanne, in 1872.

The subjects of M. Surell's study were chiefly these,—the phenomena of torrents and effects produced by them; the causes of their occurrence; means of defence which had been employed to protect the land and its inhabitants against their ravages; and measures which were more likely to prove efficient if they should be employed, which measures were plantations of trees, and herbage, and bush, over the area drained by them, combined with the erection, in subordination to this, of *barrages*, or wears, to control and regulate the flow, where this may be practicable and desirable.

Previously to the publication of the original edition—but at what date I know not—there had been published a *Memoire sur l'état des forêts dans les Hautes Alpes, les causes de cet état, ses resultats et les moyens d'y remedier*, by M. Delafont. Of this M. Surell writes,—" All the causes of the destruction and disappearance of forests are thoroughly and carefully expounded in a memoir by M. Delafont, *inspecteur des eaux et forêts*—a memoir full of well-intentioned and wise statements, which only calls forth regret that it did not inspire the Adminstration with enlarged and bold views, which alone would be commensurate with the evil; for great evils call for great remedies."

"The sad results which I am about to point out," says M. Delafont, "are deplorable on all hands. All men who have not been blinded by ignorance, or whose heart has not been withered up by selfishness, give expression to the thought that it is high time to stop the progress, ever increasing, of so fearful a devastation. They lament over the evils without number which are occasioned by the deforesting of the mountains, and seem to call us to the protection of our forest wealth. These reflections, these prayers, I have often myself heard uttered with an energy which is inspired by the profound conviction of the existence of a great evil, and of the imperious necessity which there is to stop its course. Let us listen to the cries of distress of a population alarmed by the future before it."

And M. Surell refers to this, and other statements by M. Delafont cited by him, as supplying evidence that he had himself in no way exaggerated the evil in what he said in his *Étude sur les Torrents*.

While the work of M. Surell's was passing through the press, he received a copy of a *Memoire sur la dégradation des forêts dans les arrondissements d'Embrun, et de Briançon*, which the inspector of forests, in these two arrondissements, M. Jousse de Fontanière, had shortly before addressed to the Administration. Of this he says,—" This work—prepared by a man most competent for the work, and devoted to his duties, who, after having struggled for a long time against the innumerable difficulties of his service, succumbing under the trial, took measures at last to demand aid—should have had the effect of securing the attention of the State to the frightful future to be anticipated in this department."

And he cites the following as a specimen of the out-spoken faithfulness of

the author,—" From all that has been said, it is concluded that the department of the High Alps is the one of all France in which the cultivators are most threatened in their fortune, and that they will be compelled, sooner than is supposed, to abandon the places which were inhabited by their forefathers, and this as a consequence of the destruction of the soil, which, after having supported so many generations, has given place, little by little, to sterile rocks.

"The destruction of the forests will be the principal cause of this calamity. The disappearance of these from the mountains will give up the soil to the action of the waters, which will sweep it away into the valleys; and then the torrents, becoming more and more devastating, will bury under their alluvial deposits extensive grounds, which will be for ever withdrawn from agriculture.

"The crusts, denuded of their vegetable soil, no longer permitting the infiltration of the waters, these will flow away rapidly on the surface of the ground. Then the springs will dry up; and the drought of summer being no longer moderated by their irrigations, all vegetation will be destroyed.

"The elements of destruction growing thus one out of another, we have only to observe what passes to-day to predict what will infallibly come about some ages hence. When the forests shall have entirely disappeared, fuel and water, the two primary necessaries of life, will be awanting in these desolated countries.

"The cupidity of the inhabitants of these mountains, the tenacity with which they keep to old customs, do not permit a hope that a moral conviction of this desolating future will strike their thoughts so strongly as to lead them to make some temporary sacrifice; it is, therefore, for the Administration, more enlightened in regard to the state of things, and to their consequences, to meet the evil by laws most appropriate to the requirements of the country."

Ladoucette, in his *Histoire, Topographie, Antiquités, Usages, Dialectes des Hautes Alpes*, already cited, says the peasant of Dévoluy "often goes a distance of five hours, over rocks and precipices, for a single [man's] load of wood;" and that "the Justice of Peace of that Canton had, in the course of forty-three years, but once heard the voice of the nightingale." Now the desert and the solitary plain begins there anew to flourish like a rose, and the inhabitants to rejoice with joy and singing; and there is heard the shout of children playing in the streets—a change brought about by *reboisement* and *gazonnement*, confirming the conclusion that the destruction of trees and herbage had been the occasion of the desolation.

In regard to the valley of Embrun, where a corresponding improvement has been brought about by similar means, Héricart de Thury, who has also been already cited, wrote in 1806,—"In this magnificent valley nature had been somewhat prodigal of its gifts. Its inhabitants have blindly revelled in her favours, and fallen asleep in the midst of her profusion." And Becquerel, in his work *Des Climats*, mentions also that it was once remarkable for its fertility. What it became, through the ravages of torrents, after the destruction of its trees, Surell has shown.

M. Surell cites, as in accordance with his views in regard to the influence of the climate on the formation and violence of torrents, the following remarks by Labèche, in his treatise on Geology. Writing of the geology of the Alps, M. Labèche says,—" A difference in the climate ought to produce other visible changes, as well in the superincumbent rocks as in those that were of an older formation. It is probable that the more a climate was warm, and approached that of the tropics, the greater would be the evaporation, and the quantity of rain; greater also would be the intensity of power of certain meteoric agents; consequently, according to this hypothesis, the different deposits ought to present indications of the influence of such climates, more marked in proportion as the epoch in which they were formed was more remote from the present. If rains, like to those of the tropics, have fallen on high mountains such as the Alps—even supposing that many of them had an elevation less than that of these—these rains would produce effects very different from those which we see now in the same countries; one may see that these would form all at once torrents of which the actual inhabitants of these mountains have no idea; such volumes of water would sweep away quantities of detritus far greater than those which the actual torrents of the Alps carry away, the volume of which, however, is pretty considerable.

"Thus, though admitting the correctness of this hypothesis in this, it is necessary always to take into account the differences produced on the surface of the earth by the action of meteoric agents, the which is more powerful as the climate is more warm. One ought especially to give attention to this, when from the observation of a series of the layers of the same district it appears evident that the temperature, under the influence of which they were formed, has gradually diminished. Let us examine now to what degree vegetation can, in warm climates, counterbalance the power of disintegration, and transport which atmospheric agents possess. It appears that, all other circumstances being equal, the more warm a climate is, the more vigorous is the vegetation which it produces. The question then comes to this: Does the vegetation protect the soil against the destructive action of the atmosphere? It is impossible to answer this otherwise than in the affirmative. If we want proofs of this fact we shall find them in the artifical mounds, or *barrows*, which are so common in many parts of England; they had been exposed in that climate to the action of the atmosphere for about 2000 years; and yet they have not undergone, in their form, any perceptible change, although they have, during at least a considerable portion of that time, only been covered by a light layer of turf. If now it is admitted that the vegetation protects, to some extent, the ground which it covers, it follows that the stronger the vegetation is the more efficacious is the protection which it affords, and as a consequence the ground is always defended from the destructive action of the atmosphere in proportion to the need it has of such protection. Without this providential law of nature, the softer rocks of tropical regions would be speedily carried away by the waters, and the soil would no longer be able to sustain vegetables or animals; for, although in many tropical regions we meet with vast extents of land which present the appearance of sterile deserts, but which one sees suddenly start to life after two or three days' rain, and cover themselves, as by enchantment, with a beautiful verdure, we should bear in mind that the roots of the briskly vivacious plants from which moisture causes to be produced so vigorous a vegetation—and even those of the annual plants which have passed away, of which the seeds produce leaves so verdant—interlace themselves in such a way in the soil that they oppose a considerable resistance to the destructive power of rain. In the Savannahs of America it is frequently the case that there is little vegetation, and there they experience considerable disintegration.

"I have by no means the intention to infer from what has been said that the disintegration of soil is not generally greater under the tropics than in temperate climates; it has been my desire simply to establish that in both cases the soil receives, from the vegetables which cover it, a protection proportionate to the destructive influence to which it is exposed. Let us suppose that there should occur in England one of those rainy seasons so common under the tropics. No doubt great extents of land would be washed away, and the *barrows*, of which we have already spoken, would quickly disappear. If, on the contrary, there fell there only the same quantity of rain which we have every year in the climate of England, we would find scarcely any traces of vegetation in the low-grounds, for the water produced by it would be insufficient to sustain tropical plants, and while it tended to disintegrate the soil, it would be so speedily evaporated that its destructive action would be scarcely perceptible. The quantity of rain and the vegetation are proportionate to one another; nevertheless, the disintegration of the soil increases with the quantity of rain, and the force of many meteoric agents, in such a way that, other things being equal, the greater the rainfall the greater is the destruction of the soil; and consequently, the warmer the climate, the more considerable is the disintegration of the mountains.

"In tropical regions, parasitical and creeping plants are seen in all directions, growing wherever it is at all possible to do so, and with such luxuriance as to render the forest almost impassable. The forms and the leaves of trees, and of such plants, are admirably adapted to resist great rains, and to protect the innumerable creatures which, in the rainy season, come to seek a shelter under their foliage. The noise which the tropical rains make in falling on these forests strikes strangers with astonishment; it is heard at distances which would be almost incredible to the inhabitants of temperate regions; and the rain, thus deadened and broken in its fall, is speedily absorbed by the soil; whereas, where it flows into hollows, it produces torrents, which every one must confess are rather impetuous, and cause great ravages."

M. Michel Chevalier, in his work entitled *Des Interets Matériels de la France*, writes thus :—"Besides the works executed in the river-bed, there are other measures which, according to men of experience, would exercise a salutary influence on the navigability of natural water-courses, and which concern even canals, as to feed these recourse must be had to rivers and to the smallest streams. I wish to speak specially of the replanting of mountains which have been so improvidently despoiled of their woods, and abandoned in their nakedness by a culpable indolence, or even by a fatal one, descending to niggardly interests, which the law does not recognize, but on the contrary resents, have hindered the forests from reproduction by the effort of nature alone. The rains and the snows, when they fall on the bald heights, flow away or evaporate with the greatest rapidity; in place of maintaining brooks and rivers, on the rich levels, by which boatmen may profit, and on which the proprietors of river-banks may felicitate themselves, they produce there sudden floods, inundations which suspend navigation, devastate properties, covering them with gravel, and sometimes eating into them and carrying them away; then, after these floods, there follow soon low waters, which only stop at distant points and for a short time after some storm.

Through reckless deforesting our temperate countries are thus being assimilated to southern regions, where there are nothing but torrents during the spring and autumn, imperceptible threads of water in the midst of an ocean of sand during the summer, and never smooth unmanageable rivers. The business is now to restore the soil of France to the primitive forests.

Amongst the deforestings effected within the last fifty years there is much which will be permanently profitable to the country. Deforesting is a conquest of man over nature; woods ought to disappear from the plains, and there to give place to cultivation. But, unhappily, we do not find in the valley alone ground furrowed by the plough, or lands furnishing pasturage and grass; they have plucked up the trees of sterile cantons, where wood alone should grow; they have imprudently given up to the axe the sides and the summits of our mountains; then the régime of the profitless pastures, freed from all surveillance, together with a vicious administration of public and private forests, have hindered the reproduction of wood after the felling; and the carelessness of the agents of the State in the communes have shut their eyes to the most destructive abuses. To-day the communes and the State possess thousands—millions of hectares of nominal forests, where there is just as much vegetation as there is in the steppes of Tartary, or in the desert of Sahara. The sowing ordered by the laws, or by the regulations, have been rendered illusory through the amount of the grants which were allotted to them, and a mockery through the bad faith which has too often presided over their execution. We are assured that oftener than once, and that I may say at a time not very remote from the present, the lessees of the fellings of the woods have sown sand instead of seed. About twenty years ago the evil came to a head; then the Administration established the Forest School at Nancy, which furnished workmen capable and active, and men of integrity. In 1837 the minister of finances proposed to stimulate the zeal of subaltern agents by an improved treatment, which placed them above misery, and protected them from temptation. All these improvements of the officials are doubtless to be commended, but they will be productive of little effect so long as there is not inserted in the budget a chapter in *support of replanting*. With a million devoted every year to sow and plant well-selected kinds of trees on the plots occupied by the forests, which would appear always to rebel against cultivation, the State would create in twenty or thirty years an immense capital, spread over the vast brows of the Pyrenees, of the Alps, and of the Vosges; as well as on the shores of the lands where they have applied, only on a Lilliputian scale, the ingenious and economic process of the *savant* Bremontier. In time of peace this would be an inexhaustable provision for twenty branches of industry, and notably for that in iron, which will never be wrought cheaply in France until wood shall be more abundant. In time of war this would be a resource of more ready avail than that of new taxes."

In the *Memoires de l'Academie des Sciences Morales et Politiques* for 1843, there appeared a *Memoire sur les Populations des Hautes Alpes*, by M. Blanqui, an eminent political economist, from which the following passage is cited and translated by the Hon. George P. Marsh, in his valuable work entitled *The Earth as Modified by Human Action*:—" I do not exaggerate," says Blanqui. "When I shall have finished my description and designated localities by their names, there will rise, I am sure, more than one voice from the spots themselves, to attest the rigorous exactness of this picture of their wretchedness. I have never seen its equal even in the Kabyle villages of the province of Constantine ; for there you can travel on horseback, and you find grass in the spring, whereas in more than fifty communes in the Alps there is absolutely nothing.

"The clear, brilliant, Alpine sky of Embrun, of Gap, of Barcelonette, and of Digne, which for months is without a cloud, produces droughts interrupted only by diluvial rains like those of the tropics. The abuse of the right of pasturage and the felling of the woods have stripped the soil of all its grass and all its trees, and the scorching sun bakes it to the consistency of porphyry. When moistened by the rain, as it has neither support nor cohesion, it rolls down to the valleys, sometimes in floods resembling black, yellow, or reddish lava, sometimes in streams of pebbles, and even huge blocks of stone, which pour down with a frightful roar, and in their swift course exhibit the most convulsive movements. If you overlook from an eminence one of these landscapes furrowed with so many ravines, it presents only images of desolation and of death. Vast deposits of flinty pebbles, many feet in thickness, which have rolled down and spread far over the plain, surround large trees, bury even their tops, and rise above them, leaving to the husbandman no longer a ray of hope. One can imagine no sadder spectacle than the deep fissures in the flanks of the mountains, which seem to have burst forth in eruption to cover the plains with their ruins. These gorges, under the influence of the sun which cracks and shivers to fragments the very rocks, and of the rain which sweeps them down, penetrate deeper and deeper into the heart of the mountain, while the beds of the torrents issuing from them are sometimes raised several feet in a single year, by the débris, so that that they reach the level of the bridges, which, of course, are then carried off. The torrent-beds are recognized at a great distance, as they issue from the mountains, and they spread themselves over the low grounds, in fan-shaped expansions, like a mantle of stone, sometimes ten thousand feet wide, rising high at the centre, and curving towards the circumference till their lower edges meet the plain.

"Such is their aspect in dry weather. But no tongue can give an adequate description of their devastations in one of those sudden floods which resemble, in almost none of their phenomena, the action of ordinary river-water. They are now no longer overflowing brooks, but real seas, tumbling down in cataracts, and rolling before them blocks of stone, which are hurled forward by the shock of the waves like balls shot out by the explosion of gunpowder. Sometimes ridges of pebbles are driven down when the transporting torrent does not rise high enough to show itself, and then the movement is accompanied with a roar louder than the crash of thunder. A furious wind precedes the rushing water and announces its approach. Then comes a violent eruption, followed by a flow of muddy waves, and after a few hours all returns to the dreary silence which at periods of rest marks these abodes of desolation.

"The elements of destruction are increasing in violence. The devastation advances in geometrical progression as the higher slopes are bared of their wood, and 'the ruin from above,' to use the words of a peasant, 'helps to hasten the desolation below.'

"The Alps of Provence present a terrible aspect. In the more equable climate of Northern France, one can form no conception of those parched mountain gorges where not even a bush can be found to shelter a bird—where, at most, the wanderer sees in summer here and there a withered lavender—where all the springs are dried up—and where a dead silence, hardly broken by even the hum of an insect, prevails. But if a storm bursts forth, masses of water suddenly shoot from the mountain heights into the shattered gulfs, waste without irrigating, deluge without refreshing the soil they overflow in their swift descent, and leave it even more seared than it was from want of moisture, Man at last retires from the fearful desert, and I have, the present season, found not a living soul in districts where I remember to have enjoyed hospitality thirty years ago."

And in another connection it is said by Mr Marsh,—" It deserves to be specially noticed that the district here referred to, though now among the most hopelessly waste in France, was very productive even down to so late a period as the commencement of the French Revolution. Arthur Young, writing in 1789, says,—' About Barcelonette, and in the highest parts of the mountains, the hill-pastures feed a million of sheep, besides large herds of other cattle ;' and he adds,—' With such a soil and in such a climate, we are not to suppose a country barren because it is mountainous. The valleys I have visited are, in general, beautiful.' He ascribes the same character to the provinces of Dauphiny, Provence, and Auvergne, and, though he visited, with the eye of an attentive and practised observer, many of the scenes since blasted with the wild desolation described by Blanqui, the Durance and a part of the course of the Loire are the only streams he mentions as inflicting serious injury by their floods. The ravages of the torrents had, indeed, as we have seen, commenced earlier in some other localities, but we are authorized to infer that they were, in Young's time, too limited in range, and relatively too insignificant to require notice in a general view of the provinces where they have now ruined so large a proportion of the soil."

But the voice of warning fell on deaf ears. It was like a voice crying in the wilderness—not the voice spoken of by the Hebrew seer, powerful as was that which had said,—" Let there be light," and which like it brought about its own accomplishment—but a voice crying in the wilderness, as that expression is generally understood.

Inundations in 1840, and others occurring in 1846, caused some attention to be given to the subject, and measures were about to be adopted, with a view to prevent the continued occurrence of such catastrophes, when the Revolution of 1848 took place, and forests were sacrificed right and left to provide funds required to meet the national expenditure of the day. But on the establishment of the empire the subject again commanded attention. And within the last twenty years several works, in this department of the literature of forest science, have followed each other in quick succession.

"In 1853, ten years after the date of Blanqui's memoir," says Marsh, "M. de Bonville, prefect of the Lower Alps, addressed to the Government a report in which the following passages occur :—

"'It is certain that the productive mould of the Alps, swept off by the increasing violence of that curse in the mountains, the torrents, is daily diminishing with fearful rapidity. All our Alps are wholly, or in large proportion, bared of wood. Their soil, scorched by the sun of Provence, cut up by the hoofs of the sheep, which, not finding on the surface the grass they require for their sustenance, gnaw and scratch the ground in search of roots to satisfy their hunger, is periodically washed and carried off by melting snows and summer storms.

"'I will not dwell on the effects of the torrents. For sixty years they have been too often depicted to require to be further discussed, but it is important to show that their ravages are daily extending the range of devastation. The bed of the Durance, which now in some places exceeds a mile and a quarter in width, and, at ordinary times, has a current of water less than eleven yards wide, shows something of the extent of the damage. Where, ten years ago, there were still woods and cultivated grounds to be seen, there is now but a vast torrent; there is not one of our mountains which has not at least one torrent, and new ones are daily forming.

"'An indirect proof of the diminution of the soil is to be found in the depopulation of the country. In 1852 I reported to the General Council that, according to the census of that year, the population of the department of the Lower Alps had fallen off no less than 5000 souls in the five years between 1846 and 1851.

"'Unless prompt and energetic measures are taken, it is easy to fix the epoch when the French Alps will be but a desert. The interval between 1851 and 1856 will show a further decrease of population. In 1862 the ministry will announce a continued and progressive reduction in the number of acres devoted to agriculture; every year will aggravate the evil, and in half a century France will count more ruins, and a department the less.'

"Time has verified the predictions of De Bonville. The later census returns show a progresssive diminution in the population of the departments of the Lower Alps, the Isère, Drome, Ariège, the Upper and the Lower Pyrenees, Lozère, the Ardennes, Doubs, the Vosges, and, in short, in all the provinces formerly remarkable for their forests. This diminution is not to be ascribed to a passion for foreign emigration, as in Ireland, and in parts of Germany and of Italy; it is simply a transfer of population from one part of the empire to another,—from soils which human folly has rendered uninhabitable, by ruthlessly depriving them of their natural advantages and securities, to provinces where the face of the earth was so formed by nature as to need no such safeguards, and where, consequently, she preserves her outlines in spite of the wasteful improvidence of man."

Mr Marsh adds in a foot note,—"Between 1851 and 1856 the population of Languedoc and Provence had increased by 101,000 souls. The augmentation, however, was wholly in the provinces of the plains, where all the principal cities are found. In these provinces the increase was 204,000, while in the mountain provinces there was a diminution of 103,000. The reduction of the area of arable land is perhaps even more striking. In 1842 the department of the Lower Alps possessed 99,000 hectares, or nearly 245,000 acres, of cultivated soil. In 1852 it had but 74,000 hectares. In other words, in ten years 25,000 hectares, or 61,000 acres, had been washed away, or rendered worthless for cultivation, by torrents and the abuses of pasturage.—CLAVÉ, *Études*, pp. 66, 67."

In the *Annales des Ponts et Chaussées* for 1854, is a paper by M. Belgrand, entitled *De l'Influence des Forêts sur l'écoulment des eaux pluviales*, cited by Mr Marsh as containing notices of remarkable floods occurring in different rivers in France. The Loire, above Rouen, has a basin of 2417 square miles, and in some of its inundations it has delivered 9500 cubic yards per second, which is 400 times its low-water discharge. And he gives a list of eight floods of the Seine, occurring within the last two centuries, in which it has delivered 3000 cubic yards per second, or 30 times its low-water discharge. Such is the vastness of the body of water rapidly poured into the rivers by torrents or storms of rain, by which torrents are occasioned.

In 1857 appeared *Études sur les Inondations, leur causes et leur effets* by M. F. Vallés, in which he makes several comments on the observations of Belgrand, relative to the rainfall in 1852 at Vozelay, in the valley of the Beuchat, and at Avallon, in the valley of the Grenetière. And in the *Annales Forestières*, for the December of the same year, appeared a paper, entitled *Les Inondations et le livre de M. Valles*, by A. F. D. Héricourt.

"The udometric measurement of Belgrand, discussed by Valles, constitute," says Marsh, "the earliest, and in some respects the most remarkable, series known to me of persevering and systematic observations bearing directly and exclusively upon the influence of human action upon climate, or, to speak more accurately, on precipitation and natural drainage. The conclusions of Belgrand, however, and of Valles, who adopts them, have not been generally accepted by the scientific world, and they seem to have been, in part at least, refuted by the arguments of Héricourt, and the observations of Cantegril, Jeandel, and Belland." These will be found quoted in *Comptes Rendus a l'Academie des Sciences, 1861.*

In 1856 appeared a pamphlet, published in Paris, entitled *Moyens de forcer les torrents des Montagnes de rendre une partie du sol qu'ils ravagent*, by *M. Rozet*, to which I shall afterwards have occasion to refer more in detail. And in the course of this year—whether before or after the appearance of M. Rozet's pamphlet I have not been able to ascertain—renewed inundations supplied a befitting opportunity for the Emperor to call the attention of the nation to the subject.

In the following year (1857) was published *La Provence au point de vue des Bois des Torrents, et des Inondations, par* Charles de Ribbe.

Some of the facts, historical and statistical, embodied in this work are embodied in a notice of Dauphiny and Provence, by Marsh, which will afterwards be cited.

In a work by Maurice Champion, entitled *Les Inondations en France depuis le VIme Siècle jusqûà nos jours*, a work in six volumes, published in Paris, 1858–1864, are narrated the ravages of many inundations which have devastated extensive districts. And in an erudite and able work by Alfred Maury, entitled *Les Forêts de la Gaule et de l'ancient France*, published in Paris in 1857, is collected an immense amount of statistical detail, on the extent, the distribution, and the destruction of the forests of France. By help of these the student in this department of Forest Science can carry back his studies to times that are past.

In 1858 appeared *Étude sur les Phenomenes et la Legislation des Eaux au point de vue des Inondations, par A. Monestier Lavignot*.

The same subject is discussed in a *Rapport sur les Plantations de la Solonge*, by M. A. Broignard, de l'Institut, which appeared in *Annales Forestières*, Tom. X.; and in a report to the Emperor by His Excellency, M. Magne, the Minister of Finance, relative to the planting of mountain ranges with trees, which appeared in the *Moniteur* of February 3, 1860. This will afterwards be given in full.

In this year (1860) was published a *Memoire sur les Inondations des Rivieres de l'Ardèche*, by M. de Mardigny; and in the year following (1861) appeared a pamphlet, published in Paris and Toulouse, entitled *Études sur le Reboisement des Montagnes*, par Paul Tray.

During the years which followed much information was collected through enquiries made by the Government, the substance of which was embodied in documents issued in connection with the legislation which was now employed to give effect to the suggestions which had been made, and the results were to some extent embodied in that legislation; and a good deal more was learned in connection with practical operations which were being carried on, which was embodied in reports of operations and reports of conferences held by appointment of the Administration by the officials and others employed in the work, which were published by the Administration.

Translations of most of these documents will afterwards be given. But it may be mentioned here that to meet public opinion it was deemed expedient, as the work advanced, to give more attention to *gazonnement* than was done in the commencement of the operations begun.

In the citations which have been made from works previously published, one section only of the literature of Forest Science—that relating specially, if not exclusively, to the influence of forests on torrents—has been laid under contribution. In regard to that I may say, in a word, that the French literature in this department of Forest Science is saturated with the idea that vegetation is the natural protection of the ground from the consequences of meteorological disturbances, occasioned by the destruction of forests by which a meteorological equilibrium, favourable to agricultural operations, had been established, and which may be re-established by the restoration of sylvan clothing to the mountains; and the same idea permeates much of the literature of France on subjects allied to that to which I have referred.

But, while primary importance was attached to *reboisement* and to *gazonnement*, mechanical appliances, such as Surell sought to combine, when necessary, with the extension of vegetation as a means of bridling, and stifling, and controlling torrents, did not fail to command the attention of those who were interested in the struggle, which was the more necessary that there are destructive torrents produced by the melting of snow, and the rapid melting of glaciers, or by *débâcles*, the breaking up of icy barriers confining waters, in situations in which *reboisement* and *gazonnement* are impracticable, and therefore as a remedy inapplicable; and there are other torrents of which the same thing, or something similar, may be alleged in regard to these appliances.

There is given by M. de Ladoucette an exposition of a scheme of embankment proposed by M. Delbergue-Cormant, *Ingénieur en chêf des Ponts et Chaussées*. The following is a translation of the memoir by M. Delbergue-Cormant, cited by him :—" There are two kinds of torrents, principal and secondary. The first are easily distinguished,—they always flow in the principal valleys; thus the Durance, the Guil, the Deux-Brüch, the Drac, &c., are principal torrents.

"The second descend from the lateral mountains of the valley, and come often at an angle more or less approaching 90°, to increase the principal torrent, which occupies the depth of the valley; it follows from this that the torrents of Sarrazin, of Boscodon, are secondary torrents. The means employed hitherto to control the principal torrents are to enclose them by banks faced with stones. I have shown in another memoir that one may obtain the same results more economically; but wishing to occupy myself at present only with secondary torrents, I confine myself to this.

"Before proposing means of preventing or of repairing the ravages which the secondary torrents make, it is necessary to know these torrents, and for this purpose to take them up at their birth, to examine them in their course, and in following them in the increase of their bed of deposit year by year to show the enormous extent of the damage which they may have occasioned. It is certain that a secondary torrent does but little or no evil so long as it is shut up between steep banks. It is when it leaves the lateral mountains to enter into the valley that it begins its ravages. Let us examine how this comes about.

"So long as the waters of a torrent are confined within steep banks they roll on in a great body, drawing on with them not only gravel, but even enormous rocks. Scarcely have they left the mountain, when, not being sustained and kept together by the banks, they divide themselves into a thousand little currents; and then, so far from drawing on rocks, they scarcely roll gravel along, and as their force diminishes more and more they scarcely bear along to the principal torrents some grains of sand. This explains perfectly the form taken by the deposits formed by secondary torrents. At the departure from the mountain this form is that of a portion of a cone, the summit of which corresponds to the point where the torrent comes out from the mountain. In effect, the waters, in quitting the mountain, have still an acquired force which permits them to roll the rocks on to some distance; in the second instance, this force being diminished, they deposit the rocks and carry forward only stones; in the third instance, their force being still further diminished, they abandon the stones and then carry on the gravel. Thus, then, is formed a first deposit, which will be less and less considerable in proportion to its distance from the mountain. In a second flood of the torrents the waters get freely away, and the deposits of sand and of gravel will increase less, always in this following a slope. In fine, the increase may become so inconsiderable that the sides of the cone recede from the mountain; then the torrent divides itself into two currents, and soon there comes to pass, at each of these two currents, what had occurred with the principal currents. Thus the fertile lands of the valley may disappear under the heaps of stone and of sand; as these torrents are greatly multiplied, there will come a day that their deposits, spreading out till becoming conjoined, a whole valley will become sterile, and will not be able longer to support its inhabitants.

"We have seen that the secondary torrents do not deposit the gravel and stones which they carried from the mountain; but when their waters are no longer confined by the banks—when they enter the valley—they spread themselves over a great surface, and thus lose their force; they cannot carry further the stones and the gravel, and these they abandon at a greater or less distance from the mountain. This indicates to us the course to be followed in order to control these torrents at their *embouchure*, and to prevent them covering the land with gravel.

"I would propose, then, in accordance with this principle:—First, to dig a bed for the torrent in the deposit which has penetrated to the exit from the mountain; second, to give little breadth to this bed, but great depth, in order that the waters may be there confined as they are in the natural bed which the torrent has dug for itself in the mountain, and that they may continue to sweep on the stones and gravel; third, to carry the gravel which is dug out from these cuttings to some distance from the edge, to form of them two embankments parallel to the new bed; fourth, to widen the entry of the new bed at the end towards the mountain, in order to collect the waters, and to strengthen by large stones these widened portions; fifth, to plant the embankments with willows, and other trees which grow quickly; sixth, to take care to clear away the obstructions which may form themselves in the new bed after each eruption of the torrent.

"One may see that there is no need of any building to confine these secondary torrents, and that the inhabitants of each village, with their shovels, their pick-axes, and some wheel-barrows, may secure the territory from the ravages. It is much to be desired, that being enlightened in regard to their true interest, they should lose at last that indifference which keeps them alike from preventing their ruin, and from repairing it.

"It may be observed that it is not necessary that the new bed be dug throughout all its length in a single campaign. It suffices to begin at the foot of the mountain, and to end off the open part in any year, by a more gentle declivity than that of the deposit of the ravine, to give an outlet to the waters. Thus the inhabitants would do wrong to excuse themselves by an alleged impossibility of doing all the work at one time. Further, neighbouring communities could mutually help one another.

"The advantages which the communes would derive from this work are considerable; for, not only would they not have to fear new invasions of the torrent, but the sides of the torrent, not being now exposed to the waters, might be usefully cultivated, by watering them with waters of the torrent which might be derived from the upper portion of it."

The scheme proposed resolves itself (he says) simply into digging for the torrent a straight canal through the centre of the deposit, and maintaining this canal by constant clearings. According to M. Ladoucette, whatever may be the precautions proposed by the author for strengthening the hills by means of plantations and cuttings like to continuous dykes, they will never present sufficient resistance to erosion; still less will they hold out against the undermining effects of the flood.

It is mentioned by M. Surell, that the clearing out of torrents is always a difficult operation, on account of the great size of the stones, and the hardness of the mud in which they are imbedded; and that this work, which demands great waste of muscle, and entails great expense, produces no durable result. The smallest flood suffices to overturn all, and to throw the bed of deposit into its previous disorder.

Something similar or analogous to the proposal of M. Cormant was carried into execution by M. de Ladoucette, who caused a trench to be cut in a straight line from the gorge of the Durance. He employed in this work the prisoners confined in the central house of Embrun, to the number of five hundred, and the work, prosecuted with energy, was completed in a month; but in the course of the next month there came a flood, and all was destroyed.

This scheme attributes all the ravages of torrents to the irregularity of their beds; and proposes, as a simple and sufficient remedy, to give to them a straight bed. Surell alleges that the scheme confounds cause and effect; and that torrents do not spread themselves hither and thither because they have not a straight bed; but they have not a straight bed because, continually depositing matter, they are forced to spread themselves hither and thither.

M. Cormant might justly claim to be allowed to say, in defence of his suggestion, that had the artificial bed been of a magnitude to contain the whole flood, as was evidently requisite, the success might have been complete.

In 1856 appeared, as has been already mentioned, the pamphlet of M. Rozet, entitled *Moyens de forcer les Torrents des Montagues de rendre une partie du sol quils ravagent*, to which reference has already been made. "He proposes," say Marsh, " to commence with the amphitheatres in which mountain torrents so often rise, by covering their slopes and filling their beds with loose blocks of rock, and by constructing at their outlets, and at other narrow points in the channels of the torrents, permeable barriers of the same material promiscuously heaped up, much according to the method employed by the ancient Romans in their northern provinces for a similar purpose. By this means, he supposes, the rapidity of the current would be checked, and the quantity of transported pebbles and gravel—which, by increasing the mechanical force of the water, greatly aggravate the damage by floods—much diminished. When the stream has reached that part of its course where it is bordered by soil capable of cultivation, and worth the expense of protection, he proposes to place along one or both banks, according to circumstances, a line of cubical blocks of stone or pillars of masonry three or four feet high and wide, and at the distance of about eleven yards from each other. The space between the two lines, or between a line and the opposite high bank, would, of course, be determined by observation of the width of the "swift-water current at high floods. As an auxiliary measure, small ditches and banks, or low walls of pebbles, should be constructed from the line of blocks across the grounds to be protected, nearly at right angles to the current, but slightly inclining downwards, and at convenient distances from each other. Rozet thinks the proper interval would be 300 yards, and it is evident that, if he is right in his main principle, hedges, rows of trees, or even common fences, would in many cases answer as good a purpose as banks and trenches or low walls. The blocks or pillars of stone would, he contends, check the lateral currents so as to compel them to let fall all their pebbles and gravel in the main channel—where they would be rolled along until ground down to sand or silt—and the transverse obstructions would detain the water upon the soil long enough to secure the deposit of its fertilizing slime. Numerous facts are cited in support of the author's views, and I imagine there are few residents of rural districts whose own observation will not furnish testimony confirmatory of their soundness."

He says,—"The plan of Rozet is recommended by its simplicity and cheapness as well as its facility and rapidity of execution, and is looked upon with favour by many persons very competent to judge in such matters. It is, however, by no means capable of universal application, though it would often doubtless prove highly useful in connection with the measures now employed in south-eastern France."

And he adds, in a foot-note,—"The effect of trees and other detached obstructions in checking the flow of water is particularly noticed by Palissy in his essay on *Waters and Fountains*, p. 173, edition of 1844. 'There be,' says he, ' in divers parts of France, and specially at Nantes, wooden bridges, where, to break the force of the waters and of the floating ice, which might endamage the piers of the said bridges, they have driven upright timbers into the bed of the rivers above the said piers, without the which they should abide but little. And in like wise, the trees which be planted along the mountains do much deaden the violence of the waters that flow from them.' Lombardini attaches great importance to the planting of rows of trees transversely to the current on grounds subject to overflow.—*Esame degli Studi sul Tevere*, § 53, and *Appendice*, §§ 33, 34."

In 1857 there appeared, in the *Annales des Ponts et Chaussées* and in the *Annales des Mines, Etudes sur les Torrents des Hautes Alpes*, by M. Scipion Gras, *ingénieur des mines*. Of this work the following analysis is given by M. Cézanne, in his supplement to the work of Surell :—" After having defined torrents the author divides their course into four parts—*bassin de réception, canal de réception, lit de déjection*, and *lit d'écoulement*. M. Scipion Gras distinguishes amongst torrents at the bed of deposit four classes, according to the character of the basins drained by them. The study of the laws in accordance with which solid bodies are swept away by floods leads him to the conclusion that there are two distinct modes of operation —*transport en masse* and *transport partiel*—the former effected by floods of great body and strength, the second by floods of a medium character. These different operations produce contrary effects upon the bed of deposit ; the great floods, as they exhaust themselves, deposit over this a layer of clay and gravel, over or through which the waters spread themselves in thin sheets ; the lesser floods, on the contrary, dig down into the bed of deposit and plough in it a channel for themselves, after having conveyed thither the more comminuted materials referred to."

Upon which M. Cézanne remarks that M. Gras does not occupy himself much with the basin drained by the torrent, the special subject of study by him being torrents the basins of which are not susceptible of being planted ; and impressed with the evils resulting from the dejection of detritus, he seeks to modify the natural advance of the bed of gravel, and discusses the two methods generally employed to effect this, characterized respectively by the employment of dykes and of *barrages*, or embankments and barriers. He expresses himself very decidedly in favour of barrages, and he thus sums up his opinion on the point :—" In short, the first proceeding *(l'endiguement)*, as a means of suppressing a bed of deposit, is often impracticable, or at least the success of it is dubious ; when it does succeed, it only carries the mischief elsewhere. It is, then, one which is *very defective, and which ought to be abandoned.*

" There remains the second course of procedure ; we have demonstrated its practicability, *its applicability to all torrents*, and its freedom from the drawbacks attaching to the first. And here begins the most important part of our task."

This quotation (says Cézanne) is characteristic of the method of procedure of M. Gras, which is pre-eminently systematic ; he observes natural phenomena with great accuracy and precision ; he then proceeds to distinguish, to classify, and finally to bring all his energy to bear upon a single and exclusive system. " But," says M. Cézanne, " the consideration of nature inspires one with a dislike and opposition to all systematic formulas. It may be well to run down embankments, and extol barriers ; but it is not less the case that there are circumstances in which the practical man will see at a glance that there embankments are better adapted to meet the case than are barriers. And it is necessary, in a study of this kind, to avoid all special pleading in support of the absolute superiority of one system or of another, and to confine remarks to showing clearly in what circumstances either of them should be preferred.

"M. Gras being so decidedly in favour of the exclusive use of *barrages*, or barriers, in reference to the two different categories of floods established by him, recommends, according as it may be desired to effect a complete or a partial retention of gravel, the construction of submergible barriers in the latter case, and insubmergible barriers in the former.

"The latter, insubmergible—so designated, although actually overflowed by the torrent, and expected and intended to be so at times, and it may be frequently—belong to a class of embankments which have been long in use. Erected in some favourable position in the gorges, they are designed to effect a deposit of gravel directed up the river. If the reservoir designed for this deposit be very considerable, if the transport be slow, it may tell effectively for some distance below for several years.

"Submergible barrages constitute, strictly speaking, the system which M. Gras claims the credit of originating. This system is based on a very delicate analysis of the effect of floods, which shows that high waters only acquire their full force in a narrow channel in which they are confined. If they be allowed or compelled to spread themselves out, their force is diminished, and the larger materials which were being borne along by them are deposited. To compel them to do this—to spread themselves out—it is only necessary to raise, on a widening of the bed, a horizontal sill, which cannot be washed away, worn down, or furrowed; the waters, then, not being able to concentrate themselves in any place at a lower depth than that of the whole sill, spread themselves in a sheet over the sill, and a deposit up the river follows as a consequence.

"After a great flood, such as may be of occasional but comparatively rare occurrence, floods of lesser magnitude, which are much more frequent, go over this deposit anew, and do on a lesser scale what the greater flood has done on a greater, excepting that such large blocks as could only be carried along by a great flood will remain in the places above the barrier in which they had been left. And the effect of the whole will be, that great floods will be less disastrous, the work done by them being effected by a great number of floods, the consequences of which are innocuous.

"It is not necessary that these *barrages* should be of great height, nor, consequently, of great solidity; it is enough that their upper surface sustain the friction of the pebbles carried down by the flood, and that their base can sustain the slight water-fall which they occasion.

"From this it may be seen," says M. Cézanne, "that the system of operations proposed by M. Scipion Gras is the very opposite of the course formerly followed, in so far as formerly, when a dam or barrier was to be erected, a narrow depth in the bed of the current was selected, that the structure being short there might be given to it, at little expense, the thickness necessary to enable it to resist the violent action of the water. He recommends to select expansions in the bed of the current, and even proposes to erect, on the cones of dejection, works of the same kind, which he calls *barrages radiers*. To secure the plain of *Bourg d'Oisans*, in the basin of the former lake *Saint Laurent*, against ravages by the *Romanche* and the *Vénéon*, which debouche each by a different gorge, he proposes the erection of such barriers, spread out horizontally, the length of which should be not less than 763 mètres, or 2500 feet, upon which he supposes that the two torrents, uniting their floods, will pour out the mass of water in a regular sheet, 32 centimètres, or 12 inches, in fall."

Of this locality it is mentioned elsewhere by M. Cézanne, that in 1157, after a storm of rain, two torrents of the Oisans, which look directly across from one bank to the other of the Romanche, the *Vandaine* and the *Infernay*, raised a barrier across the principal valley; a lake formed itself immediately behind this dam, which was known under the name of the *Lake Saint Laurant*, because the storm had burst on the day of St Laurant. This lake stood for sixty-eight years, but in the night between the 14th and 15th September 1219 the barrier gave way, the waters laid waste the lower parts of the valley, and two towns, Vozille and Grenoble, were almost entirely destroyed. Since the thirteenth century onwards there has often been a threatening of the formation again of this barrier, but in despite of this there has sprung up, in the dried basin of the Lake Laurant, the Bourg d'Oisans, which M. Gras proposed thus to protect.

M. Cézanne states in detail objections to which the measure was deemed by him to be open; and referring to two practical applications which had been made of the system proposed by M. Gras—one on the Roise, near Grenoble, the other on the Riou-Bourdoux, in the Lower Alps—he cites observations made by Professor Culmann, who visited the former some three years after the publication of the *memoire*, and reported of it thus:—"At the time of our visit (October 1860), we found that a strong *débâcle* had just passed over a *barrage*, and that a great mass of rubbish had been stopped behind the upper barrier. The little wooden bridge a little above it had evidently had too weak a channel, and it was carried away, and the barrier itself could not resist more. . . .

"It is clear that the work had maintained its resistance until the deposit above it had attained the top of its slope, and that so soon as blocks of even small size began to roll over the inclined plane the links of the binding chains, formed of iron bars ·02 métre, or four-fifths of an inch, in thickness, yielded to the shock and opened."

"Beyond this," says M. Cézanne, "M. Culmann criticises the mode of constructing rather than the theories of these barriers, but he does not appear to attribute to them other effect than to determine a deposit in the same way as does every other kind of barrier."

M. Cézanne visited La Roise in 1869, and he says,—"In point of fact, the bed of La Roise presents to a visitor the ordinary appearance of the bed of a torrent. The repaired barriers are surmounted by deposits, and the old state of things appears to be exactly reproduced at a higher level.

"According to M. Marechal, *Ingénieur des Ponts et Chaussées*, the experiment tried on the Riou-Bourdoux has not been more successful; the barrier has perished through defective or vicious construction."

M. Cézanne states, in concluding, that notwithstanding failures, which have followed a practical application of it, which have been made, in some of which the failure was attributable to unsatisfactory workmanship, engineers who have to do with torrents, but who have not had much personal experience in connection with torrential phenomena, will read with much profit the memoir by M. Gras; they will find a great many facts carefully noted, and will learn how to make observations themselves. And others who have written upon the subject go, I may mention, far beyond this in their commendations of the measure proposed by M. Gras.

In regard to *triage*, selection or successive deposit of materials of different bulk or gravity, a subject underlying the proposal of M. Gras, M. Cézanne writes,—" The *triage* of the matters borne along is very strongly marked in torrents which tend to extinction, or only, if the case be so, to take a *régime* of greater constancy. It happens even that the lesser stones, &c., being all borne along, there remain only the larger; the bed is then furnished with a self-created rockery, which energetically resists erosion, and as a necessary consequence the torrent cannot deepen the channel in which it flows. It is then necessary to give some assistance to the torrent, and the larger blocks are removed and ranged along the bank. The water re-collected between these rude embankments digs away anew. This system is much used in Switzerland. The course to be followed is this: replant with woods these parts of the basin in which this can be done; and when the torrent shows a tendency to cut a bed in the dejection, facilitate the process by removing the self-formed rockwork of blocks denuded by the *triage*."

In 1865 was published *Memoire sur les barrages de retenue des graviers dans les gorges des Torrents*, by M. Philippi Breton, *Ingenieur des ponts et Chaussées*. Of this M. Cézanne says,—" This treatise may be justly characterized a treatise on torrential geometry; the author demonstrates in it, with beautiful clearness and distinctness, the principal theories which relate to the transport of gravel,—to the profile or outline of the bed of deposit,—to the different kinds of cones thus formed, the troncature or section of which, and the reproduction of which, are explained by beautiful sketches taken from nature."

Of the design of the work, M. Breton writes,—"Different questions connected with the establishment of *barrages*, or barriers, for the retention of gravel, have been raised and discussed. But, notwithstanding all that has been done, it appears to me that ideas in regard to what results are to be expected from these *barrages* are still vague, varied, and undetermined; there is still a great want of decision in regard to selection of location, to the number of *barrages* to be employed, to the best or most suitable means of constructing them, and to the duration of their efficiency. After having reflected long on these subjects, I have come to be of opinion that, to preserve a plain from invasion by a torrent which debouches on it, it is necessary to establish, in the first place, a single *barrage*, situated at the outlet of the gorge, or very near to this; then a second *barrage* at some mètres [or yards], and not more, above the first, when that one shall cease to be efficacious; then a third at some mètres above the second, when this in its turn shall have completed the service it can render; and so on. Such is the subject of this memoir."

"From this it appears," says M. Cézanne, "that the proposal of M. Breton is the very opposite of that of M. Scipion Gras, submitted eight years ago; he speaks not of barriers but slightly raised above the level of the bed, or of silts stretching across expansions in that bed, but of solid massive walls, carried up as high as possible by successive stages into the throat of a gorge, and constructed, not of blocks bound to one another by chains, but of hydraulic masonry of the strongest that can be obtained.

"*Barrages* in which wood is employed to meet the want of cohesion in gravel, last (says he) but for a short time,—for the wood, buried half of its bulk in the gravel, often dry and often wet, will quickly rot, as quickly as do the *Cabrettes*, and more quickly than do the coffers known under the name of arks *(arches)* in the mountains of Dauphiny and Provence. *Barrages* constructed entirely of rockwork, and those constructed of dry stones, never cost much less, and they sometimes cost more, than those built with Roman cement, and these have a great advantage over the others in their greater cohesion. As soon as a breach occurs at any height in a *barrage* of rocks or dry stones, the violent current, passing through the breach, begins at once to enlarge it, and it soon effects a great destruction. In the hydraulic masonry any opening can only enlarge itself slowly, and the flood will have exhausted itself before the destruction has become serious.

"In saying what I have done I am only extending to barriers retaining gravel the practical rule adopted in the department of Isère for longitudinal dykes. M. Picol, and the engineers under his orders, have often remarked that a dry-stone dyke is rent from the bottom to the top when a small breach has been made in the foundation. Wishing to make these observations complete by comparison, they made the experiment of building with stones set in good hydraulic mortar. The experiment was not long in revealing—first, that the dykes so constructed did not cost much more than did those built of dry stones, as they could build with smaller material, and they did not require to give the same thickness to the wall; and then, what is of primary importance, that a wall built with good mortar can sustain a considerable destruction at its base without being instantly rent to the top, for the part above sustains itself in the condition of an arch or vault; and thus time is afforded for assistance."

After having discussed in detail the different questions which are connected with *barrages*, M. Breton thus meets an objection which is often brought up:—"I have frequently heard educated and intelligent men object against the system of retaining gravel by *barrages* the danger of a rupture in the works. When these works shall have amassed a great mass of gravel behind them, if a rupture should occur, that entire mass, so retained above its level, would, it is said, suddenly begin to move, and would produce a frightful catastrophe below. And as a proof in support of this fear they adduce the effects attributed to the sudden emptying of the Lake St Laurent, which, escaping from the plain of the Oisans, laid waste the valleys of the Romanche, and of the Drac, as far as to Grenoble. They might adduce, in like manner, the lamentable disasters produced in a single night by the rupture of the reservoir at Sheffield! But they forget that in these two cases, as in all others which may be cited in which the rupture of a reservoir has caused a sudden catastrophe at a lower level, the statement refers to a reservoir of water, and not to a reservoir of sand, and earth, and gravel.

"It is thus that I have no dread of this objection, if the work be judged of only by builders accustomed to see the movements of water, and of sand and gravel, and know the difference between them; never will an engineer bring himself to believe that gravel will flow as does water."

Numerous cases illustrative of the effects of the rupture of a *barrage* are then given. But M. Breton, while writing thus, is not unmindful of the importance of the *boisement* or *gazonnement* of the basin drained by the torrent. He admits distinctly that it is vegetation which has the power to extinguish torrents; he only proposes *barrages* as a temporary expedient against torrents which cannot be prevented, as are sometimes those connected with glaciers, or as temporary appliances where, through the strength of prejudice or legal difficulties, the forest treatment must be for a considerable time postponed.

In the same year, 1865, there was published in Lausanne, a *Rapport au Conseil Fédéral sur les Torrents des Alpes Suisses, inspectés en 1858, 1859, 1860,* and *1863, par M. le Professeur Culmann.* Of this work M. Cézanne writes,—" Switzerland is a land privileged indeed ; the philosopher, the artist, the humble foot-soldier—in a word, every one, whatever may be the tendency of his mind, finds there numerous subjects of study. By hundreds of thousands, tourists, from both worlds, annually visit this classic land of noble landscapes, of natural science, and of freedom." He mentions in a foot note, that he was informed, by the monks of the great Saint Bernard, that they lodge upon an average 40,000 visitors annually, and sometimes 800 in a single day at the height of the season. And he goes on to say,— " Looked at from the point of view of our study, Switzerland is seen to be a protuberance, like a boss on a shield, which rises above the lofty plateaux of Europe; it is a reservoir, whence water is distributed; it is also a laboratory, whence issue many thousand torrents—working away, in combination with the glaciers, to level down the rough and rugged back of our planet. All of these waters, flowing from the eternal snows, precipitate themselves in cascades to the depths of the valleys below ; they keep on, ever sowing anew with their alluvial deposits the basins of a hundred score of lakes ; thence, partially clarified, they escape towards the four points of the compass to throw themselves into four seas, after having watered Germany by the Danube and the Rhine, France by the Rhine, and Italy by the Po and the Adige.

"The engineers of this country, brought up within the sound of the torrents, and accustomed from infancy to the thousand caprices of the mountain streams, quickly acquire a special experience in this matter. They are little given to generalizations, to systematic theories, to geometrical definitions; they give themselves more to the study of particulars, and seek out for each case a special solution adapted to the local circumstances. And such is the character of the work of M. Culmann.

"In 1856, the rainfall which devastated France did not spare Switzerland ; the Federal Council bestirred itself and commissioned M. Culmann, one of the most distinguished students of hydraulics, to go through the whole of the cantons, and to report, in regard to each torrent, on the evil and the remedy. And at the same time, to meet the public demand, which attributed justly to the destruction of forests the ravages of the torrents, a commission was organized and appointed to report at the same time in regard to the forests. The two reports have been published in German and in French. They agree on the conservation effected by forests. That of M. Culmann relates more especially to those water-courses connected with which the mechanical appliance of the engineer is required to come to the aid of *reboisement.*

"The report of M. Culmann passes in review many hundreds of torrents ; it is a repertory of isolated facts, well observed, calmly stated, with simple demonstrative sketches." . . .

With regard to *boisement* and *gazonnement,* he says,—" In Switzerland, as elsewhere, the evils produced by torrents is not a necessary evil ; it takes birth often from the waste and recklessness of the inhabitants. The principal remedy, and the only one which is decisive and definite, is the *boisement* or *gazonnement,* which stifles the evil at its source, *principiis obsta.* The cantons which have given attention to their forests have been least attacked ; those which have devastated them—in particular the Italian cantons—are threatened, as are the HighAlps of France, with complete ruin." And he gives a great many examples of cases in which, in consequence of the grubbing up of a wooded place, a torrent, which till then had been inoffensive, became all at once dangerous.

M. Culmann attaches greater importance to the initiative being taken by the people, than to interventions by the Government; and he cites facts in support of his opinion. But he attaches, I may say, primary importance to securing connected action by all interested; of the advantages of which, and the disadvantages arising out of the want of it, he gives facts in illustration.

Proceeding to the consideration of mechanical works of engineering, he recommends—(1) The clearing of the water-courses of all large blocks resting there, in which he is supported by M. Cézanne; (2) The erection, when and where it may be expedient, of *barrages* and dykes or embankments.

"In regard to such works, the theory of M. Culmann," says M. Cézanne, "may be stated thus :—*Barrages* are but a temporary expedient to be employed while awaiting *reboisement*; it is necessary to construct them in a series, commencing from below; when the first barrier is filled to the level with gravel, then should be constructed a second behind it, and so on continuously."

Barrages are the preferable structures to erect against torrents, dykes or embankments against rivers. With these, as with *barrages*, it is necessary to proceed from below upwards.

Barrages constructed of facines or of wood, &c., when but a temporary effect is to be produced, are often preferable to those of stones or of masonry, because they accommodate themselves to movements in the bed. Structures of facines form a moveable enclosure, on which vegetation easily establishes itself; *barrages*, constructed with hurdles, are very useful in ravines, and even on sinking slopes; but in general stone-works are preferable to those constructed of wood; these, however, are very serviceable in cantons in which the population employ them. "All the *barrages* in torrents constructed hitherto in Switzerland," says he, "have proved beneficial; all the people who have made use of them have showed themselves satisfied with them."

Epis, or stakes, avail nothing against torrents, or against mountain rivers; they are available only against peaceful rivers, bearing along but little solid matter suspended in their waters, and this composed only of sand and small gravel, and not of blocks, which are able to attack the bank and change the direction of the liquid stream. "The stockade of these," M. Culmann says, "should form a continuous line; for if some do not reach to this, and others go beyond it, the current, thrown from one bank to the other, may do greater damage than if there had been no works of enclosure. The *epis*, or stakes, should be sufficiently close to the bank to prevent any loop or expansion of water being formed between them; and the less the banks the closer should they be. In mountain banks the space is so limited that continuous dykes are less costly.

"In Bavaria, for example, the lower Danube is too small to allow of the system of stockade being applied; they are under the necessity of adding a more or less extended wing to the back of each spike—that is to say, to construct immediately a portion of the future bank. They have also abandoned the system of spikes along the Rhine, where they were greatly attached to it, and where they now construct continuous embankments of facines. Stakes can no longer be employed along the Lech; and since the alteration of the course of the *Linth*, they have become satisfied that parallel dykes are much better suited for the enclosure of this small water-course. And on all the lower channel of the Linth, they have little by little replaced with these the stakes wherever these were not absolutely necessary."

M. Culmann then reports in detail in regard to the location and construction of bridges; and on the phenomena and effects of glaciers, torrents, avalanches, and landslips.

The former subject is of local importance; and the information communicated may be utilized, to some extent, by any employed in making surveys for roads and bridges; but in every case local circumstances have such an effect in determining operations that it is deemed unnecessary to cite the views advanced. With regard to glaciers, torrents, avalanches, and landslips, the case may seem to be similar. But avalanches, at least, are not confined to Alpine regions; and though woods may prevent the formation of a landslip, they cannot arrest its progress when once in motion. There is not a little in the graphic details of engineering operations given by M. Culmann in this chapter of his work which commands attention and illustrates the importance of the work.

"Torrents issuing from glaciers," says M. Cézanne, "are numerous in Switzerland; they are subject to formidable *débâcles*, or outbursts of water, when the glacier in its movements of going and coming, after having dammed up some secondary valley, gives free passage all at once to its waters. To prevent such evils is for the engineer a formidable undertaking, and a difficult problem. How contend against a glacier? What physical force can he bring against the mass which is being unceasingly renewed by the ever recurring winters, and which, making use of the hardest rocks, transporting blocks of stupendous size by a movement almost imperceptible, would annihilate the most irresistible work of man? Here are two cases reported in which a simple idea sufficed to vanquish the inert Colossus:

"The glacier d'Aletsh, an affluent of the Rhine (Valais), dammed up a small lateral valley, situated behind the Eggishchorn, and created thus the lake of Mærjelen. 'This lake,' says M. Agassiz, in his *Études sur les Glaciers*, 'was formerly more extensive than it is now; and when it happened that the melting of the snow and ice became excessive, it would often happen that the whole of this body of water would with violence eat away an outlet under the glacier, and occasion the greatest destructive ravages in the bottom of the valley. To obviate this they dug, in the direction of the glacier of Viesch, an artificial channel to this lake, which could no longer rise above the level of its orifice. The ice did not rest immediately on the water; there was, on the contrary, between the bottom of the glacier and the surface of the water a space of some centimètres, perhaps an inch or two, occasioned by the temperature of the lake being always during summer higher than that of the glacier. By means of this space, enormous blocks of ice often detach themselves and float on the surface of the lake, imitating exactly the floating icebergs of northern regions.'

"But the most characteristic example is that furnished by the glacier of Giétroz—the assault made against which is somewhat dramatic, and exceedingly interesting:

"At the bottom of the valley of Bagnes, one of the branches of the Drause, at sixteen kilomètres, or about twelve miles from Chables, there rises vertically a high wall of rocks, surmounted by the glacier of Giétroz. The moving mass protrudes itself, projects beyond the support, and falls at the foot of the precipice; the broken fragments congeal anew and form a cone-shaped glacier, which pushes before it its moraine. What ensues must be given in the narrative of *Guide Joanne*:—' In those years in which avalanches are very frequent the heat of summer does not suffice to melt a quantity of ice equal to what the mountains cast down. The enormous block which then forms a bridge on the Drause becomes always larger and larger, and as the arch of this bridge, dug in summer by the torrent, closes up in winter, it happened in 1597, and in our own times, in 1818, that the early months of spring sufficed not for the Drause to open for itself a passage, and a lake was formed behind the ice.

"'When this became known (wrote M. Simond, some months after the event), alarm spread at once, not only throughout the whole valley but in Le Valais, and on so far as Italy. Travellers feared to take the route of the Simplon; it was felt that when this dyke should come to break up there would be there a sudden débâcle which would sweep over the country to a great distance. The preceding winter had been severe; the ice had even then cast a dam across the valley, but without stopping the water, which had eaten out a passage for itself; but a second severe winter had produced such a fall of ice that the obstacle had become insurmountable and impervious.

"'The Government sent an engineer (M. Venetz); he found that the dyke was 110 toises (nearly 700 feet) in length from the one mountain to the other, 66 toises (or about 400 feet) in height, and 500 (or 3000 feet) in thickness at its base. The lake was 1200 toises (or upwards of 7000 feet) in length, and had already risen to half the height of the dyke, that is to say, was from 30 to 40 toises (from 180 to 240 feet) in depth. The engineer determined to cut a gallery or tunnel through the thickness of the ice, beginning 54 feet above the actual level of the lake, to give time to complete the work before that height should be reached by the accumulating waters, which were rising at the rate of from 1 to 5 feet per day, according to the temperature; and he began the work on the 11th of May at both ends of the tunnel. Fifty men in relays, relieving one another alternately, wrought there night and day at the peril of their lives,—one and another of the avalanches which were falling every moment threatening to bury them alive in the tunnel; many were wounded by lumps of ice, or had their feet frozen, and the ice was so hard that it frequently broke the pick-axes used. In despite of all these difficulties the work advanced rapidly. On the 27th of May a great piece of the dyke broke off from the bottom with a fearful crash; it was believed that the whole was about to break up, or to rise in a mass, and the workmen fled; but soon they courageously resumed their work. Similar accidents occurred repeatedly; some of the floating masses, calculating from the distance at which they stood above water, must have had a thickness of 70 feet submerged. On the 4th June the tunnel, 608 feet long, was cut from end to end; but as it had an elevation of 20 feet or more in the centre it was necessary to level it. The weather had been cold, and the lake had not yet risen to the level of the mouth of the tunnel, so they continued to lower this till the 13th, the day on which the flow commenced, at ten o'clock at night. The lake still rose for some hours; but next day at five o'clock in the afternoon it had sunk 1 foot; on the morning of the 15th, 10 feet; on the morning of the 16th, 30 feet; at two o'clock that day the length of the lake had shrunk 325 toises (nearly 2000 feet), for the tunnel, being continually eaten away, lowered itself as quickly as the lake. The Drause flowed, filled from bank to bank, but without overflowing, and a few days more would have sufficed to empty the immense reservoir.

"' But detonations in the interior of the dyke announced that *glaçions*, blocks and pillars of ice, were detaching themselves from the mass, through their low specific gravity, and were thus diminishing the thickness of the dyke on the side towards the lake, while the current out of the tunnel was eating away this dyke on the outer side, and was threatening a sudden rupture; the danger increasing, the engineer despatched from time to time expresses to warn the inhabitants to keep themselves on the out-look. The water began to make way under the ice, sweeping along the stones and earth at its base under the tunnel; the crisis appeared inevitable and close at hand. At half-past four o'clock in the afternoon a tremendous crash announced the rupture of the ice-work; the water of the lake shot along with fury indescribable; it formed a torrent 100 feet in height, which traversed the first 6 leagues, or 18 miles, in forty minutes, although kept back in many places by narrow gorges through which it had to pass, carrying off in its course 130 chalets or cottages, a whole forest, and an immense quantity of earth and of stones. Debouching over against Chables, the chief place of the valley, the water was seen pushing before it a moving mountain of all kinds of debris of 300 feet in height, from which was rising a thick black cloud like the smoke of a conflagration. An English traveller, Mr P., of Lausanne, accompanied by a young artist, and a guide, was returning from seeing the works, and going towards Chables; happening by chance to turn round, he saw advancing with fearful rapidity the moving column, the distant roar of which he had not heard through the noise made by the Drause. He hastily warned his two companions and three other travellers who had joined them; all leapt from their mules, scrambled up the mountain, and got safely beyond the sweep of the deluge, which filled in an instant the whole gorge beneath them. But Mr P. was nowhere to be seen; for some hours they believed him to be lost; but then they learned that his mule, shying at an overturned tree which she saw on the road, wheeling round, saw all at once an object far more dreadful close upon her, and, darting off towards the mountain, had carried him far away from the scene of danger.

"' From Chables the débâcle arrived at Martigny—4 leagues, or 12 miles, distant—in 50 minutes, carrying off, as it advanced, 35 houses, 8 mills, 95 barns, but only 9 people, and no cattle, the inhabitants having all been warned to be on guard. The village of Bovernier was saved by a jutting rock turning off the flow of the torrent; and the people saw it pass like a shot by the side of the village without touching it, although much higher than their heads. The rocks and stones were dropped before it arrived at Martigny, blasting with sterility extensive meadows and fertile fields.

"' There it divided, but 800 of the houses of this town were carried away, many others were damaged, and the streets were strewn with trees and earthen debris; 34 people only appear to have lost their lives there, the inhabitants having betaken themselves to the mountains.

"' Below Martigny the débâcle, finding a great plain, spread itself out and deposited a great deal of mud and wood, and that to such an extent as to render healthful, as was hoped, a great marsh there. The Rhone received it little by little, and at different parts of its course, without overflowing; it reached the lake of Geneva at eleven o'clock at night, and was lost in the great extent of that lake,—having traversed a course of 18 leagues, or upwards of 50 miles, through Switzerland, in six hours and a-half, by a movement gradually retarded.

"'All the bridges having been carried away, the inhabitants on the two sides of the Drause could have no communication for some days, or inform one another of their respective losses, but by throwing across the river notes attached to stones; and the putrifying slime threatened them with an epidemic. It is somewhat remarkable that an old man of ninety-two saved himself by getting on a hillock supposed to have been formed by a débâcle in ancient times; the new one followed him to the very summit, where he maintained his footing by the aid of a tree which was not carried away.

"'M. Escher estimated at *eight hundred millions* of cubic feet the mass of water which had accumulated at the time it began to flow out by the tunnel. This mass had been reduced to *five hundred and thirty millions* in the course of the three days following, and the level of the lake was lowered by 45 feet. If the tunnel had not been made the lake would have risen 50 feet higher, and the mass of water would have attained a measurement of *seventeen hundred and fifty millions* of cubic feet when it began to flow over the dyke, instead of the *five hundred and thirty millions* to which it had been reduced when it began to pass across the tunnel, and would have spread its ravages over the whole of the lower Valais.'"

M. Culmann goes on to say,—"When, in the course of the winter of 1821-22, the dyke of ice threatened to form again, and had already covered about 400 mètres, or upwards of 1300 feet, of the bed of the Drause, M. Venetz undertook to destroy this mass of ice, the face of which measured 22,300 square mètres.

"He succeeded completely in doing this by the help of wooden aqueducts, leading on to the glacier streamlets of water from the mountain Alia, heated in some measure by passing over the rock; by these means were made great gashes, which detached blocks of 800 and 1000 cubic feet in measurement. In falling down, these broke in pieces, which were carried away by the Drause.

"After having destroyed the cone of the glacier, from 1822 to 1824, M. Venetz undertook precautionary works to prevent the blocks of ice precipitating themselves anew to the depth of the valley. He constructed simple barriers across the valley over against the glacier. The summits of these are perfectly straight and horizontal; they produce thus so great a lateral extension of the surface of the water, that the ice-work cannot make a vault across it. The blocks of ice fall, then, always into the waters, remain constantly in contact with this, and melt away by degrees. Thus the stream can never be covered up, and the blocks of ice cannot precipitate themselves further. From the time that the cone overhangs, by 2 or 3 mètres, 7 or 10 feet, the stream which has dug away its base, the portion in front detaches itself, and is borne away. These sometimes fall beyond the stream, and form a small glacier at the side of the moraine, on the left bank. And these masses maintain their ground sometimes for a pretty long time, but they can never cover up completely the stream.

"From that time onward—that is to say, from 1826—these *barrages* have sufficed to prevent the ice-work from covering up the Drause, and thus damming up the valley.

"In acknowledging the great merit of these works, we may express the wish that the engineers of this canton could be enabled always to avail themselves of the means necessary to maintain such useful structures, so as to

I

erect similar ones in other valleys—amongst others, in those of Saas and of the Massa."

These extracts, from the report of M. Culmann, will suffice to show how interesting this report is for engineers who have to do with torrents.

As yet comparatively little had been done to carry out *reboisement* and *gazonnement* in France. M. Culmann visited the High Alps, having had his attention directed to several of the works published in regard to the torrents of that region, and the remedies proposed by Fabre and Surell, and advocated by others. And he thus reports on what he saw,—" Our expectations were disappointed. One torrent alone had been subdued, and that not one of the most formidable of them; it was not in the basin of the Durance, so cut up with ravines, but in the comparatively peaceful one of the Isère. In what is, properly speaking, the domain of the torrents, they have made an experiment in *reboisement* by a plantation of pines, of some thirty or forty acres in extent, in the *bassin de réception* of the formidable torrent of Chorges.

"These, and some few others, on the smaller mountain banks, are the only practical results which all the studies of the engineers have produced since the close of the last century. . . In no country is the *Administration des Ponts et Chaussées* so centralized and so well organized as in France; but in whatever direction we look we are saddened by the painful impression that a state of things far superior, previously existing, has been brought to nought. It may be asked, perhaps—Why then devote so much time to it? And what has the condition of a foreign land to do with Switzerland?

"We were convinced that our general description of torrents could not be closed more advantageously than in showing how a country has, little by little, been brought to a state of ruins, when its population did nothing to maintain it—did nothing but consume the products of the soil, and sought not by any natural or artificial process to repair their losses, or to preserve its power of production.

"Let this state of things be considered by us while it is not yet too late; and let no one reply,—We shall never sink so low as that; if the country be more and more neglected—if its condition be allowed to go on becoming worse for an indefinite length of time—it will end, as will also its population, in differing so little from what we have just described that their conditions will be identical."

M. Cézanne remarks on this,—"It is humiliating to meet with such a testimony in an official document, published in two languages, by a foreign Government, and spread over the whole of Europe. It is a canton of our own France which has thus been pointed out to all as an example of the evils to which inertness of administration may lead. It is in vain that eloquent appeals have been made since the commencement of the century; nothing has been done, and the ruins of the valley of the Béouse, in Devoluy, described by Surell, are still there to supply a subject for heart-rending pictures."

If these severe observations be now no longer true—if anything has been done—thanks to the law of 28th July 1860 on the *reboisement* of the mountains, and above all to that of the 8th June 1864 on *gazonnement*. The last, I may state, was passed in the year following the completion of M. Culmann's tours of inspection; and translations of the text of both laws, with documents connected with them, will be given in a subsequent Part of this compilation.

I found in a paper which appeared in *Revue des Eaux et Forêts*, for

April 1866, the following striking illustration of the effect of woods on torrents :—" The State possesses, in the department of Vaucluse (says the forest conservator, Labuissière), a forest of more than 3000 hectares, situated on the portion of the mountain Luberon nearest to the valley of the Durance. This region is very much cut up, and traversed in all directions by very narrow and deeply embanked ravines in the midst of masses more or less dense of Aleppo pines and green oaks.

" These ravines are almost the only outlets for the transport of wood, in consequence of the difficulties which would be encountered, and the expense which would be incurred, in making more practicable ones on the rapid declivities, strewn with enormous masses of rock. There exists one so situated, called the Ravine de Saint-Phalez. The direction is from north to south, in the midst of a mass of Aleppo pines in a state of growth more or less compact.

" Its length, and for four kilomètres, or from the road from Cavaillon to Pertuis, to the domain of Saint-Phalez, of an area of about 50 hectares, forms the *bassin de réception* of the torrent.

" This land is well cultivated ; there are no declivities too steep for cultivation ; it comprises vineyards, meadows, and arable land ; the soil is argillaceous.

" The ravine of Saint-Phalez receives many affluents, the most important of which is that of the Combe-d'Yeuse, which joins it near the summit, where are some hundred mètres of the cultivated grounds of which I have spoken.

" The ravine de la Combe d'Yeuse is of much less considerable length than that of Saint-Phalez ; it is scarcely two kilomètres. It is strongly embanked, surmounted by steep declivities, covered with green oaks of eight or ten years' growth, and with Aleppo pines of different ages. Its *bassin de réception*, of about 250 hectares, or 113 acres, comprises the whole slope, precipitately inclined, with a general south-west aspect ; it is closed at the top by a deep bed of rock cut into peaks of the most imposing aspect.

" The geological formation is absolutely the same, as are all the other conditions, at all the points which I have examined. In no part is to be seen either spring or appearance of humidity ; no water is seen excepting at the time of the storms or great rains, and this water soon passes away, with the differences which will afterwards be mentioned. At all other times these ravines are of a desolating aridity.

" In the night of the 2d and 3d of September 1864 there fell a rather abundant rain over all this portion of the mountain. In the morning the argillaceous grounds of Saint-Phalez were saturated, of which evidence was found by anyone attempting to cross them. The ravine of Saint-Phalez, the receptacle of the surplus water, had flowed but slightly ; that of the Combe-d'Yeuse remained dry.

" The day of the 4th September was very warm ; a water-spout borne along by a south-west wind struck on the Luberon. Its passage did not last more than forty minutes ; but scarcely had it come when the torrent of Saint-Phalez became awful. Its maximum deliverance was about two cubic mètres. It did not flow more than fifty minutes, but with an average delivery of half a cubic mètre ; it had then passed in all 15,000 cubic mètres of water. Its height had been 0·04 m. ; each square mètre had received 40 litres, and the 50 hectares of Saint-Phalez 20,000 cubic mètres. The ground had only retained 5000, which is sufficiently explained by their argillaceous character and their state of saturation the night before. While the torrent of Saint-Phalez flowed, filled from bank to bank, seizing and carrying off rocks which had been employed to form a road which was believed to be safe against all contingencies, that of the Combe-d'Yeuse and all those traversing wooded lands remained dry, or gave only an insignificant quantity of water.

"On the slope opposite to that of which I have been speaking, in the valley of the Peyne, a carriage-road newly formed did not experience the least injury throughout the whole of the portion of it passing through the forest of the domain; but at its issue, on the lands of the Libaude and of the Roquette, it had been, so to say, destroyed. A cart loaded with faggots was upset and smashed by the waters, which flowed from all the cultivated slopes, and tore along, with the noise of thunder, at the bottom of the ravine.

"My good fortune secured to me another subject of study on the same ground.

"On the 25th October following I went to the sale of the fellings of the Tarascon, where there fell an abundant rain. The next day (the 26th) the weather was clouded. I set off for the Luberon in the hope of arriving there at the same time as would a storm of rain, which I saw approaching. I arrived first; the ravine of Saint-Phalez was still moist, from the passage in small quantity of the waters of the night before; they had served, as appeared, to saturate the lands of the domain, as had previously happened on the 7th [3rd?] September.

"I had scarcely gone over two kilométres in the ravine when the water began to rush with great violence; ten minutes later it precipitated itself in its ordinary *canal d'ecoulement*, completing the work of destruction begun in the month of September. The lands of Saint-Phalez had then absorbed but little or none of the water that day.

"The storm was not of long duration—an hour at most. The time was unfavourable for collecting on the ground exact measurements, but I reckon that the torrent delivered, at its maximum, somewhat less water, perhaps, than on the 4th of September. The flood, however, was more frightful; it swept away rocks with so much the greater ease that nothing had been repaired since the first storm, which had left the stones dug out, and without bond of cohesion among themselves.

"To gain the forester's house, which was on the slope of the left bank, it was necessary to make a long circuit—to go round the domain of Saint-Phalez, and to cross the grounds belonging to it, in which one sank to the depth of 0·30 mètres, or 12 inches. Before arriving at my home I had still before me the ravine of the Combe-d'Yeuse, and I feared I should be stopped there by a new obstacle. I was agreeably surprised to find it dry. An hour after the storm the ravine of Saint-Phalez had ceased to flow.

"It rained throughout the whole of the 28th without there being anything to remark similar to what had happened on the preceding days. The only effect of this was that, on the evening of the 30th, near the forester's house, and at 200 or 300 mètres from the ravine of Saint-Phalez, there was seen coming down, in that of Yeuse, a small fillet of clear water; its volume increased perceptibly during three days, to diminish in like manner during the two which followed; its passage broke down a little of the foot-path which goes along the valley, but caused only a damage easily repaired. This foot-path did not present the same solidity of structure as that of the Combe de Saint-Phalez, built on enormous blocks of rock which had stood for several years, and which had allowed of passage with a waggon some days before its destruction by the storm in September. If the Combe-d'Yeuse had yielded as much water as that of Phalez, and if these two masses of water had come at the same time, the damage caused in the plain would have been considerable, and the Durance, which received these waters, would have been so much the larger.

"Thus we have two torrents very near and under the same conditions—except that the basin drained by the one comprises 50 hectars of cultivated lands, that of the other 250 hectares of woodlands. The first receives, and allows to flow away, the waters of the greater part of a storm in a few hours at most, causing thereby considerable damage; the second, which had received a greater quantity of rain, stores it—keeps it for two days—evidently retaining a portion of it, and takes three or four days to yield up the surplus, which it does in the form of a limpid and inoffensive stream.

"The day on which I took the notes which I have copied in part I did not think they would ever be of use to me; what I had seen taught me nothing new; my old convictions had been simply confirmed, and I remained anew persuaded that it was imprudent to deny what the inhabitants of the country—better observers and more clear-sighted, when acting for their own interests or their own property, than is generally supposed—have long affirmed, supporting their allegations on abstract theories. What I had just seen in the Combe-d'Yeuse, however, had roused my curiosity. I wished to know if it had been remarked before, and that invariably. With this view I held a kind of inquest; I interrogated an old warden and woodman, and wood-merchants of the country.

"I will tell in a few words what I learned from them.

"Before 1840 the fellings of the Luberon were sold with power to bud the Aleppo pine; the prescribed period of exploitation was ten years.

"From 1823 to 1833 the whole of the Combe-d'Yeuse was exploited. The growth was composed principally of green oaks; the Aleppo pine was only found in clumps, often very sparse of trees, scattered over the whole surface.

"In 1829, the year of the building of the forester's house, the Combe-d'Yeuse yielded such a great quantity of water that enormous trunks of pine lying in the ravine, or on the slopes, were carried away by the torrent. The mason who built the house has confirmed to me the correctness of this last statement, telling me that the day after the storm the purchasers ran over the plain below the road from Cavillon to Pertuis to seek out their timber, scattered about and half-buried in the ground. It is probable that at that day the basins of reception of Saint-Phalez and of Yeuse being in pretty much the same conditions, the waters arrived at the same time in the *canal d'écoulement*. It is easy to conceive what damage they occasioned to the rich cultivated lands on the banks of the Durance.

"From 1829 to 1840 the Combe-d'Yeuse only twice yielded a little water.

"At this time the woods were on an average of twelve years' growth; the green oak-lopped and well-exploited had sprouted again with vigour, and were covering the soil, to which they already gave protection. It must not be forgotten that it was in 1840 that there occurred the great inundation of the Rhone, which drove the sheep from the Crau-d'Arles; and the forest of Luberon afforded them shelter, which saved them from certain death.

"Some time before there had been made a *barrage*—barrier, or wear—at the outlet of the Combe-d'Yeuse where the passage is straitest; it stopped a considerable quantity of water, estimated at 24,000 cubic mètres by M. Cournaud, land surveyor; the warders bathed in it, and it was of such a depth that they could swim in it. Unhappily, this water could not be retained, in consequence of so many fissures in the rock and the rapid percolation through the soil; and they broke down this *barrage*, which had been built with such great hopes, to facilitate the bringing out the timber of the last exploitation in 1856.

"In 1843 there was still water in the Combe-d'Yeuse; there was no more seen until 1856, but from this time onward the water flowed almost every year only in small quantities. But the year 1862 must be excepted; although much less rain fell than in 1864, it delivered much more water into the ravine; but this it did without causing damage worth speaking of.

"The comparison of all these dates seems to me to supply valuable instruction. If the woods be young the ravine flows every year, often causing thereby considerable damage. As their age augments it flows at intervals more and more remote, and ends in being almost completely extinguished.

"These conclusions will not astonish foresters who have been accustomed to the exploitations of copse wood. For, slight as may be the slope of the ground with a light soil, the annual fellings are cut into ravines by a single storm, whilst nothing like this is to be seen in the felling at its side, which has grown for 20 or 25 years, according as the one or other of these rotations of the fellings has been adopted."

In 1849 there appeared a pamphlet by A. Marschand, entitled *Ueber die Entwaldung der Gebirge*, which was published at Bern; and in 1872 was published, at Arbois, *Les Torrents des Alpes et le Paturage*, par M. L. Marschand, *Garde Général des Forêts, Ancien élève de l'Ecole Forestière*.

The preparation of this treatise was undertaken at the suggestion of M. Faré, *Directeur général de l'Administration des Forêts*. It embodies the results of observations made during a residence of seven years in the valley of Barcelonette, and during a tour of observation in the Austrian Alps, and observations made in Switzerland, whither M. Marschand had been commissioned to go to complete his study of the subject.

The attention of M. Marschand was given, primarily and chiefly, to torrents and the means to be employed to arrest and counteract them; attention was also given to pasturing of flocks and herds on the mountains as the original cause or occasion of the destruction of forests, which destruction of forests had been followed by the appearance of the torrents in the regions in which they are so numerous. Every facility was given to him by the forest authorities, officials, and subordinate *employés*, in the prosecution of his studies; and he states that by them were furnished many of the documents and ideas embodied in his treatise.

M. Marschand appears to have been led to conclude that the effects produced by trees, observed by Surell and others, was, primarily and principally, if not exclusively, produced by their roots; and by these modifying the hydroscopicity, capillarity, and permeability, of the soil and subsoil; and that this they did even when this ground was rock. And from this standpoint he deals with the subject.

Surell and others had given an exposition of what may be called the mechanical effects of the roots of trees in preventing the formation of torrents; he, while accepting this, was led to conclude that there was more in this than had been evolved.

After glancing at the natural history of mountains; at their primary condition; and at modifications of this effected by aqueous influence, including disintegration by frost; at the arrest of these by vegetation, and the resumption of the operation of these, which occurred consequently on the clearing away of forests, &c., he says,—" There is an action but little observed, but one which goes on with very great activity, in the decomposition of rock—it is that effected by roots. This influence has been studied by Julius Sacks, and reported in his *Manuel de Physiologie végétale*.

" In twelve days the roots of the *phaseolus multiflorus* [the scarlet kidney-bean] has produced, on polished white marble, great markings, a demi-millimètre in depth, like the traces of an engraver's tool; experiments made with other plants, and on other kinds of rock, give similar results. Of these twelve days, six were taken by the root in reaching the marble, and in the remaining six days these markings were made. It may be inferred that the presence of forests, which develope a great many roots, deeply penetrating the ground, will have for its effect considerably to increase the riches of the soil, by expediting the decomposition and disintegration of the rock. If we think of the results obtained in a few days by the experiment in question, we may form some idea of the influence of forests acting throughout hundreds of years; and we may be prepared to admit that the rocky subsoil of the forests, although protected against extreme atmospheric influences, may be disintegrated, at least as rapidly as if it had been exposed to the direct influence of the atmosphere, through the influence of the roots of trees. It is to this operation we may attribute the gently rounded forms of calcareous rocks covered by the soil of forests."

Citing next experiments by Thurmann, in which cubes of different minerals, thoroughly dried, weighing each 100 grammes, were immersed in water for five minutes, he states that these gave the following results:—Liassic triassic, compact jurassic, liassic triassic and oolithic limestones, granite, serpentine, basalt, dolerites, trachytes, &c., gave a mean absorption of 0·50 gramme of water. Similar minerals, including gneiss and compact marl schist, somewhat disintegrated and changed, gave a mean absorption of 1·50 grammes; limestone still further decomposed, ferruginous oolites of Mt. Jura, liassic schists and grits from the Vosges, and eruptive rocks perceptibly changed, a mean absorption of 4 grammes; variegated grits, green coloured grits, calcareous chalks, gravelly clay, and sands, 7 grammes; and clays, Oxford marls, kaolin, an absorption of from 10 to 30 grammes.

These observations he considered indicative of the absorption of water being proportional to the state of subdivision of the material composing the rock; and the effect he resolved into their hydroscopicity and their capillarity—the former, the power of each molecule of the rock to retain around it a layer of moisture difficult to withdraw—the latter, the property possessed by many molecules of earth, to retain, in interstices by which they are separated, small globules of water.

Apart from these, he treats of the permeability of soils as something quite distinct, and existing in very different proportions—as, for example, in oolitic limestones, which absorb and retain very little water, but which are very permeable by water, through abounding cracks and chinks, and vertical fissures, by which they are subdivided, in consequence of which water falling upon the surface of them does not remain there, but disappears in innumerable fissures. To such chinks the name *lésines* has been given.

Thus is it with extensive plateaux on the Jura chain of mountains—and thus is it with those of the Karst, near the Adriatic. In such places there are no fountains, and there are no torrents; but after denudation by the destruction of forests, the fertile layers of soil do not the less disappear, being swept away with the waters into the *lésines*. He cites, in proof, the Karst, which was formerly wooded, but it is to-day only bare limestone, cut up by *crevasses*. And referring to a progressive impoverishment of certain parts of the Jura, he states it is attributable to the same cause.

In general, rocks which are highly hydroscopic are not very permeable, for the molecules, once moistened, cohere and present the appearance of a compact mass impermeable to water, as may be seen in clay.

On a permeable soil or subsoil, trees create and maintain on the surface a layer of humus of considerable hydroscopic and capillary properties, retaining water, and modifying the general permeability. While, on an impermeable rock, the roots would break up this and increase the permeability.

The principle which he seeks to establish is, that forests have the effect of modulating the properties of rocks, giving to them what they have not; and he alleges it is in this way, more especially, that their action is salutary in the control of waters on the mountains. Of this view of the subject he gives the following illustration:—" When the rain falls on a denuded brow of a hill, composed of argillaceous earth, the water moistens the surface—this absorbs a great quantity of it, through its hydroscopicity and capillarity—but when once this surface is moistened, the transmission of water goes on, only very slowly from particle to particle, for the permeability is almost nothing, in consequence of the minute subdivision of the molecules which are brought into the closest contact; that which is absorbed remains on the surface and dilutes the superficial layer, which is soon thus transformed into a thin clay devoid of cohesion. A layer more or less deep will then detach itself from the surface of the mass, and will flow to the bottom as mud more or less fluid, according as the rain may have been more or less violent. By a very gentle rain, a superficial layer is moistened; but the water falling slowly may be able to penetrate it completely, in virtue of its hydroscopicity and capillarity. In this case there will be only a superficial flow, for the greater portion of the water will penetrate the soil.

" But suppose that this same argillaceous land, or other unstable ground, were wooded, the trees in spreading the fall of the rain over an expanded surface, that of their foliage, would moderate the rain-fall, and would at the same time augment the absorbent power of the soil, as well as its permeability, and as a consequence augment the quantity of water retained superficially. The mobility of the surface thus softened would, undoubtedly, be increased, but the roots imprisoning it would retain the softened ground to such a degree that no amount of water falling upon them from the heavens alone could cause it to slip away. Wherever landslips occur on wooded grounds they can be otherwise accounted for.

" If, in conclusion," says he, " we examine a forest situated on a land permeable *en masse*, as are the plateaux of jurassic limestone, the first effect of the forest would be to cover the soil with a thick layer of humus and of moss, which combine in a very high degree hydroscopicity and capillarity. The quantity of water retained thus in the upper layer of the soil will be much greater than it would have been had there been no forest there, for on the rocks referred to the destruction of woods is almost immediately followed by a denudation of these rocks of soil.

"It follows from this that on these lands the forest arrests the descent of the waters to the bottom of the valleys, for it is only very slowly that water retained by hydroscopicity and capillarity quits the substances which they penetrate. Moreover, as the greatest storms of rain never do saturate completely the layer of humus on wooded soils, it is impossible to form torrents on these."

In successive chapters M. Marschand treats of soils, and the hydrological influences to which they are subjected, of atmospheric influences, transport by waters, decomposition by roots, of the condition of disintegrated rocks; the hydroscopicity, capillarity, and permeability of different soils thus produced; the influence of vegetation in *gazonnement, buissonement,* and forests; the *boisement* of the Alps; the meteorology of the Alps; Alpine torrents, limpid and muddy; the extinction of torrents—general principles of procedure, preparatory works; *barrages,* or barriers of stone, and of wood, and *barrages vivants; clayonnages,* or wears of hurdle, and estimate of cost; *saucissons,* or complementary works; *pérés continus, murs en travers, rigoles en clayonnage;* works on the mountains of the torrents, *clayonnages,* or hurdle barriers, of different forms, with estimate of the cost; landslips, *assainissement; reboisement* in the Alps, preparation of the soil, sowings, plantings, selection of plants, selections of kinds of trees employed.

In the second part of the work he discusses at length the subject of pasturages, treating in successive chapters of the pasturage of cows on the Swiss Alps, and the effect produced on the ground not always injurious, but impoverishing; the pasturage of goats on the Alps, and the devastations committed by them; the pasturage of sheep on the French Alps; the pasturage of sheep on the Swiss Alps; the devastations occasioned by sheep, and comparison between the advantages connected with the rearing of cattle and the rearing of sheep.

He states that the effect of *gazonnement* is to augment the hydroscopicity and capillarity of the surface of the soil, but that this is not sufficient to secure the absorption of all the water that falls upon it in a storm of rain, and he cites facts in support of the allegation. The same good effects, he states, are produced to a greater extent by *buissonnement,* or the planting of bushes, while a layer of humus of great hydroscopicity, produced by the decay of their leaves, co-operates in the production of these good effects; but he cites evidence that even *gazonnement* and *buissonnement* combined have failed to prevent erosion and the formation of torrents.

But forests produce in a surpassing degree each and all of the effects referred to, as produced in a minor degree by meadows and bush. (1) They form by their detritus a very hydroscopic layer, and in consequence augment the quantity of water retained; (2) They augment the expansion of surface on which the water falls; and (3) They augment the capillarity and permeability of the subsoil.

"I have," says he, "in treating of the permeability of the soil, explained the influence of forests on this. In retaining for some time the water at the surface they augment considerably the quantity which is absorbed, particle by particle, by hydroscopicity and capillarity, for this absorption is slow; and thus, in a word, the forest tempers the action of rain dashed downwards in a storm, and leads the water gently on to the soil, as if it had fallen in a gentle shower; and further, it augments, in fine, the permeability of the soil, by keeping the surface unhardened and in some sort always open to receive the water which comes slowly from the heavens.

K

"I make no mention of the influence of forests in regard to evaporation—in regard to the direct absorption of water—and in regard to the humidity of the atmosphere, &c. I take up one point of view alone of the torrential management of waters in the high mountains, and these relate to this only indirectly.

"If a storm of rain beat upon a forest the whole of the water which falls is temporarily retained, all penetrates more or less deeply the soil without flowing on the surface; and, it may be objected, if the subsoil is impermeable the result will be the same. But the objection is without foundation. I shall suppose, what is frequently the case, that there is impermeable rock underlying the *humus*: all the water should arrive at this bed of rock and flow down, but the hydroscopicity and capillarity of this *humus*—of the ground—of the foliage—of the branches of the trees—in a word, of the material of which the forest is composed—will arrest the water to such a degree and measure as to regulate temporarily the delivery.

"In support of what has just been said," says he, "I shall cite an observation made quite recently in the canton of Appenzel, in Switzerland. The torrent of Weissenbach formerly appeared in a swollen state at Weissenbach about three hours subsequent to the bursting of the storm on the mountain; but since the woods have been destroyed—and this has only been done to a partial extent, and those destroyed have been replaced with a fine *gazonnement*—the floods appear at Weissenbach within an hour after the storm. In this we have a very striking illustration of the influence of forests, and of the *gazonnement* which has taken their place. And whoever has resided in the mountains will understand that a delay of two hours in the appearance of the flood, and in its protraction (which augments by four hours the period of flow), may suffice to prevent the most serious disasters; for there everybody knows that the great danger from Alpine torrents arises from the suddenness, amounting almost to instantaneousness, of their flood.

"I have glanced rapidly at the action of forests, in view simply of their effect on the water which falls on their surface; but their function is by no means limited to this, for they serve also to arrest the waters which come from the pastures above them. They constitute in some measure a kind of immense and powerful *barrage*, or barrier, placed between the summit and the bed of the valleys.

"In support of this allegation, I shall cite personal observations which seem to me conclusive. Never have I seen, during the most violent storms of rain, superficial flowings of water in the forests situated under pastures, though such flowings may have existed in the meadows at a greater elevation than the forests; all the waters which these supplied were literally absorbed and retained by the forest soil. I except, intentionally, well-marked ravines, which coming from above traverse forests, for the question here is only of *slopes somewhat uniform, or but slightly undulated*; it is evident that the soil of the forest will not absorb the water of a stream which traverses it encased in a bed.

"I take, for example, a valley which rises to a summit line somewhat elevated. The end situated at a great height is formed entirely of pasture lands, which stretch out equally on the summits of the brows of the mountains; at a lower level beneath these are the forests. The waters which fall into the cistern formed by the head of the valley rapidly accumulate, and give birth to a torrent which traverses the forest. On the contrary, **that which falls on the pasture lands above the brows do not commonly reach** the depth of the ravine: descending to the forest zone uniformly extended over the soil, they are there absorbed.

"In a word, the zone of the forest absorbs generally the water flowing from the zone of pasture lands which correspond to it. In support of these observations, I appeal to all who, in the Alps, have observed storms of rain in the forest. I except water accumulated in ravines or depressions, which are in another condition.

"But the beneficent action of the forests does not limit itself to this; the flow in the ravine may also, if it be not completely absorbed, be by them rendered less injurious if it should come to spread itself over a *cone de déjection* in a forest otherwise covered with wood. I have observed, in connection with this, numerous muddy floods in ravines which, spreading themselves out in the middle of a forest, come out thence very limpid, depositing in it their slime, and leaving in it also almost the whole of the water.

"The great forest of the Ofen, in the Grisons, has supplied me with many instances of this. The soil, composed of the dolomite limestone of the triassic period, is somewhat unstable; in the middle of the pasture lands which surmount the forest there are formed every year numerous torrents, which to an enormous extent carry off the small pebbles, which are characteristic of the dolomite. All these torrents arriving in the forest, then expand and diffuse themselves, and very rarely do they penetrate to the bottom of the valleys. In the upper portion of the Munster-Thal, I have seen on the right-hand side an enormous ravine, the muddy torrents of which are arrested by the forest. And the waters of the Munster, so well enclosed at this point, are a proof of the beneficial action of the forests. In fine, from the moment that the forests begin to retain the mud they retain also temporarily the greater portion of the water in which this was suspended, which are arrested by the enormous absorbent powers they possess."

Facts in accordance with some of these latter statements have been observed and recorded by others.

M. Marschand makes the following remarks on *The Influence of Vegetation on the Flow of Water :*—

"GAZONNEMENT.—Many people suppose that on the steep parts of the Alps a good *gazonnement* would be enough to keep up the soil and put an end to torrents. Experience has shown me that *gazonnement* alone is nearly always powerless to moderate sufficiently the action of water flowing over steep declivities.

"I have been surprised at storms when passing through meadows fit for being mowed, situated at 2200 mètres altitude—that is to say, above the forest region. After some minutes, if the storm was pretty violent, the water ran off the turf, collecting in the depressions of the ground, and forming small clear torrents. On the 17th August 1869, in particular, I observed in the upper basin of the Tincé, in the Maritime Alps, a storm of wind and hail which hardly lasted half-an-hour, but which gave rise in the meadows to a number of these little torrents, the junction of which would produce a very considerable rise in the Tincé.

"A storm observed at the same point in October 1868 threw immense masses of water into the same river in spite of the perfect *gazonnement* of its upper basin; the same storm caused great havoc in the upper basin of the valley of Abriès, among the pastures on the hill of Grange-Commune. Two of my friends had great difficulty in crossing the meadows situated near the summit, so large were the torrents which had suddenly formed.

"All the places mentioned are covered with very good turf, and the soil is formed of grey schist.

"The canton of Tessino is destitute of wood in most parts, but, as compensation, it possesses magnificent pastures which, in spite of the maddest mode of depasturing, preserve (thanks to the wonderful soil) their thick and perennial mantle of verdure. The inundations there are terrible, although at the lower end of the torrents are to be found lakes which retain the alluvial soil and moderate the rapidity of the rise in the rivers; the effect produced by these lakes is very great but insufficient; in 1868 the level of Lake Majeur rose 7 mètres at Locarno, and in the public square the water rose to the first storey of the houses.

"It would be interesting and useful to ascertain the quantity of water which, falling with the rapidity of a thunder shower, would be sufficient to saturate a turfed surface, but the quantity is very small, and depends on the steepness of the descent. This phenomenon is easily explained. Alpine turf, beat down by cattle, is formed of small plants growing close together, the interlaced roots forming a sort of felt. When rain comes it makes the rootlets swell, which, pressed together, imprison the soil and form a scarcely permeable covering, through which the water gradually passes only by means of capillarity and hydroscopicity. If the rain is slow and continuous these two properties are enough to permit all the water to pass through. If, on the contrary, it is violent, the water runs over the surface without being absorbed.

"But, supposing the surface to be horizontal, the effect just described is also produced; the excess of fallen rain, however, lies on the herbage to be gradually absorbed, for the quantity of rain retained by the herbage is in inverse proportion to the slope of the ground, and varies continually.

"Turf, from the special point of view which we occupy, is chiefly useful in consolidating the soil; this end is partially attained, in so far as any surface whatever, when turfed, will always resist the direct action of the rain, but as a whole it is not: the excess of the water absorbed unites, forms little streamlets, and, according as the inequalities of the ground on which they recur are steep or narrow, the turf is attacked by the running water, the soil is laid bare at some one point, and in a few minutes there is the beginning of a ravine, which will always grow larger after every new storm if a remedy be not promptly applied.

"To return to the subject, I would say that turf increases the capillarity and hydroscopicity of the surface of the ground, but these two properties are unable to absorb spontaneously all the rain which falls during storms, and the excess runs down the surface if the ground be steep, or lies on the top if it be flat.

"These observations lead me to conclude that all flat surfaces may be turfed without inconvenience; and that the turf on steep declivities will moderate very slightly the rising of the floods.

"There is plenty of opportunity for making experiments on this point. Places for making observations should be chosen on the same kind of soil, the surfaces of which have been examined, and where there is no water but from the sky. It can be easily done in the mountains. So soon as there are signs of a storm those employed should repair to such spots, there to measure the quantity of rain with a pluviomètre; and to measure, by means of *barrages* with rectangular sections, the quantity of water issuing from the basin where it is received; and also to note the duration and nature of the storm, whether it be accompanied by snow or by rain, &c. These observations are so simple that the guards who live on the mountains will be able to make them.

"What I have mentioned will, I trust, make it very plain that on the Alps *gazonnement* alone is not enough; this opinion is no longer disputed in Tessino, where, as I have already said, the herbage is exceedingly good; observations on the rise of torrents in the ravines which descend from the pastures of our own Alps give the same results.

"LE BUISSONNEMENT.—I have often spoken of *buissonnement*, or planting with shrubs, as being enough to put an end to Alpine torrents; this opinion I now believe to be erroneous. Shrubs may undeniably be of great use— they may be able to cope with purely local accidents, but in no circumstances can they be substituted for a zone of forest. I think I have explained before that Alpine forests create on the surface a bed of humus possessing great hydroscopicity. Shrubs do not supply similar results—it is in this they are inferior. They consolidate the earth well enough, but with rare exceptions there is not found under them the thick mobile layer which carpets large forests, and thus the soil receives atmospheric influences too directly. In conclusion, I may cite in support of my opinions the mountains of Tessino, which surround Bellinzona; which are well turfed, and covered with beautiful shrubs, but amongst which are found ravines and erosions.

"Looked at in regard to the regulation of the water flow, there is not much difference to be remarked between places which are turfed and those which are covered with bushes; whilst in the upper part of the valley of Tessino, towards Aïrolo, there is proved to be an enormous difference between the rise of torrents in the wooded valleys and in those which are not wooded.

"LES FORETS.—Forests are on a grand scale what meadows and shrubberies are on a small one; their effects are—(1) The formation by their detritus of a highly hydroscopical bed, and in consequence of this augmenting the quantity of water retained by the soil; (2) The augmentation of the surface of the dispersion of the water; (3) The augmentation of the permeability and capillarity of the subsoil."

In 1872 appeared the Supplement by M. Ernest Cézanne, *Ingénieur des Ponts et Chaussées, Representant des Hautes Alpes a l'Assemblée Nationale*, to the work of M. Surell, published conjointly with a second edition of the work, and containing a review of treatises which had been published, and of works which had been executed, in the interval which had elapsed subsequently to the original publication of that work.

In this work, while holding that *deboisement*, or clearing away of forests, is not always and everywhere to be condemned, but is in many circumstances necessary for agriculture and the promotion of civilization, and that the general *reboisement* of the Alps would be the ruin of the country, M. Cézanne states that the great service rendered by Surell was the disengaging and treating apart from the general question of forest science the special problem, of local importance, relative to the effect of forests on water-courses, which, being carefully defined, was thus prepared for treatment according to scientific method.

In an introductory chapter he gives a condensed history of *deboisement*, or the destruction of woods in France. In a second chapter he gives a careful discussion of the question,—Has the *deboisement* of France modified the mean annual temperature of the country either one way or the other, as is maintained by opposing parties, which discussion he concludes with the statement that he deems it would be wise, in existing circumstances, to hold by these words with which Gay-Lussac replied to Arago, before the Commission of Enquiry of 1836 :—"According to my opinion there has not yet been obtained any positive proofs that woods have by themselves an actual influence on the climate of a great country, or of a special locality, or that they have an influence different from that of vegetation of every kind. The questions involved are so complicated, when we face them under the climatic point of view, that the solution of them is very difficult, if we may not say impossible."

The third chapter is devoted to the consideration of the effects of forests on the rainfall. In this he alleges that in so far as this is effected by mountains it is less the local elevation than the local inclination of a place by which the effect is produced, and that the effect of this is different according as the pluvial cloud may be ascending or may have commenced its descent; and this, he contends, is an important element to be employed in the correction of pluviometrical observations. He considers that the effect of forests on the quantity of rainfall must be infinitesimally small, and that numerous corrections, some of which he specifies, must be made upon pluviometrical observations before they can be made available for a satisfactory solution of the question raised.

In succeeding chapters are discussed evaporation, infiltration, *ruissellement*, or the source and flow of water in water-courses, with the action of forests on each of these, and the result of such actions, which he sums up thus :—" This action depends on circumstances peculiar to each watercourse, and even to each affluent. This action is proportionally more certain and more energetic according as the water-course is more torrential.

" But what it is impossible to deny, and is beyond all dispute, is the influence which forests exercise in conserving the soil of the mountains against being washed away. In doing this, and preventing the formation of ravines, forests modulate the flow of the waters. And this supplies the only certain criterion we have of the utility of forests in this connection."

The chapter which follows reports what was done in the ten years following the passing of the law of 1860, including the passing of the law in regard to *gazonnement*, with details of the circumstances which led to this; which chapter he concludes with the remark—"After such testimony one cannot feel free to doubt that the operation is good, seeing that it satisfies everybody—the Administration, men of science, and the people."

In another chapter he reports the work done, and results obtained in connection with the artificial extinction of torrents.

Then follows an analysis of a memoir by M. Guiny, sub-inspector of the exploitations of the mountains thus redeemed, which appeared in the *Revue des eaux et forêts*, for 1865, with remarks of his own in support of the proposal to substitute cows for sheep, and more especially for the immigrant sheep from Provence, the pasturing of which is destructive and unremunerating.

This is followed up by similar analyses, with remarks of *Études sur les torrents des Hautes Alpes*, par *M. Scipion Gras*, and of *Memoire sur les barrages de retenue des graviers dans les gorges des torrents*, par *M. Philippe Breton*, and of *Rapport au conseil fédéral sur les torrents des Alpes Suisses, inspectés en 1858, 1859, 1860, et 1863*, par *M. le Professeur Culmann*—works relating chiefly to torrents to which *gazonnement* and *boisement* are inapplicable as means of extinction, as is the case with many which derive their floods from glaciers,—and which treat of the absolute and the relative advantages of dykes or embankments, of *barrages* or wears, and of artificial channels for drawing off the excess of waters,—while the last of them supplies not a little detailed information in regard to Swiss torrents similar to what has been cited in regard to torrents in the French Alps.

And in a concluding chapter the information obtained by induction through the study of the torrents of the Alps is applied to geological phenomena which find, or do not find, a satisfactory explanation in deductions made from what has there been seen.

In this chapter he shews that extensive districts of the country, some of them far away from the Alps, show indications of torrential and glacier action, upon which, when this has once been seen, it is as impossible to look without this being seen, as it is to look upon the remains of extinct torrents in the Alps, referred to by M. Surell, without perceiving them to be such, when once they have been seen to be so in the light of M. Surell's observations.

The expansion of the theory is so very great that some preparation of mind may be desirable before taking up his views, and the more advanced views of others upon the subject, whether this be done with a view to accepting, or comparing and weighing, or rejecting them. This may be pleasantly obtained by a cursory perusal of the following little fancy sketch, embodied in a defence which he makes of graphic details of physical geography, embodied in the memoir by M. Breton, analyzed in his work.

"M. Breton," says he, "almost apologizes for pausing to describe effects so well-known in the mountains. But, apart from the circumstances that the delightful character of his demonstrations secures for him the favourable consideration of his readers, do not many pass by the most interesting phenomena of nature without observing them? And is it not delightful for a traveller when, enlightened by the instructions of a master, he knows how to account to himself for all the peculiarities of those distracted surfaces, and to decypher at a glance in these archives of stone the ancient history of the mountain?

"This steep declivity is a cone of crumbled down earth which descends from that gap; this one here, less inclined, has been produced by an avalanche; that other presents the subdued slopes of a torrential cone. This small hill leaning its back on the mountain is an ancient cone which would fill up the valley; near to the gorge a village conceals itself, the vane of the clock peers out from above massive domes of walnut trees, towards the base the river has lately opened a troncature or section of the cone by a rush upon it, and then she has thrown itself against the other side of the valley; a recent cone has engrafted herself on the older, a little in advance of the exposed section; not far from that a *moraine*, more ancient still, almost buried in the cone, carries back the thoughts to the times long past, when these fields, to-day so rich and animated, were like to the desolate fiords of Greenland, and slept enshrouded in a mantle of ice."

All this seems natural and sound; we feel that it is not a mantle of fiction, but a mantle of fact which is being thrown over the scene, and we find pleasure in the reproduction of what was in the olden time, and in time much older than that to which that designation is generally given. But he (Cézanne) takes us over extensive districts of France, and shows us the same kind of things every where. Nor does he in doing so recede into the inaccessible, **where we cannot test the correctness and verisimilitude of what he says.**

"A journey of a few hours (says he) may enable any one, from the window of a carriage, to verify the greater part of the observations made. From the railway station of Hendaye, which is on the shore of the Ocean, may be seen on adjacent rocks the covering of lœss, and the torrential pebbles.

"In the cuttings towards Biarritz and Bayonne, the bank of gravel is well marked, and from place to place very deep and extensive. From Peyrehorade to Pau, the railway follows closely the foot of a terrace, the slope of which often presents a remarkable regularity. It is the base of a cone cast up on the tertiary deposit by the *Gave de Lourdes*. The town of Pau is built on the edge of this terrace. From the Place Royale we look down upon the valley of the Gave, some 30 mètres, or 100 feet beneath; opposite, on the left bank of the river, the undulating knolls of Jurançon are remains of the glacier deposits of the *Gave d'Ossau*; on the right bank towards the east, and on to the mountains stretches in dimishing perspective the valley, divided in its primary plan by a small chain of low hills, crowned with villas and small umbrella-like pines; these are *testimonies* to the work of erosion committed by the Gave when it opened up its channel and valley through its own deposits. The horizon is bounded towards the northeast by a straight line of regular inclination, which is the culminating ridge of the cone of Lourdes.

"From Pau to Nay the plain is sown with rolled and water-worn pebbles. From Nay to Saint-Pé the grounds show, from time to time, the unmoved rocks covered with lœss and glacier pebbles, and once and again terraces cut up by the Gave in its own deposits. From Saint-Pé onwards appear *moraine* blocks, which continue to appear until Lourdes is reached, and the spaces between these *moraines* are filled with pebbles rolled by the torrents. Lourdes is the highest point of the railway, corresponding at once to the summit of an angle which formerly divided the glacier into two branches, and to the summit of the glacier-deposited cone; the latter is on the hills which rise to the left of the railway station. The Bléout, with its erratic boulders, rises on the right of the station, and partially encloses the valley of Argelès.

"On leaving Lourdes the road descends towards Tarbes by a riverless valley; and between Lourdes and Adé there have been counted, in the railway cuttings, seven separate and distinct *moraines*, partially buried under the argillaceous lœss and the torrential deposits.

"The line proceeds for some way between the two cones of the Gave and of the Adour; but thereafter the plain expands through the erosion effected by two parallel water-courses. The Echez, pretty far to the right, is still eating away a scrap of the cone of the Adour; while on the left bank the Mardaing is attacking the cone of the Gave, the fine regular ridge of which may be seen after passing the station of Ossun.

"In approaching Jullian the railway passes through a cutting in descending a terrace cut up by the Echez in the dejection of this torrent, which took its rise towards Adé from the eastern branch of the glacier. The strength of this torrent, now no more, is still testified by the dimensions of the blocks of stone which it has rolled down and spread over the plain.

"From Tarbes to Tournay the tunnels and cuttings are cut under the lœss of the Adour, and we traverse several open valleys following the crest of the cone. The rails at Tournay pass over the Arros, which flows between the glacier dejections of the Adour and those of the Neste, and rises, as the cone of the Neste, by an inclination of ·034 along the valley of the Lene.

"At Capvern the railway debouches on the plateau of Lannemezan, where the view extends over a plain of varied contour, which is bounded to the south by the lofty amphitheatre of the Pyrenees, and which sinks away towards the north and is lost in the horizon. We are then under the col by which the loess was spread out, and a momentary glance may be had of the valley of the Neste, whence the glacier degorged.

"The culminating point is near the station of Lannemezan; it is there, near this summit, and in accordance with the torrential character of the phenomena, that we see the largest sized pebbles of the kind seen in travelling thither from Tarbes, some of them larger than a horse's head. And it is necessary to re-descend so far as Montréjeau to find in a *moraine*, brought to light by a cutting, blocks of a size comparable to that of those found in this culminating ridge.

"From Montréjeau to Toulouse, and more especially in the plain of Muret, the plain is bounded towards the left by the regular formed ridge of the cone of the Neste. The hills are cut in terraces, which become less important as they recede in distance from the cone. Montréjeau and Saint-Gaudens are built on the edge of a slope of water-rolled blocks, some of the last traces of which may be recognized at the gate of Toulouse.

"If, quitting the main line at Portet-Saint-Simon, we go up the valley of the Ariège, we shall not be long in finding unequivocal evidence of torrential action. In the environs of Pamiers the plain is completely covered with blocks of stone, perfectly rounded, which the husbandmen have collected into heaps all around the cultivated spots, seeing which one could almost imagine himself on a recently grubbed cone of a recent torrent like that of Embrun. These blocks are found of increasing size as we get higher, and in the station of Foix may be seen, in the garden of the station-master, most beautiful chips of glacier blocks, with streaks and other indications of friction produced by glacial action. The terminal moraine was in these *barrages* or dams, and the minute study of the environs permits, if it do not suggest, the supposition that in the glacial period the beautiful elliptical basin, at the bottom of which flows the Larget, which at Foix falls into the Ariège, was repeatedly filled and emptied, forming an important lake and formidable floods."

In a foot-note it is stated,—" The town of Foix and its picturesque chateau are situated on what looks like the gate of a sluice closing the strait gorge by which the Larget debouches; but behind this gate the basin opens up and ramifies on a large scale; and this basin, closed by a barrier near the extremity of the glacier, was in circumstances exceptionally favourable to the production of a lake and of débâcles, or breakings up, emptying it in a great measure of its contents."

From the study of such and such like phenomena, M. Cézanne has been led to conclude that there must have been a period when torrential action has been much more stupendous and much more extensively diffused than at present.

But M. Cézanne alleges that there are indications, no less marked, of glacial action in deposits underlying some of these torrential deposits. He remarks that the theory of glacial action, considered as a chapter of geology, presents this peculiarity—that all the phenomena embraced by it, and all the circumstances in which they occur, may be observed in our own time. There is no difference, excepting in the scale of magnitude of the phenomena, which has been greatly reduced. L

The action of glaciers has been studied in the Alps; results obtained have been applied to the phenomena presented by and in connection with the parallel roads in Glen Roy; these have been satisfactorily shown to have been produced by glacial action, exhibiting on a grand scale the same phenomena as are to be seen in the valley of Bagnes, marking out what must have been the banks of the ancient lake of Giétroz.

He quotes a paper by M. Ch. Martins, which appeared in the *Revue des Deux Mondes* for March 1867, in which it is said,—" Throughout almost the entire length of the valley of Glen Roy—that is to say, for ten miles and upwards—there may be traced on the opposite declivities three terraces or parallel banks, perfectly horizontal and corresponding exactly on the two sides of the valley. From a distance they are distinctly visible; when reached they are found to be a pebbly surface from 10 to 60 feet wide, the slope of which is less steep than that of the mountain above and below. The lowest of the terraces is 750 feet above the level of the sea, the second about 210 feet higher, the third upwards of 80 feet above the level of the second,—all terminating at the head of the valley on the col which separates it from the valley beyond.

" In 1840 Buckland and Agassiz visited Glen Roy, and perceived that temporary barriers to the flow of water could alone account for the formation of these level lines. Glaciers coming successively to close up the one or the other issue of the valley, the stream which ran through it would form a lake, which would flow by the col towards which the terrace inclined. Agassiz recognized polished and striated rocks and the ancient moraines which he had learned to distinguish in the Alps; and subsequently Mr Jamieson has given a chart and details confirming completely the view of the illustrious Swiss naturalist.

" Mr Jamieson carries back the formation of these terraces to the close of the second glacial period, when it was due to an oscillation of the glaciers descending from Ben Nevis and the surrounding mountains. These barriers have barred up, one after another, the valley of Glen Roy and the neighbouring valleys. The waters, stopped in their flow, have formed lakes at different levels, determined in each case by the height of the col which closed the extremity of the valley opposite to that barred up by the glacier. The entireness of the terraces prove also that subsequent to the formation of these Scotland has never been submerged."

And M. Cézanne goes on to say,—" If this theory be correct, it follows that we may expect to find at the *débouché* of each of the great valleys of the Pyrenees and of the Alps masses of lœss, deposited in accordance with the characteristic forms of torrential deposits; and the dimensions and composition of which may be in accordance with the immense duration of the glacial period, and so in accordance with the greatness of the phenomena to which this has given rise.

" At the foot of the Pyrenees there exist such deposits; if not most considerable, they are at least most perspicuous. Let us ascend some height such as the *Pic du Midi* of Bigorre, or the *Cap* of the Col d'Aspin, the ascent of which is very easy; from this elevated point the observer, turning to the north, can freely cast his eye from the west to the east, over a vast plain which rounds itself off at the horizon, as does the sea, and the bleak fawn colour of which contrasts strongly with the green and sombre wall of the Pyrenees.

" This general effect is pretty well rendered by the chart of the *État-*

major, on which one may, besides, recognize the smallest undulations of the ground, which to the observer are flattened by the aerial perspective. According to the chart it is not a compact plain which stretches from the foot of the mountain, it is a series of vertically rounded plateaux, or of flattened cones, the summit of which is at the gorge of each important valley.

"The two sheets of this chart, representing the districts of St Gauden's and Tarbes, placed side by side, represent in a striking manner, to whoever may have seen well-marked torrents, three vast cones of *déjection* which *débouche* from the valleys of the Gave, at Lourdes—of the Adour, at Bagnères —and of the Neste, at Hèches."

The cones thus represented are furrowed by numerous water-courses from the mountains; and they themselves to some extent intersect or cover one another. Full details are given, with tabulated measurements, and references to a coloured geological map of the district. And having referred to difficulties which had been experienced by others in attempting to account for all the phenomena, he goes on to say,—" That these cones have come out from the gorge's entrance, which are guarded by them, a single glance at the map suffices to show; and an examination of the places themselves leaves no doubt upon the subject.

"The glacial origin of these vast deposits is not less certain. In each valley it is possible to follow, from the moraines which remain intact on the extremities of the cones, step by step, the progress of the rocky fragments which little by little lose their glacial characteristics, become rounded, diminished in size, reduced to ordinary gravel, or even to clay or glacial mud. In the valley of the Adour, for instance, facing the village Santa Marie, are two conjoined gorges, descending, the one from Tourmalet, the other from the Col d'Aspin. Between these two gorges, and overlooking the confluence, is a terrace or bank, the slope of which, seen from below, recalls by its irregularity a gigantic railway embankment. If we trace this embankment, following it along the road which leads to Luchon, we soon discover, on the slope above this, a wood which imperfectly conceals a confused mass of enormous blocks called the *Moraine de Grip*, recognized at once as a moraine, such as may be seen near a glacier of the Alps. In this picturesque spot the Adour has cleared for itself a passage among the blocks, some of which, from their forms and size, may be compared to houses; the *Pic du Midi* and the *Pic d'Arbizon*, each at the bottom of one of these gorges, look down from their azure pyramid on this scene of disorder, where their ruins lie confounded.

"On leaving this, on to Bagnères, there may be seen, on the two sides of the valley of Campan, the traces of a glacier: it must have gnawed, on a former time, at the vertical wall which rises on the right; while on the left a series of terraces mark the different levels of the moraines, and the torrential alluvia deposited along the glacier.

"Even at Bagnères we are still in a country full of glacial ground: half-rolled blocks, in size to be compared to a sheep or sack of corn, lie about everywhere; they encumber the bed of the Adour, form heaps along the highways, and enter into the construction of the walls. Towards the summit of the cone—that is to say, on the hills of the right bank—they are very numerous, and also quite as large as in the valley; but on leaving this point they diminish rapidly in size, following the same law of decrease in the clay of the hills as on the floor of the valley: twenty kilomètres below, at Tarbes, they have scarcely the size of a man's head; and towards Mont-de-Marson we meet almost exclusively with clay covered over with the sand of the Landes, borne thither probably by the wind.

"The clay itself presents alternating colours: it is yellow, ochre coloured, or bluish; and near Bagnères we can in the trenches recognize some sort of stone which would furnish some one or other of the colours. By digging out blocks in all stages of disintegration, we may be said to see in actual operation the manufacture of clay. At some places a cutting in the ground presents the appearance of mosaic work, in which granitic pebbles, perfectly recognizable in their rounded forms, but softened by time, may be cut like butter, or rather like *nougat* [a cake made of almonds and honey], each of them leaving still recognizable, in spite of its decomposition, the rock from which it had been torn.

"The cone of the *Gave* supplies similar facts; it is isolated on all sides; its head may be said to be in the air. The glacier which produced it would meet at Lourdes, on coming out of the valley of Argelès with a small mountain of schist, which would necessitate it to divide itself; the exterior branch would direct itself towards where Tarbes now stands, extending as far as to Adè; the other branch would descend towards the position of Pau, reaching as far as Saint-Pè. Between the two branches would be turned off the loess; so the summit of the glacial cone would rest on the schistose mountain in the angle formed by the two branches of the glacier. At the time of the retreat of the glacier all the waters of the valley of the Argelès were united towards the west in the Saint-Pè branch; and the other branch, that of Adé, is still a void valley without a river—the railway from Tarbes to Lourdes has been constructed there; but this valley, devoid of a single considerable water-course, is full of torrential indications.

"The valley of Argelès has been the subject of special study by MM. Ch. Martins and Collomb, indefatigable explorers, who, after having dwelt on the glaciers of the Alps in the *Hotel mouvant des Neuchatelon*, have sought, from Spitzbergen to the Sahara, the traces and the causes of the glacial period. These *savants* thus sum up their memoir, published in *Bulletin* of the *Societe géologique de France*, 2 série, t. xxv., p. 141, seances sur, 18th November 1867, and *Mémoires de l'Académie des Sciences de Montpelier*, t. vii., p. 47 :—"To sum up these observations, we have ascertained in one of the principal valleys of the Pyrenees—the valley of Argelès—the existence of an old glacier of the extent of 53 kilomètres, which shed its terminal moraines on the sub-Pyrenean undulating plain, and extended to within 15 kilomètres of Tarbes, with an altitude there of 400 mètres, about 1350 feet,—its point of departure being at a mean altitude of about 3000 mètres, 4000 feet, the mean slope of its surface being 1 of 0.039.

"This glacier, including its affluents and its higher *névés*,—in a word, its hydrographic basin—would cover an area of about 1·400 square kilomètres, or 140· hectares.

"The thickness of the glacier reached, at Gèdres, 850 mètres; at Saint-Sauveur, 800 mètres; at Pierrefite, 675 mètres; at Argelès, 600 mètres; at the Pic de Jer, near Lourdes, 412 mètres.

"The summit of the Beout, a conical mountain which rises above Lourdes in the middle of the valley to the height of 792, was covered by the glacier; and even from the railway station of Lourdes may be seen distinctly, in profile against the blue sky, the erratic boulders, scattered over the ridge of the mountain, at an elevation of 450 mètres, 1350 feet, above the Gave.

"In the cutting from which has been obtained the material for the embankment on which stands the railway station have been found, with their characteristic fossils, limestones conveyed undamaged from the Cirque of Gavarnie, and, side by side with these, blocks torn from the granite summits of Cauterets. The scientific explorers cited have given with their memoir a longitudinal profile of the glacier, and a chart of the terminal moraine.

"'In studying,' say they, 'the traces which the glacier has left upon the soil, we have seen that it comported itself as do all the glaciers known; it transported materials of great bulk, and at the same time minute debris, which we find in the form of moraines exactly in the place which is assigned to them by the accepted laws of the movement of translation of glaciers, and taking in these an arrangement or disposition which excludes every thought of other mode of natural transport.

"'At the same time the glacier has polished and scratched the resisting rocks with which it was in contact,' (and it should be admitted that with the thickness given above [I am quoting M. Cezanne] that the rubbing and friction of the glacier, with a pressure approaching to 1000 tons per square mètre, prolonged throughout some hundreds of ages, would suffice to account for the erosion of a valley many hundreds of mètres in depth). 'Then, in the third place, the mud produced by the continual friction of the ice against the rock, finally ejected by the waters produced by the melting of the glacier, and by the glacial torrents, have contributed to form the principal material of that loess which covers the place far beyond the périmètre of the ancient glacier.'

"One might, were it not necessary to avoid repetitions, give proof as demonstrative in regard to the cone of the *Neste* which forms the *plateau de Lannemezan*. This plateau is a vast deposit of loess brought out from the valleys of the *Neste*; but a noteworthy circumstance is, the less important of these two valleys, that of the *Neste*, has supplied the uppermost dejections; its cone has partially covered up that of the Garonne, and the great river has been turned out of its course by its affluent. All the strange windings of the Garonne, and of Saint-Bertrand de Comminges at Montréjeau, explain themselves at once on the spot by the strongly characterised moraines which the Luchon railway has exposed.

"Much less ramified are the other valleys of the Pyrenees—those, for instance of the Nivelle, of the Nioi, and of the Joyeuse; those of the Gaves, of Mauléan, of Aspe, and of Ossau—that more especially of the Ariege—descending from less elevated summits, have had their glaciers and their torrential débâcles proportionate to the local circumstances."

These are but specimens of numerous details given, most of them with measurements and other indications of precision such as science demands, of the district thus traversed, of the districts beyond, of the Alps, of the basin of the Rhone and its affluents, with references to what is to be seen in the basins of the Po, the Danube, and the Rhine,—producing an impression that the whole of these districts, together with much of the intermediate region, is covered with torrential deposits of a magnitude and extent far surpassing those of any which the engineer of modern times is called to treat—reminding one of the statement, "There were giants in those days."

And in view of the whole he (M. Cézanne) is led to conclude that this which he is disposed to designate the torrential geological era must have been immediately posterior to what is known as the glacial period. To the consideration of the phenomena illustrative of this point, and of phenomena occurring during the alleged era, is devoted the penultimate chapter of the work.

More immediately connected with the subject of *reboisement* is the following *résumé* given by him of the whole series of phenomena brought under review :—

" There may be given in a few words a *résumé* of the whole series of these phenomena.

" The mountains are the result of a series of upheavals following one upon another in the same region. A final agitation gave to the different chains of these the existing elevation ; it elevated the summit and opened up deep fissures or divisions, which have become the valleys of the present time. From the time this occurred the waters began to fashion the *thalwegs*, following the line which best suited them ; wearing down outlets and filling up basins. It is necessary to admit, or to assume, that the depth or thickness of the alluvial deposits in the bottom of certain valleys—for instance, those of the Isère in the Graisivaudan, or of the Rhine in Alsace,—is to be reckoned by hundreds, and perhaps by thousands, of mètres or yards ; for even yet certain lakes existing in depressions of the Alps have their bottom below the level of the sea.

" After a long series of ages the mountains assumed the leading features which they now exhibit, when, the climate changing, great glaciers carried on actively the work of erosion ; these have planed away escarpments, and fashioned into something like horizontal lines the rocky belts of the valleys.

" *Débâcles*, or inundations, from the escape of the waters of pent-up lakes, and deluges resulting from the tremendous rains of summers on the extensive fields of ice, have carried away and deposited in the principal valleys in certain favourable places, but more especially at the *débouchures* of lateral gorges, the masses of loess which have formed cones in the higher plains, and in which the water-courses have subsequently dug out the secondary valleys.

" At a later period, after the melting away of those glaciers, the torrents seized upon the bared mountains ; and without restraint they have dug out their basins, and have again taken up the materials disintegrated by the glaciers, and deposited these in the gigantic cones which give to certain regions a physiognomy peculiarly their own.

" But after a time the forests, spreading by degrees, stifled the waters under a mantle of verdure ; the torrents became extinct,—an era of peace and of comparative quiet supervened in the mountains ; then the tribes of men, who during the glacial period rambled over the low-lying plains, in company with the reindeer, the aurochs, and the bears, began to spread themselves in the high-lying valleys. The most ancient settlements were made at the gorges of the torrents, towards the summit of the cone ; in point of fact, there are to be found in the mountain valleys very few of these gorges in which we do not meet either with an existing village or with an ancient ruin.

" In this location, which was then one favourable to their pursuits, the primary inhabitants could profit by the exceptional fertility of the cone of deposits; they had nothing to fear from the principal river, which flowed through the lower-lying lands, nor from the torrent, which was then extinct; they commanded the plain, and found themselves at the gate of the mountains; the adjacent gorge supplied them with water, the forest supplied them with wood, the rock supplied them with stone, and their flocks spread themselves over the verdant ridges around them.

"Little by little, a reckless use of the forests and of the pasturage disturbed the equilibrium of the natural forces; and now the old sore is re-opened, and anew, by man's deed, the mountains are inoculated with the leprosy of the torrents. The evil has gone on increasing during prolonged ages of disorder and recklessness; the position of the cultivated grounds, and of the villages established at the *débouché* of the torrents, has now become critical in the extreme; and unless we go back, as we have done, to the olden times, we are unable to account for men having taken up their dwelling in the spots, of all others, which at this day appear to be those which are more immediately threatened.

"But at last an era of reparation begins; and, thanks to the eminent men who have in byegone years given their mind to the work, the next generation may hope to see the final decline of the modern renewed Torrential Era."

In 1874 was published *Les Torrents leur lois, leur causes, leur effets: Moyens de les reprimuer et de les utiliser: leur action geologique universelle, par* Michel Costa de Bastelica, *Conservateur des Eaux et Forêts*.

This work treats of another aspect or of another department of the subject than any discussed in the treatises already mentioned, which the author designates,—*Le phénomène torrentiel*, or *la torrentialité*; and thus is opened up another chapter of the natural history of torrents.

"The question raised by torrents," says M. Costa, in the introduction of the work, "is a very complex one. Behind the technical questions embraced by it, there are others which connect themselves with the forest economy, and with the pastoral occupations and the agriculture of the inhabitants of the mountains, and which involve serious difficulties of administration and of legislation. To operate on the basin of torrents brings one in contact, and sometimes into collision, with the requirements and the customs of the population. The two spheres of thought are quite distinct. The technical element of the question is admitted to be the more important of the two, and it is made the basis of the system of operation. I shall, therefore, confine myself exclusively to it. It will be easy to remove the difficulties of another kind, which beset practical operations under the requirements of the case, when it shall be demonstrated by science, and established by experience, that it is possible to put a stop to the outburst of water-courses by a combination of simple works, comparatively inexpensive, and wisely-devised conservative measures.

"With this view I desire to give synthetically the fundamental idea of the new torrential theory.

"It has struck me in all preceding discussions, in regard to hydrology in general, that they relate almost exclusively to the *débit*, or quantity of water passing or delivered; it is admitted that they take into account the materials borne along by the currents at the time of floods; but on the whole the supposed cause of inundation is always spoken of as the excess of the delivery over what it is at other times. All the discussions which have taken place, and in which the most eminent savants have taken part, have been confined to that of causes which could act on the delivery; and the whole discussion has come back to that of the permeability or impermeability of the soil. Even the effect of forests has not been studied beyond what the consideration of them from this point of view required. All of the researches which have had for their aim to enable us to combat inundations, have had no other object but an action to bear on this delivery. All who have written on the subject have reasoned and made their calculations as if, at the time of an excessive flood, nothing was occurring but an augmentation of the volume of the current, without any variation in the hydraulic law by which it was regulated. I have no hesitation in saying I consider this way of looking at the subject erroneous; and it is at this point that I take my departure from those who have preceded me—on a new enquiry.

"From my point of view there is seen to be something more than simply a variation in the delivery. At the time of a great flood, when a current—be it great or small—bears along considerable solid masses, consisting of earth and stones of all sorts and sizes, a peculiar phenomena of special importance is evolved. This is a perturbation, more or less marked, in the progress of the current, and in the laws by which it is regulated; and this it is which I call the *torrential phenomenon*, or, if a word must be created under which to speak of it, the *torrentiality*—an action of perturbation which is the greater in proportion as the secondary causes by which it is produced—namely, the solid matters borne along—are the more considerable.

"From this point of view, the most furious torrents of the Alps are seen to be only extreme cases of a general phenomenon, which is produced more or less imperceptibly, or more or less distinctly marked, in all currents of water which are not perfectly tranquil in their flow.

"The characteristic effect of this perturbation is an instability in the course of the stream.

"When a current of water does not bear along solid matter, whatever may be the volume of water, the flow is effected with great stability in accordance with hydraulic laws. Sudden variation in the delivery, in raising or lowering the level, produce variations in the rapidity of the flow; from this there is thus a certain consequent perturbation; but the action of gravitation, in its omnipotence, being constant, and this accommodating itself to the resistances due only to friction, the stability of the stream tends uninterruptedly to maintain itself. To a rise of level there being a corresponding increase of rapidity of flow, it is rarely the case that such waters rise higher than the banks.

"The perturbation produced by solid material borne along is, on the other hand, very serious. If the substance of the current be very greatly changed in consistence—for example, if for a limpid water, possessing all its fluidity, there be substituted a viscous liquid—if, further, the torrent be required to perform the mechanical work of conveying a certain quantity of solid matters—the conditions are greatly modified. In the first place we have no longer simply pure water, but water subject to every degree of variation in so far as fluidity is concerned; and thus the work of transport imposed on the current developes resistances which are subject to every degree of variation. From this birth is given to an extreme instability in the current, or in other phrase, to *torrentiality*.

"Experience shows that this perturbation, produced by second causes, exercises on water-courses a much more powerful action than that proceeding from simple variations in the quantity of water delivered.

"In the great torrents of the Alps, which bear along at the time of great floods enormous masses of material, from the grain of sand to the largest blocks of rock, and which, moreover, are extremely muddy, the perturbation is such that the laws of hydraulics would appear at times to be entirely reversed, and to produce effects diametrically opposite to what are produced in a normal condition. For example,—the bed, instead of being concave, is convex; the current, instead of following such depressions in the soil as offer the most rapid declivity, tends to raise itself, and to follow the prominent points in the ground. The surface of the water itself is convex; the most extraordinary dynamical effects are produced; and the water-course —a prey to a veritable revolutionary state of things—becomes the picture of the maddest instability.

"We have there, I repeat, an extreme case of the torrential phenomenon, and one the study of which is pre-eminently adapted to reveal to us the laws by which it is regulated; for, though less remarkable, this perturbation is nevertheless perceptible in the currents of ordinary streams which bear away solid matter when in flood. This formidable phenomenon betrays itself by certain indications. The surface of the stream tends to assume a convex form; it is furrowed with currents which change their position with great mobility and varying rapidity. The principal current, instead of establishing itself in the deepest part of the bed, tends, on the contrary, to follow the line of the highest parts of this, and to invade the banks of gravel, if there be such there. In rebound from the normal state the greatest rapidity of flow is along by the banks, and this is one cause of the erosion of these.

"It is evident that these are effects which cannot be other than the product of secondary disturbing causes, since it is physically impossible that variations in the quantity of water passing along could be the producing cause of any such instability.

"A trained eye, morever, can judge at once, by the appearance alone of a water-course, what is the degree of torrentiality to which it is subject.

"First, when the banks are, through a stretch of some length, covered with verdure to the water edge—or when the willows allow with impunity their branches to be borne along by the current—it is a certain sign of great stability and tranquility of flow. If, on the contrary, the banks are despoiled of vegetation, and show traces of erosion—and further, if there are to be seen here and there banks of gravel—this is symptomatic of the first stage of torrentiality.

"These indications become more and more pronounced, according to the special *régime* of each water-course; and when, as in the Durance, the torrential phenomenon attains a great degree of intensity, the water may be seen straggling over immense plains of pebbles, and dividing into many branches, which change their position on the smallest increase of flood.

"The condition and appearance of the islands formed by these branches present also a certain characterestic of greater or less stability in the *régime* of a water-course. When these islands are covered with old trees, and better still, if people have made up their mind to dwell on them, although there be occasionally great floods, it is a sign of great stability. If, on the contrary, these deposits are devoid, or despoiled, of vegetation, or have not

M

even acquired that dull tint which prolonged exposure to the air gives to them, it is indicative that the instability is very great.

"The existence, then, of a torrential perturbation, attributable to matters borne along, is demonstrable. But more than this, this perturbation is subject to laws as constant as those which regulate the flow of water. The *a priori* proof of this is the form taken by deposits which are the products of this action.

"Nothing is more irregular, to all appearance, than the floods of the great torrents of the Alps. Those who have read the impressive descriptions of them given by M. Surell, know that they look like chaos: blocks of stones rolling along with powerful crashes, knocking one upon another, and a current black as ink, bounding over all obstacles, and spreading itself with extreme mobility over a widely extended surface without being able to fix itself any where. One is ready to believe that this enormous body of stones, borne off by the waters, is about to be scattered abroad at hap-hazard, and to form a confused mass, setting at defiance all rule; on the contrary, it is a curious fact of immense compass, that though the torrent, in pursuing for ages its work of clearing away in the mountain, and of embanking in the plain, may have multiplied indefinitely its floods and its transports of material, the constant result of this continuous action—the one completed result of all these elementary embankments—that which is designated the *lit de déjection*—has assumed a geometrical form of the most perfect regularity! The determining of the geometrical law, by which the contour of these deposits with its numerous distinct characteristics has been regulated, presents considerable difficulties. I shall afterwards state what is my opinion on this point; but, whatever that law may prove to be, there evidently is some such law; and it is enough, at this stage of the discussion, that this has been established."

Alleging, then, that we may conclude with certainty that that work of the torrent, in appearance so irregular, has been governed by laws, and these the laws of torrentiality, he goes on to say, that it is desirable to determine what these laws are, as they are likely to throw some light upon the problem of inundations, and to indicate a rational solution of this, while the solution of the problem is of no small interest to the science of terrestrial physics, and even to that of universal cosmogony.

In accordance with what is stated by M. Costa in regard to the convexity of the flood of the torrent, when charged with earthy matter, are the observations of M. Surell in regard to the convexity of the *lit de déjection*, the last form taken by the suspended earthy material as the water subsided, though this convexity may be otherwise accounted for. Something similar may be observed in a flow of treacle, or of tar, or of quicksilver, or other molten metal; but something similar may be seen also in a very rapid flow of water comparatively pure.

By Marsh it is stated, in a foot-note appended to a passage in his treatise on *The Earth as Modified by Human Action*,—"Many physicists who have investigated the laws of natural hydraulics maintain that, in consequence of direct obstruction and frictional resistance to the flow of the water of rivers along their banks, there is both an increased rapidity of current and an elevation of the water in the middle of the channel, so that a river presents always a convex surface. Others have thought that the acknowledged greater swiftness of the central current must produce a depression in that part of the stream. The lumbermen affirm that, while rivers are rising, the water is highest in the middle of the channel, and tends to throw floating objects shorewards; while they are falling, it is lowest in the middle, and floating objects incline towards the centre. Logs, they say, rolled into the water during the rise, are very apt to lodge on the banks, while those set afloat during the falling of the waters keep in the current, and are carried without hindrance to their destination; and this law, which has been a matter of familiar observation among woodmen for generations, is now admitted as a scientific truth."

A phenomenon similar to that reported by the lumbermen of America may be observed in the rising and falling of mercury in a barometer tube. When rising, the surface of the mercury is convex; when falling, it is concave; and so constantly is this the case, that directions have been given to observe whether the surface be convex or concave, to determine, irrespective of the pointer, whether the mercury be rising or falling. The explanation is to be sought for in the relative strength of the attraction of cohesion keeping the particles of the fluid mass together, and the attraction of adhesion attaching them to the surface of the confining body, together with a third element, that of velocity of movement, which may be relatively different in its effect upon the two attractions named.

This explanation of how the phenomena reported are brought about, taken in connection with phenomena which are cited by M. Costa, enables us to see how it may come to pass that destructive effects on the banks of rivers are frequently produced by the floatage of timber. In many cases the injurious effects produced upon lands by the clearing away of forests are increased by measures adopted in bringing the felled trees out of the forest, and in sending the timber to its first destination. By Marsh, in speaking of a common practice followed in America and elsewhere, it is said, —" The lumbermen usually haul the timber to the banks of the river in the winter, and when the spring floods swell the streams and break up the ice, they roll the logs into the water, leaving them to float down to their destination. If the transporting stream is too small to furnish a sufficient channel for this rude navigation, it is sometimes dammed up, and the timber is collected in the pond thus formed above the dam. When the pond is full a sluice is opened, or the dam is blown up or otherwise suddenly broken, and the whole mass of timber above it is hurried down with the rolling flood. Both of these ways of proceeding expose the banks of the rivers employed as channels of floatation to abrasion; and in some of the American States it has been found necessary to protect, by special legislation, the lands through which they flow from the serious injury sometimes received through the practice described."

And, in reference to the bringing of felled trees out of the forest, he says, in an appendix,—" The methods of transporting timber employed by the lumbermen in the Alps are often more destructive than the baring of the soil. Forests frequently grow in Alpine glens, or other mountain localities, inaccessible to wheeled vehicles or even to sleighs. In such cases the timber is sent down by slides, which, if long used, become the beds of new torrents, or it is conveyed to larger streams by the method of floatation described.

" The *Rapport au Conseil Fédéral sur les Torrents, des Alpes Suisses inspectés en* 1858–63, published at Lausanne in 1865 [that commented on by M. Cézanne] gives a great amount of information respecting this scourge and its causes, among which the practice of floatation is particularly noticed. The amount of damage done to the commune of Campo, on the Rovana, (a tributary of the Maggia, in the canton of Tisino) in great part from the effect of floatation, is most striking (*Rapport* I., pp. 7–13). The force of the torrent Rovana has been augmented to such a degree by baring the soil, and by suddenly opening the dams near its sources, that in the course of four years it excavated below the village a new channel one hundred feet deeper than its ancient bed, and of course undermined the left bank, which was composed of comparatively loose materials, for a long distance. Deprived of its original support, the steeply inclined soil of the commune to the extent of twenty-five hundred acres, including the village of Campo, began to slide downwards in a body. The movement still continues (1875). Many of the houses have been carried off, some overthrown, and the walls of most of the remainder dangerously cracked. Unless costly measures of protection are soon adopted, the whole of this vast moving mass will be washed by the Rovana into the Maggia, and by that river into Lake Maggiore. So insecure is the soil considered at Campo, that as I was lately told on the spot, meadow and pasture grounds, which if safe would be worth 100 dollars (£20) per acre, cannot now be sold for 10 dollars (£2)."

In the first part of his work, M. Costa treats of the phenomena of transport of solid materials by running water, and the laws regulating these in different states of the current—from that of a tranquil flow and the first movement of sand and stones through acceleration of the flow, through various degrees of speed, to the deposit of these in consequence of a diminution of this—and having shown that these phenomena include two modes of transport—one appropriately characterized as *triage*, or selection, bearing onwards lighter or smaller material, while heavier or larger is left, or only rolled along, or dropped, and another in which the whole appears to be borne along *en masse*, water and stones and mud commingled, but keeping their relative position while being borne onward—a section is devoted to the discussion of the laws of viscosity, of which this is a form, and of density, as this is effected by immersion in a fluid.

An opportunity will afterwards present itself for stating somewhat in detail the phenomena he has observed in connection with the transport of solid materials in both of the modes described.

Proceeding, in the second part of his work, to treat of the torrents, he calls attention to two different typical forms of torrential floods—the comparatively limpid floods of the Vosges and the Pyrenees, and the floods of the High Alps, loaded with earth and stones, which they are sweeping along; and looking upon the former as virtually *extinct torrents*, to employ the phrase introduced by Surell and now consecrated by use, he confines his remarks to the latter class of torrents, and discusses in connection with them what he considers the essential parts of these—the basin and the deposit. These are treated of at length, and more especially so the geometrical form of the deposit and the laws regulating its increase; also certain remarkable incidental phenomena connected with torrents, and the phenomena attending extinction of torrents.

One of the remakable phenomena of which he treats is the bounding of stones before the mountain wave, of which mention has been made.

In regard to this he writes,—"Some of the effects of torrents have appeared so extraordinary that, the law of torrentiality not having been ascertained, the imagination set to work to seek out fanciful explanations of what was seen.

"Thus, for example, has it been with the generally alleged fact, that at the moment of flood large stones set off of themselves, rolling in advance before the current had touched them, under impulse from a current of air preceding the advancing head of waters.

"Eye-witnesses, and these grave men, have affirmed this fact to myself; and M. Surell has collected numerous testimonies of this phenomenon, and has sought to account for it theoretically. In reality, the fact as reported is absolutely impossible. Resistance increases as the square of the velocity. Let a calculation be made, from the velocity necessary to a current of air to displace a stone no larger than an egg; what velocity would require to be imparted to a current of air capable of displacing a stone such as some of those of which this has been told, which must have been at least 50 centimètres, or 20 inches in diameter! The thing alleged is physically impossible; and it must be remarked that the people who allege they have seen those things occur under their eyes, at a few paces from them, do not dream that if they had been caught in such a current of air they would, at least, have felt it!

"When these witnesses are cross-questioned, they all declare that they have seen the stones rolling dry before them; but no one says he has seen these stones *begin to move*. These witnesses are trust-worthy, in so far as it is true that the stones were seen rolling before their eyes; but the point in which they deceive themselves is the explanation of the phenomenon. They have attributed this to the force of the current of air; there is the mistake.

"The fact is a very simple one, and easily explained in accordance with what has been evolved by the study of the effect of a sudden retardation on a current of matter. Through the velocity acquired, and the upward direction given to their movement, the stones, detaching themselves, are projected forward from the water by which they were borne along.

"We find that it is towards the contraction of a water-course, occasioned by a bridge, that the phenomenon manifests itself with most intensity. It is, moreover, at such points that it ought to be most easily observed. At the time of a great flood, there are few spectators in the deserted gorges of the torrents.

"From the moment that we are in possession of principles, nothing is more easy than to account rationally for all the effects, and all the accidental incidents which they may produce."

And in a similar way does M. Costa account for other remarkable phenomena which have been observed in connection with torrents.

In a third part of the work, M. Costa discusses at great length the extinction of torrents. In his statement of this question, he says,—"According to the opinion of all the engineers who have had to contend with torrents, with a view to the protection of valleys against their ravages, works simply defensive have been acknowledged to be in most cases, if not useless, at least altogether insufficient, and often dangerous, intensifying at times the evil.

"M. Surell more especially has put this fact in a clear light; and he has established beyond all controversy that it is necessary to carry into the basin the works designed for the protection of the land, that the evil may be attacked at its source.

"He has demonstrated, not less triumphantly, that *boisement* is the most potent means of extinguishing torrents, as by vegetation we can act, at one and the same time, on the delivery of water, and on the consolidation of the soil.

"M. Surell did not confine himself to preaching the *reboisement* of the mountains, he pointed out at the same time the advantages to be derived from *gazonnement*, and from small artificial works of consolidation formed of facines properly disposed in the ravines.

"His logical mind perceived the advantages which might be derived from more extensive and costly artificial works, but he did not believe it possible to guarantee their solidity and their durability under the circumstances in which they would be erected.

"MM. Scipion Gras et Phillipe Breton have also loudly proclaimed, in a way the most explicit, that the *boisement* of the valley appeared to them the most efficacious measure which could be adopted against torrents, and that it was only in default of proceedings with a view to extinction being adopted —the application of which, when they wrote, was still surrounded with obscurity and uncertainty—that they proposed the measures they did, as means of diminishing, at least provisionally, the danger.

"I do not feel called upon to relate here the difficulties and vicissitudes, moral and administrative, which the foresters had to encounter and overcome in the commencement of the operation.

"The alarms of the peasants, in regard to their pastoral interests, were such that they rose in open rebellion. The ferment was extreme in all the mountain regions, more especially in the region of the Alps; and, as always, political passions and local animosities mixed themselves up with the question at issue and envenomed the discussion.

"Now this agitation is almost calmed down, and it is but right to acknowledge that this happy result is due in a great measure to the spirit which presided over the direction of the operation.

"The means at our command form three categories: *boisement, gazonnement*, and artificial works of consolidation. In order to determine precisely to what extent, and in what circumstances, each of these means should be adopted, it is necessary to study apart their respective actions, and afterwards resume, in a general discussion of the question, the system to be adopted in a plan of extinction." And he proceeds accordingly.

In speaking of the good done by forests on the face of mountains, forming a basin drained by water-courses, he says their beneficial action is manifold; and though this manifold action it may be difficult to unfold, the attempt to do this will place beyond all question that their beneficial action of the water-course is at once most marked and considerable.

"In the discussions which have taken place on this subject," says he, "the point which has engrossed attention to some extent has been almost exclusively the permeability or impermeability of the soil, and the proportion borne by the water absorbed to that which flows off. This is certainly an important question, and no difficulty is found in showing that forests diminish to an enormous extent the amount of water which flows away; but the service which they render is perhaps greater still in regulating, as they do, the flow, and in securing the delivery of only water of perfect fluidity.

"The study of torrents has shown that the evil done consists not so much in the greater or less volume of water discharged as in the disturbances or perturbations of the flow connected with this. The principal causes of these are sudden changes or variations in the delivery and in the degree of fluidity of the flood. And if it be shown that the forests have, in relation to both of these, a regulating power superior to that of any other force operating on the torrent, it will be proved that they are the most potent means of extinguishing torrents.

"If we could expose, by a vertical section, a wooded slope, it would show in the upper portion a layer of varying thickness, but most frequently of from 30 to 40 centimètres (12 or 15 inches) of humus, in which the fibrous rootlets are so developed that the whole has the appearance of a woolly material. This layer is at once a sponge and a filter. The large roots of the trees penetrate more or less into the subjacent rock.

"When the rain falls on ground covered with wood a considerable portion of the water is restored to the atmosphere by evaporation; another portion is absorbed by the immense expansion of foliage and boughs. If the rain be prolonged the water comes at length to the ground, which again is capable of absorbing an immense quantity. A flow from this is slow to establish itself; it is necessary, first, that the saturation of the sponge-like layer be complete; and when this is effected—when the water has been able to make a passage for itself by an infinite number of imperceptible channels—the flow, like that of a charged syphon, maintains a certain uniformity of flow, and this it continues for a long time after the rain has ceased.

"So much is this the case that opponents have alleged that forests are more hurtful than beneficial, as they tend to prolong floods. The flood is prolonged, it is true, but the delivery is regulated—diminished at the commencement and increased at the close: the total quantity of water drained away takes a longer time to flow; it flows during the whole of that longer time; and, what is of more importance, it flows uniformly and equally, with no sudden variations, and thereby much evil is avoided; and, what is of more importance still, the forest acts at the same time as a filter, delivers no water but what is of perfect fluidity, scarcely even discoloured by the washing away of organic matter, and unable to wash away the earth of the subsoil protected against erosion by its thick covering of humus.

"When, on the contrary, the rain falls on a soil stript of vegetation, it tends to cut this up into ravines, and it does so if the tenacity and resistance of the ground be not sufficient to withstand it; and the flood is subject to great variations in its current, carrying off here and there the earth and other debris of the soil.

"Forests have, then, a double action; on the one hand they consolidate the soil, on the other hand they reduce and regulate the flow of the current —acting at once both on the delivery and on the perturbation,—in other words, on the primary cause and on the secondary causes of the overflowing of water-courses.

"It has been tried to subject to experiment and observation the meteorological and hydrological influences of forests. And doubtless studies so interesting are by no means lost to science. They cannot be too much encouraged; but it should be borne in mind that they can have comparatively little value in this question, seeing that they cannot take cognizance of this modulating and regulating action.

"In regard to any flood which we may wish to make the subject of study, it would avail comparatively little to know what quantity of rain falls annually in the basin drained by it. What is necessary to be known is— In what way did the flow of the flood operate during the duration of the flood, taking into account the quantity of water discharged, and all the causes or sources of perturbation operating; which is a much more difficult problem.

"And in resolving the whole question into the permeability of the soil, and its capacity of absorption, it appears, importance is attached exclusively to the reduction of the volume of water which flows away. It seems to be forgotten in this that water-courses, if *steadily supplied*, constitute it may be said the principal riches of a country, and the most potent of all instruments of labour.

"By their modulating power forests act as vast reservoirs, not only in preventing sudden variations of delivery during a flood, but in feeding the water-courses and raising their level during the period of exhaustion. In what relates specially to the torrents of the Alps, it has been demonstrated that the renewed devastating power which they have exhibited, and which has assumed such portentous magnitude in the course of the last forty years, is a consequence of the disappearance of the woods. When one goes over these lands—cut into ravines and despoiled of all vegetation—he meets with numerous stumps of pine and of larch, which testify that at a period as yet still recent they were covered by vast forests.

"M. Surell cites, as an example of the action of forests, the torrent of Savines, now completely extinguished, and the basin of which is everywhere adorned with a magnificent forest of firs and pines. The forest has effectively contributed to the extinction of the torrent, but at this point the following observations may be made:

"This natural extinction of the torrent goes back to ages most remote. The cone is of a perfect geometric regularity. At its base, opposite the Durance, it presents a troncature or section, produced by the erosion of the river, and the escarpment of which is about 30 mètres (100 feet) in height at its culminating point. This section of the ground lays open the interior of the torrential deposit formed of rolled pebbles.

"The whole surface of the cone is cultivated, and on one portion has been built the large village of Savines, the chief place of the canton.

"This enormous heap of deposit is situated at the foot of a high mountain called Morgon, in the flanks of which are dug out a profound gorge surrounded by a vast basin, the work of the water. All the upper slopes of the mountain are hung with a beautiful forest, producing firs more than 30 mètres (100 feet) in height, and 3 mètres (10 feet) in girth.

"The lower slopes are deeply ravined, but wooded to the very edge of the *thalwegs*. A pretty strong stream rises from the principal gorge, but it swells but little; it carries down no materials, and it flows into the Durance by a bed deeply enclosed in the left bank of the cone. Extinction and stability are complete; but it is certain that if the forest should be made to disappear, anew would disorder revive, and this with the same intensity as before.

"In going over the basin with attention, I satisfied myself that everywhere the bed of the *thalwegs* of the gorges and the ravines, formed of the hard rock, were absolutely incapable of being undermined. From this it may be inferred, that during the activity of the torrent, when the basin was being deepened more and more, the surface could not have been wooded. But from the time that the waters had everywhere reached the hard rock, and that these could no more be undermined and washed deeper, their *thalwegs* in the upper slopes tended to consolidate themselves, taking their natural stable declivity; and from that time vegetation could begin to take hold and complete the extinction.

"This remark is important in this way, that if the disappearance of a forest always gives birth to torrential disturbances, it does not always hold true that one can put a stop to them by the planting of a forest alone.

"Much as an unstable ground is protected by being wooded—though it maintains itself and behaves in a hydrological point of view as do the most solid lands, if the wood come to disappear, if the ground be deeply ravined, if the bottom of the ravine continues to be easily undermined and washed away—it becomes extremely difficult to establish vegetation on the mountains, which continually crumble away, and which with this instability no longer retain any trace of vegetable soil.

"In the Alps there are numerous cases of old mountains which crumble away when the foot of the slope is undermined by the water. And one is thus left, if he desire to effect a radical and prompt extinction of a torrent, to give, artificially, to the bottom of the ravine a power of resistance to undermining and washing away, by appropriate works of consolidation.

"But be this as it may, the potent action of forests is beyond all question. Whatever be the character of the woods—timber forest, coppice-wood, or simple shrubbery—all contribute to give firmness to the soil, to retard and to regulate the flow of the water drained off.

"In comparing the different kinds of woods, it may be said that lofty timber forests, with their vast apparatus of foliage at a great elevation above the soil, are of most use with a view to meteorological and hydrological effects; and that young trees serve perhaps better to insure the consolidation of the soil on steep declivities. But as generally, on poor land, the soil of timber forests covers itself with branches, &c., it follows that a mixture of the two kinds of woods accomplishes best the end which it is sought to effect."

In regard to *gazonnement*, he says,—" To report efficiently the influence exerted by a bed of turf or herbage covering the soil, it is necessary to follow the very interesting natural process which goes on when pasturing is suppressed on land, till then, given up to the abuse of pasturage.

"The facts which I am about to state are not exclusively theoretical, they are confirmed by numerous experiments of enclosures which have been made during a great many years.

"I ought first to make an important remark on the subject of the different disintegrations of the soil which occur. Some are simply superficial, and in no way compromise the stability and the solidity of the bulk of the ground. The surface is more or less disintegrated, but the subsoil is unmoved. With others, on the contrary, if they occur on unstable grounds, or on grounds badly poised, breaking up the mass, they tear it deep and move it to its very foundation.

"It is apparent from what is said that the influence of turf or herbage, even if it could be produced there, can be of almost no effect in this latter case. There is nothing but woods, with their strong and deep roots, which can render firm and protect a soil so unstable, and often on such the wood itself requires the assistance of artificial works of consolidation to sustain it. But the action of turf or herbage is, on the contrary, very powerful if it be employed only to repair a superficial evil by removing traces of erosion. Let us take up, then, the description of the work of nature.

"When the soil is no longer trampled, and the few herbs which it is still capable of producing are no longer gnawed to the root, there is a real awakening of the forces of nature. All the buried seeds spring to life. From the very first year the ground changes its aspect, it begins to show a green hue. In autumn the plants shed their seeds. From year to year the vegetation spreads and takes possession of the place more and more extensively.

"In proportion as this change is produced in the state of the surface, the water arrives less rapidly and in less abundance in the ravines and in the hollows of the soil. Its power of carrying off material is weakened; from the first it has no longer the strength to carry off the larger stones, which roll to the bottom of the ravines and stop there.

"To the former work of erosion, and of carrying off material, succeeds the opposite action of *colmatage*—depositing mud, &c., in its course—and of levelling. This action, at first slow, increases with a rapidly advancing progression. The tendency to effect a general levelling extends throughout the whole section of the ground over which the water flows; and a retardation of swiftness succeeds to acceleration. The vegetation promoted by the fertilizing *colmatage* takes on an energetic development; it invades more and more the bottom of the ravines. It is there a characteristic sign of victory being assured to the vegetation.

"When the upper slope is surmounted by crests formed of crumbling rocks, these summits, more exposed to the destructive action of the elements, continue to produce masses of minute debris, which sustain the action of *colmatage* on the lower slopes. Ravines, and all depressions of the soil, tend more and more to efface themselves. The soil goes on rising in these unceasingly. In this new permeable and minutely subdivided layer the turf developes itself with more and yet more vigour, and it finishes by reaching a considerable thickness—it is often 30 centimétres (or 12 inches) thick. One may estimate from this the influence produced by the thick layer of turf or herbage.

"In regard to the consolidation of the soil, the protection is complete.

"In regard to hydrology, the absorption of the water by this permeable layer is so much the greater as—be it in consequence of a greater levelling of the surface, or be it through the effect of the long herbage—the flow of it is subjected to a very great retardation.

"The levelling, in extending indefinitely the section, reduces the mass of water to an extremely thin sheet; and then each shoot of herbage easily breaks this sheet, so that the water which can only acquire velocity by a certain concentration is broken up to such a degree that all flow is impossible, excepting in some extraordinary case as when a water-spout breaks and pours out itself on a single point.

"Woods also induce *colmatage* on the higher slopes which they occupy when the crests are denuded and formed of crumbling rocks, throwing off not only fine debris, but also very often stones and large pieces of rock.

"In certain forests all the trees are severely grazed on the upper side by the shock of stones which roll from the upper part of the mountain. When these projectiles are launched with very great velocity they roll to the foot of the mountain; but most frequently they stop on the lower slope, and form, by their accumulation, a layer of variable thickness.

"When the *colmatage* acts slowly and regularly it is extremely favourable to the vegetation. It is then one of the causes of the beauty of the woods on the upper slopes.

"In a perfected forest culture it would be possible, by light works, to enrich the soil by favouring this natural *colmatage* on steep lower declivities, and if it be desired to fix voluminous materials, wood is preferable; but on gentle slopes the turf and herbage, which act on the small gravel and the finest sands, secure a *colmatage* more complete and more compact.

"It is a fact, ascertained by experience, that lands so covered are more equably levelled than are wooded lands.

"One may then draw this consequence—that, in the given cases, even for lands which it is desired to cover with woods, it is often preferable to subject them previously to the treatment of simple enclosures, that they may be subjected to this natural preparation which levels and fertilizes them.

"When, in consequence of the bad state of the soil, and the too advanced state of the ravine, action of this kind would be too slow to heal over these deep sores, it is necessary to aid them by artificial appliances. It often suffices to put some facines across the ravines, to induce the process of *colmatage*, and to give to this great energy. It is impossible on this point to lay down any fixed rule. The principle is this—when it is by its concentration that the water acquires its velocity, and its power of destruction, it is necessary, as much as possible, and at all points, to diminish the velocity by extending the section.

"Simple enclosure does not produce everywhere a pure *gazonnement*; there is required a certain altitude favourable to the turf forming plants of high mountains, and also certain conditions of the soil. In the lower districts of some countries, from the time that a piece of ground is no longer open to flocks and herds, vegetation revives, and all the plants of the locality, the seeds of which have been preserved in the soil, or have been borne thither by the wind, develope themselves. These are the lavender, the broom, the fescue grass, and very often forest trees, especially the oak, the seeds of which are very tenacious of vitality.

"All this natural vegetation, whatever it may be, is valuable when it acts to restore an impoverished land, and to combat the redoubtable effects of ravines.

"In conclusion, vegetation, under all its forms, is the most powerful means of the restoration and consolidation of the soil, and through this also the most active and most valuable agent in the extinction of torrents; but there are certain cases in which the evil has made such fearful progress that nature, left to its own powers, would be powerless to repair it. It is absolutely necessary to come to its aid if we wish to protect the valley effectually, and above all if we wish to do it quickly."

Referring to the use of *barrages* advocated by Surell, by Scipion Gras, and by Philippe Breton, M. Costa discusses the whole theory of such appliances, viewed both as designed for the consolidation of the soil in danger of being washed away, and as designed to retain or collect gravel— the former the purpose for which they were proposed by M. Surell, the latter the purpose for which they are prescribed by MM. Gras and Breton. He discusses both at considerable length, and also the diversion of torrents into new courses, and what combinations of each and of all of them with *reboisement* and *gazonnement* were likely to be most efficient in different circumstances; and in illustration of the success in consolidating ground which had followed proceedings such as he advocates, he cites what had been accomplished in the extinction of the torrent of Saint-Martha, near

Embrun, to which I shall afterwards have occasion to refer; and he proceeds in the next division of the work to the consideration of the torrential phenomena in great rivers.

He says, that with the knowledge which has been attained in regard to mountain torrents by observation and experience, the question, How can they be controlled and stifled? may be considered as settled. But the same cannot be affirmed in regard to rivers, which throughout their course are governed by the same laws—both those governing or regulating the flow of water and those governing or regulating the torrential phenomena—any apparent modification being attributable to the greater quantity of water, its greater fluidity or lesser viscosity, the lesser rapidity of its flow, and to the more extended reaches throughout which this maintains a uniformity. All that has been done hitherto in regard to rivers has related to the delivery or quantity of water in flow. Attention has not been given to the perturbations of torrentiality and to effects produced by these.

In the regulation of river currents it is desirable that the delivery or water in flow should be equalized, and all perturbation in that flow reduced as much as possible; and he says,—" This double result is obtained by the *reboisement* of the mountains, but it may be brought about in two different ways. When the object to be accomplished by the planting of forests is to equalize the delivery or quantity of water it is necessary to extend the *boisement* over extensive areas, comprising the greater portion of the basin. If, on the other hand, it is the perturbations in the flow upon which it is desired to act, it is necessary to concentrate the *reboisement* on properly selected points, and, it may be, to strengthen the action of these by a series of the artificial accessory works employed for the extinction of torrents.

" It is this latter system which is alone efficacious and practicable in acting on a great water-course.

" In a basin such as that of the Loire, for example, there might be planted a hundred thousand hectares of land without perceptibly modifying the *régime* of the river, if the lands were not selected with intelligence with a view to the consolidation of the soil and to the accomplishment of the end desired. A study of the whole course of the river, and a comprehensive plan of operations founded thereon, is absolutely requisite as a preliminary measure.

" It is necessary from the first thoroughly to know the *régime* of the particular water-course, and to ascertain its torrentiality. This may be accomplished by a general reconnoissance.

" All the affluents should be classified in a hydrological chart, according to their degree of torrentiality.

" Most frequently an inspection of the state of the confluence suffices to reveal the *régime* of the affluent. When such is torrential it will be found to straggle over an extended bed before flowing into the main stream.

" By subjecting, then, every one of the affluents to such an examination, and following out this in all the upper ramifications of the river, it is easy to determine what are the main centres of production of the stone or clay materials borne along by the river, which are the causes of the perturbation which have to be fought.

" By this procedure the evil is localized, determined, and circumscribed; and it is often astonishing to find how limited in extent, compared with the area of the basin, are the whole of the sources whence the gravel is obtained.

" By such a procedure the operation is not left to chance. All is done rationally, with an adaptation of means to the end. From the time that the extent of the sore which has to be cicatrized is known and defined, it is easy to report beforehand on the importance of the work to be done, on the expense it will entail, and on the time which will be required for its execution.

"It may be necessary to limit the operations to bring them within the means at command, but what is done is done in accordance with a fixed plan and with the assurance of success.

"I do not conceal from myself that I expose my remarks to the charge of being premature. I state them more for the future than for the present. A work so colossal cannot be improvised. Every new idea requires to be matured before it be accepted. It has got, when true, to pass through the sieve of contradiction and opposition, but it issues in triumph.

"The *reboisement* of the mountains, looked at from this point of view, has already overcome obstacles; and it has stood the test of public calamities. It is making good its position day by day, and in proportion as it becomes better understood, more and more will the necessity of developing it be felt.

"To state my opinion in a few words, this is the necessary solution of the matter: it is an efficacious one, and there exists not another for a problem which we cannot elude, and which presents itself in a more and more threatening aspect.

"I shall esteem myself happy if by this treatise, which is imperfect, but which is expressive of deep convictions, I may contribute to hasten on the time when our beautiful rivers shall no longer inspire dread or bring danger, but become magnificent highways of navigation."

The title of the work of M. Costa bears that he treats not only of the laws, causes, effects, and means of repressing torrents, but also of the means of utilising them. Means of doing this are indicated again and again in the course of the work, but the suggestions thus given exhaust not his views of what may in this way be effected.

"The great perturbations in the order of nature which leave often behind them saddening traces of their occurrence," says he, "fulfil, nevertheless, a useful, and it may be a necessary, function in the work of creation. The storms which create a turmoil in the atmosphere purify the air. Without the cyclones of the Indian Ocean, the latitudes in which they occur would not be habitable. And storms on the sea help to prevent a tainting of the waters, by commingling with the superficial layers waters from deeper layers more nearly saturated with salt.

"The inundations of water-courses, against which we seek to protect ourselves now, have served to create fertile alluvia on gigantic deltas, and on many rich valleys, some of them the most beautiful parts of the earth, in which human society has been able to develope itself, and to bring forth its marvels.

"Even in our own times, beneficent inundations—natural or artificial—by depositing in certain valleys an earth which is repairing an exhausted land, are the means of generating wealth. We have, then, in inundations a force or power which sometimes occasions ruin and devastation, but which sometimes becomes a valuable instrument of good, according as its action may be chaotic or controlled.

"Having seen how this force may be controlled and kept within bounds, it is reasonable to suppose, and I cannot but believe the supposition to be in accordance with fact, that in accordance with that unity which pervades every thing, it is possible to indicate a way by which it may be utilised; and this all the more that such a way there is, based on the very laws regulating torrential phenomena which have been brought under consideration.

"It is this consideration which has determined me to devote some pages to that interesting agricultural operation known under the name of *colmatage*, or warping, and practised by the Egyptians from time immemorial with great skill.

"To transform deserts into meadows—stony ground, absolutely sterile, or producing only a sorry pasturage, into alluvial lands, capable of bearing a covering of the most luxuriant and richest vegetation—is certainly not only one of the most lucrative enterprises, but also one, in every aspect of it, most interesting. Everywhere, where it is has been tried in favourable circumstances, it has produced results surpassing all expectations which had been entertained.

"There exist in France extensive districts, especially in the south, in which this operation might be carried out advantageously.

"The immense plain which extends from the town of Arles, in Provence, to the sea on the left bank of the Rhone, known under the name of the *Craw*, is in its central part a veritable desert of *forty thousand* hectares, covered with pebbles, thoroughly burned up by drought in summer, but where, during the rainy season in winter, there grow some stalks of herbage on which the flocks of transhumant sheep feed.

"The fertilization of this plain by *colmatage*, by means of the waters of the Durance, would be of immense benefit not only to Provence but to the whole country. It would be there a creation of enormous agricultural wealth, which would, without fail, have a reaction on the national wealth and the general well being."

He states that the credit of first entertaining this idea does not belong to him; that it has again and again engaged the attention of men given to the study of natural phenomena, and of great ameliorations of which terrestrial conditions are susceptible; and he gives great credit more especially to M. Scipion Gras for what he had done and was doing to promote the enterprise. Having done so, he proceeds to expound his views of what might be effected.

Next in importance to preventing the devastations occasioned by inundations, by the washing away of earth and earthy materials from the higher-lying basins drained by torrents, and by the deposit by these on fertile fields and valuable lands of a covering of sand and gravel and stones, the detritus of mountains washed away by the torrent in its rage, he seems to have deemed the plan of so constructing water-courses that, where practical or desirable, these should be made to make some compensation for the mischief done by them, or by others of their kind, by covering barren plains with fertile soil.

As the result of the study of natural phenomena, he states, that on pure clay a *gazonnement* of herbage is not produced, but that it developes itself with great vigour on miscellaneous deposits, and this gives rise to the speedy formation of an upper layer of vegetable soil, and that thus, in a very short time, there is produced there fertile grounds requiring only in addition a little manure to promote their fertility.

Leaving the subject of torrents, except in so far as their phenomena might serve to illustrate his subject, he prepares for the discussion of the possibility of preventing inundations by a discussion of the torrential phenomena in great rivers. And considering, as he did, that attention had been given too exclusively to variations in the delivery of water-courses—no previous writer, so far as was known to him, having even admitted the idea of perturbations in the flow giving rise to a confusion of the so-called hydraulic laws, of something like a revolution breaking out in a water-course and producing an instability which mocked all provisions and precautionary measures alike—to these perturbations and their phenomena he gives special attention; and he again brings under consideration what it is which constitutes torrentiality, with a view to showing that it is seen in rivers as well as in what are designated torrents.

According to M. Costa, his definition of a torrent embodies the idea of its bearing along earthy matter in suspension; and he states that it does so both in a mass and in what is known in France as *triage*, dropping some and carrying on others of the materials in question; in the former case all the rocks, pebbles, and lesser fragments are carried along in something like their relative positions, as would be the case in a viscid mass or in a glacier; in the latter the weightier materials are dropped first, and this going on more or less continuously, the matters in a state of extreme comminution are carried furthest. The difference in mode of transport appears in connection with difference in the velocity of the flow. When this is so great as to bear the whole along in a mass, the stones, whatever their size, do not come into collision, and if any were withdrawn they would be found to be as little rounded as are the stones falling from a glacier and forming a *moraine*.

But when the velocity is being impaired, as this goes on the stones begin to roll, *suspended in the water*, and they may come into collision one with another; and the heavier sinking, these are for a time rolled along the bottom and subjected to collision and friction. At length they rest, and where they rest the collision of others following and proceeding further subjects them to continued abrasion; and what happens thus to the heavier masses happens there or further in advance, in succession as the velocity is reduced, to others of lesser weight.

In view of these phenomena he makes valuable suggestions in regard to the structure of *barrages*. He suggests that, by submerged *barrages* of little height, if these be properly disposed, the velocity of a torrent might be so reduced as to secure a deposit over a great extent of ground of the impalpable mud borne down by a torrent. He states that much of the mud thus carried along would be infertile, but that much fertile vegetable mould is thus buried in the sea; and he proposes that in certain circumstances in which this may be practicable and desirable this should be so secured; he points out places of great extent in regard to which he proposes that this should be done; and he cites what has been effected in the High Alps in evidence of the practicability and advantages of the measure.

M. Costa looks, in the light of these observations, at the geological phenomena which led M. Cézanne to conclude that the so-called glacial or drift period was succeeded by what he considered a torrential era; the different appearance of stones found in some heaps from that presented by these in others—these being in the one angular and in the other rounded, the former like the stones forming moraines deposited by glaciers, the others like those found in *lits de déjection*, and attributable to attrition, escaped in those, having befallen these, having been suggested or been called in to support the theory or hypothesis. M. Costa, looking to the torrential phenomenon of transport of material *en masse*, differs from M. Cézanne only, or chiefly, so far as to attribute the whole, or by far the greater part of the deposits in question, which are extensively distributed over some parts of France, to torrential action alone ; maintaining and citing in support of his views phenomena of torrents established by his previous observations which go to prove that torrential action is equal to the production of all the phenomena of these deposits,—the transfer of the blocks of the greatest magnitude seen, and the transfer of these and of lesser stones without damage to their angular outlines, and to the deposit of them where they are, and in the form in which they are found.

He does not deny that the effect of glaciers is what it is believed to be, but he alleges that torrents, charged and surcharged with earthy matter, bear off rocks and stones, and such earthy matters, in certain circumstances, in a somewhat viscid mass, in which each constituent part may be conceived of as retaining its relative position very much as such matters do in a glacier, and therefore with their angularities unbroken. But in regard to the composition and contour of the beds of deposit, he cites observations of M. Cézanne and of his own which seem fully to warrant a conclusion drawn by him from them, that all moraines—deposits chiefly associated with glaciers—have not been produced by these, some having been produced by landslips and avalanches, if not otherwise, but that all cones of dejection are the products of torrents. But he goes further, alleging that while the melting or breaking up of the ice is only an accidental and local phenomena, torrential phenomena are common and universal, and are so to such an extent as to make the term torrential era objectionable ; torrential force being a force which not only has manifested itself in a permanent and continuous manner within and throughout the historic period, in modern alluvia, and in the geologic period, in ancient alluvia—buried, some of them, to the greatest depth ; but also in what may be called the cosmogonic period, at every instant of the earth's life exercising an influence on the very contour of the globe, if not also acting in the sun and in the planets.

Such are the views entertained by M. Costa of such deposits as are described in my citations from the work of M. Cézanne.

And with all his enlarged and comprehensive views of torrential action, he appears to have held the same views as those I have cited as the views of Marschand and of Cézanne in regard to the later history of torrents in France, to the extinction of these by the spread of vegetation, to the resuscitation of them by the destruction of forests, and to the re-extinction of them by *reboisement* and *gazonnement*.

Thus do all concur in pointing out to us the stage of the process in South Africa, and in other newly settled countries, indicated by torrential floods, when looked at in connection with the destruction of grass and herbage and bush and trees witnessed in and beyond the portions colonized by Europeans.

In regard to the means to be employed to secure the extinction of these, he says,—" When the torrentiality is feeble, and the evil consists mainly in the quantity of the water, it is by *boisement*, and the spread of vegetation, that it should be sought to effect the restraint or extinction of it. If, on the contrary, the torrentiality is extreme, and the devastations produced by it proceed principally from the perturbations in the flow being now violent, now feeble, this should be rectified by the extensive application of works of consolidation, such as Surell has recommended.

"The works of *reboisement* and *gazonnement*, to be effectual, require to be extended over large areas. Works of artificial consolidation, on the contrary, may be confined within a limited space, and the evil may in some cases be stifled by attacking it in its principal source.

"Sometimes it may happen that, through the pastoral and agricultural operations carried on, it is impracticable to give to *boisement* the whole extension necessary to meet the evil. Every case must be decided on its own merits. And, from the general considerations adduced, it is apparent that it is impossible to lay down invariable rules of procedure applicable to every case.

"When there is no special urgency for securing immediate results, it is preferable to employ at once vegetation. By enclosing a space, it is found that the spontaneous work of nature exercises a most favourable influence on the soil. Cover, then, with woods all the lower slopes, where there is no fear of the earth crumbling away, and where the spontaneous work of nature is not likely to cover them with vegetation. With this done, the delivery of water will insensibly diminish; losing bulk and velocity, it retains no longer the same power of undermining and washing away; the hills are less frequently and less powerfully attacked; and where it is reckoned that the torrent is sufficiently enfeebled, there may then, if it be thought necessary, be established in the gorges, with greater ease and at less expense, works of consolidation deemed useful.

"This order of procedure is more sure, and more economical, but more slow than is the reverse.

"With the vegetation there may be combined, on the upper slopes and in the lesser ravines, a great many small works of consolidation, the design of which is to effect this by retarding the velocity of the flow, and the substituting of *colmatage* for the undermining and carrying away of the soil.

"The time for undertaking works of consolidation in the gorges must be determined by the degree of urgency for a speedy extinction of the torrents, and by administrative considerations, of which the superior authority is the judge."

In illustration of what may be done, M. Costa cites the extinction of the torrent of Saint Martha, already referred to.

Besides the works which have been cited, the following have been published in France :—

BELGRAND, membre de l'Institut, inspecteur général des ponts et chaussées. *Hydrologie et météorologie du bassin de la Seine.*

BELANGER, ingénieur en chef des ponts et chaussées. *Essai sur le mouvement permanent des eaux courantes.*

COLLIGNON, cours d'hydraulique professé à l'Ecole des ponts et chaussées.

COMOY, inspect. gén. des ponts et chaussées. *Mémoires sur les ouvrages de défense contre les inondations.*

DARCY ET BAZIN. *Recherches hydrauliques.* Première Partie—Recherches expérimentales sur l'écoulement de l'eau dans les canaux découverts. Deuxième Partie—Recherches expérimentales sur la propagation des ondes.

DUMONT, ingénieur en chef des ponts et chaussées. *Les eaux de Lyon et de Paris,*—projets, tracés et détails d'exécution suivis d'une pratique des distributions d'eau.

O

Dupuit, inspect. gén. des ponts et chaussées. *Des Inondations, examen des moyens proposés pour en prévenir le retour.*

Dupuit, inspect. gén. des ponts et chaussées. *Études pratiques et théoriques sur le mouvement des eaux courantes, suivies de considérations relatives au régime des grandes eaux, au débouché à leur donner, et à la marche des alluvions dans les rivières à fond mobile.*

Fargue, ingénieur des ponts et chaussées. *Étude sur la corrélation entre la configuration du lit et la profondeur d'eau.*

Fournié (V.), ingénieur des ponts et chaussées. *Résumé des expériences hydrauliques exécutées par le gouvernement américain sur le Mississipi, et remarques sur les conséquences qui en découlent relativement à la théorie des eaux courantes.*

———, ingénieur des ponts et chaussées. *Progrès récents de la météorologie.*

———, *Amélioration des rivières torrentielles.*

Graeff, inspect. gén. des ponts et chaussées. *Théorie des réservoirs.*

Krantz, ingénieur en chef des ponts et chaussées. *Murs de réservoirs.*

Lamairesse. *Hydrologie du département du Jura.*

Malézieux, ingénieur des ponts et chaussées, professeur à l'Ecole des ponts et chaussées. *Rapport sur un voyage aux Etats-Unis exécuté par ordre de S. Exc. le ministre des travaux publics.*

Mangon, ingénieur en chef des ponts et chaussées, professeur à l'Ecole des ponts et chaussées. *Instructions pratiques sur le drainage.*

———, *Expériences sur l'emploi des eaux dans les irrigations.*

Monestier-Savignat, ingén. des ponts et chaussées. *Études sur les phénomènes, l'aménagement et la législation des eaux au pont de vue des inondations, avec application au bassin de l'Allier, rivière à régime torrentiel, affluent de la Loire.*

Nadault de Buffon, ingénieur en chef des ponts et chaussées, professeur à l'Ecole des ponts et chaussées. *Des submersions fertilisantes, colmatage, limonage, irrigations d'hiver.*

———, *Des irrigations, canaux d'arrosage de l'Italie septentrionale.*

Partiot, ingénieur des ponts et chaussées. *Etude sur le mouvement des marées dans les parties maritimes des fleuves.*

De Passy, ingénieur en chef des ponts et chaussées. *Etudes sur le service hydraulique.*

Plocq, ingénieur des ponts et chaussées. *Etude des courants et de la marche des alluvions aux abords du détroit du Pas-de-Calais.*

Poirée (M. A.), inspect. gén. des ponts et chaussées (en retraite). *Quelques mots de réponse à la brochure de M. Dupuit, intitulée: des Inondations.*

Vigan. *Irrigations des Pyrénées orientales et phénomène dit: Production des eaux.*

Several of the subjects embraced in this department of the forest science of France have engaged the attention of students of nature in other countries. Copious extracts from French works, with copies of official documents issued in France relative to *reboisement* and *gazonnement*, are embodied in an official report, issued in the province of Luxemburg, entitled *Die Reboisement des Terraines vagues Rapport présenté au Conseil provincial par la Députation permanente Session de 1867.* But being a compilation and report, made with a view to work being undertaken, it communicates no accounts of hydrological results obtained.

Strofflour, in a paper, *Ueber die Natur und die Wirkungen der Wildbäche*, which first appeared in the *Ber. der M. N. W. Classe der Kaiserl. Akad. der Wiss.* for February 1852, maintains that all the observations and speculations of French authors on the nature of torrents had been anticipated by Austrian writers; and in support of this assertion he refers to the works of Franz von Zallinger, 1778, Von Arretin, 1808, Franz Duile, 1826,—all published at Innsbruck,—and Hagen's *Beschreibung neuerer Wasserbauwerke*, published in Konigsberg in 1826. And M. Cézanne, in his continuation of the treatise of M. Surell, says, and says unhesitatingly, after speaking of the importance of utilizing, taming, and domesticating torrents,—as beasts and birds have been tamed, domesticated, and utilized,—" France and Switzerland are not the only countries in which the struggle against devastating running waters is being carried on with alternate triumphs and defeats. And we may conclude from the works now analyzed, and from the numerous publications which there are of the same kind, that the time is still remote when man shall have completely subdued, and, if the word may be used, *domesticated*, tamed, and utilized the wild waters of the mountains. But there is one happy land, the picture of which, contrasting with these gloomy sketches, may be offered to inquirers as a model and as an encouragement. It is the German *Hartz*.

"This mountainous mass, almost isolated on all sides, and but lately divided amongst four Governments, raises its highest summit—the Brocken—to a height of 1250 mètres, upwards of 4000 feet; steep slopes and deep *thalwegs* are not awanting, nor are abundant rains—the rainfall ranging from 600 to 1500 millimètres (from 24 to 60 inches). The ground is very diversified; granitic eruptions have dislocated schists of all kinds; all circumstances and conditions favourable to torrential phenomena are there in combination; but the mining industry, in quest of motive power, has seized upon the water—a force supplied without money and without price, and renewed unceasingly by nature; and it may be said that there there is not a single drop left to follow its natural course; from the highest slopes the rain is collected in furrows forming gutters; all the ravines are closed up, and numerous ponds store up their supplies; collected in canals the waters make the circuit of the brows of the hills, are carried across valleys, bury themselves in projecting spurs, and, conducted to the gate of the factories, move the hydraulic wheels placed one below another at all the descending levels of the mountains; and, coming at length to the *thalweg*, the waters are not yet freed,—they are made to descend into the mine and there to work underground.

"Seventy ponds or reservoirs of the Ober Hartz have an area of 240 hectares; they store up fifteen millions of cubic mètres, which put in movement above ground 180 water-wheels, and underground 23 wheels and 2 hydraulic presses."

From the report of MM. Belgrand and Lemoine, in the *Annales des Ponts et Eaux*, for 1868 (II. p. 307), it appears,—" There are in all 200 kilomètres (upwards of 150 miles) of canals employed to bring the waters to the ponds, and to lead them to the manufactories and to the mines.

"From the highest situated pond (Hirschler Teich) to the *Lautenthal*, there is for wheels above ground a total fall of 292 mètres (well-nigh 1000 feet).

"For the mines the available fall is still greater; it is about 370 mètres (nearly 1250 feet). These waters underground give motion to draining pumps and other machinery, are re-united in different galleries or tunnels—

finally in the Ernst-August-Stollen—situated at about 370 mètres (1250 feet) under the plateau of Chausthal. This canal, in which are collected all the waters of the subterranean sheets, conveys them to the open air at Gittelde; it is not less than 23·600 mètres in length, and its other dimensions are considerable, for this long subterranean passage can be made by boat. Its breadth is 1·90 mètre (6 feet 4 inches), and its height 2·70 (or 9 feet). This magnificent work was completed in 1864, and cost a little more than three millions of francs."

M. Cézanne goes on to say,—"Many of these works date from the beginning of the eighteenth century. It is a hundred and fifty years since the inhabitants of the German Hartz have subdued, tamed, and turned to use their running waters; it is only ten years since we began to give ourselves to the attacking and mastering of the torrents of the Alps!"

In the German literature of Först-Kunde are not a few treatises on torrents, on their destructive effects, and on the means of preventing and counteracting these.

Streffleur refers to a brochure by Franz Duile. It was published in 1826, under the title *Ueber Verbauunng der Wildbäche in Gebirgs Ländern*, and in it the author gives an exposition of the principles which should be applied to all works of this kind.

"He studied successively," says Marschand, "the construction of stone dams and of wooden erections having the same object to accomplish. The former, described by him, are dry-stone dykes, and are composed of a horizontal vault with the arch directed up the stream, and sustaining walls forming kinds of butments where the hills are not of rock, and to prevent underminings a *radier* or screen of stone retained by wood-work. The summit of the dam is lowered somewhat in the middle to facilitate the flow of water, and it is covered with a wooden plank designed to maintain the solidity of the summit. The wooden *barrages* described by him are similar to those in use in many parts of the French Alps.

"Duile superintended numerous works on torrents, but through neglect, or perhaps through the force of the current, they all crumbled into ruins; and at the close of his life he expressed to Professor Culmann regret that he had undertaken works against torrents.

"He treated also of *reboisement*, assisted by *clayonnages* or hurdles, and, in a word, of everything relating to the extinction of torrents.

"Great works undertaken by his advice leave no doubt of the efficacy of his system."

In 1844 was published at Darmstadt *Das Verdrängen der Laub-Wälder*, in *Nordlichen Deutschlande durch die Fichte und der Keifer*, by Edmond von Berg. In 1852 was published at Erlangen *Das Verhalten der Waldbaume gegen Licht und Schatten*, by Gustav Heyer; both of which have reference to the subject under consideration.

In a *Handbuch der Physischer Geographie* by Klöden, referred to by Marsh, it is stated by the latter that the author, "admitting that the rivers Oder and Elbe have diminished in quantity of water—the former since 1778, the latter since 1828—denies that the diminution of volume is to be ascribed to a decrease of precipitation in consequence of the felling of the forests; and states, what other physicists confirm, that during the same period meteorological records, in various parts of Europe, show rather an augmentation than a reduction of rain."

The statement made by Klöden is in accordance with observations made by others elsewhere. The effect of forests, and of the destruction of forests, on climate, both as regards the water supply and the temperature, has received great attention from Dr Draper, Director of the Meteorological Observatory in the Central Park, New York, with results, to some extent, in accordance with those stated, deduced from observations in the United States of America.

I have not seen the statement made by Klöden, which occurs in his *Handbuch der Physischer Geographie* (p. 658), but taking the import of it to be as given by Mr Marsh, the phenomena may be susceptible of explanation. There may have been a general increase of the rainfall, but a diminution of the drainage of the surplus moisture of the land. The ground may have become more desiccated, and that to such an extent that even an increased rainfall does not maintain the rivers at their height.

Of other German works bearing on the subject of torrents I may mention the following:—*Die Oesterreichischen Alpenländar und ihre Förste*, by Joseph Wessely—published in Vienna, 1853. *Ansiehten uber die Bewaldung der Steppen des Europæschen Russland*, by J. Van den Brinken—Braunschweig, 1854. *Die Gebirgsbache und ihre Verherungen*, by Franz Muller—Landshut, 1857; the author was a Bavarian engineer, and the work treats of the construction of *barrages*, more especially those of masonry and wood, but it treats also of the fixation of mountains, by means of hurdles, with a view to *reboisement*. *Der Wald samt dessen wichtigen Einfluss auf der Klimat, &c.* —published in Vienna, 1860. *Die Alpen in Natur und Lebensbildern*, by H. Berlepsch—Leipsig, 1862; a work of which an English translation, by —— Stephens, has been published.

In Italy much attention has has been given to irrigation, and the utilization and economising of the water supply; the Italian literature on subjects connected therewith is very voluminous; and the effect of vegetation on the humidity of the climate, and the supply of moisture for the promotion of vegetation, has not been overlooked. Of Italian works relating thereto the following may be noted:—*Dell' Immediata Influenza delle Silve sul Corso delle Aeque*, by Castellani—Torino, 1818-1819. *Dell' Impiante e Conservazione dei Boschi*, by Guisippe Cereni—Milano, 1844. *Necessita dei Boschi nella Lombardia*, by Antonio Giovanni Batti Villa—Milano, 1850. *Connisulla Importanza e Coltura die Boschi*, by Pietro Caimi—Milano, 1857. *Le Condizioni de Boschi de fiumi e de Torrenti nella Provincia de Bergamo*, by G. Rosa, in Politecno, Decembre, 1861, pp. 606-621. *Studii sui Boschi*, by the same writer, in Politecno, Maggie, 1862, pp. 232-238.

The subjects of *colmatage* or warping, alluvian drainage, and defences against inundation—all of which come within the scope of the French treatises I have cited—have also found a place in the Italian literature of hydrology. To this chapter belong the following works:—*Memorie sul bonificamento delle Maremme Toscano*, by Fernando Tartini; *Sulle Paludini Pontine*, by Eustachio Zanotti; *Relazione e voto sopra il deseccamento delle Paludi Pontine*, by Gaetano Rappini; *Sopra la distribuzione delle alluvione*, by Vittoria Fossombroni; *Richerche idrauliche relative alle Colmate*, by Pietro Paoli; *Intorno al ripararo delle innondazioni dell' Adige la citta di Verona*, by Antonio Lorgna.

In our own language has appeared in a first edition, entitled *Man and Nature*, published in 1863, and in a second edition, entitled *The Earth as modified by Human Action*, published in 1874, by the Hon. George P. Marsh,

Minister of the United States of America at Rome,—a work in which there is embodied a great mass of valuable information on the subject of torrents, or the extinction of these, and on subjects closely related to these.

I shall afterwards have occasion to quote at length statements by Mr Marsh in regard to the provinces of Dauphiny and Provence, to the valley of the Rhone, and to the department of Dévoluy. His own remarks on subjects connected with the occasion, phenomena, and control of torrents are not less deserving of consideration. His position as Minister of the United States at different courts, with a perception of the importance of such matters, have given him exceptional advantages for the study of this matter, as of much besides, of which he has conscientiously availed himself, and embodied the results in his more comprehensive treatise.

By the information supplied by such works as *Les Inondations en France depuis le VIe Siècle jusqu'à nos jours*, by Champion, and *Les Forêts de la Gaule et d l'ancient France*, already cited, the student in this department of forest science can carry back his studies to times that are past.

Of these Mr Marsh writes:—"The remarkable historical notices of inundations in France in the Middle Ages collected by Champion are considered by many as furnishing proof that, when that country was much more generally covered with wood than it now is, destructive inundations of the French rivers were not less frequent than they are in modern days. But this evidence is subject to this among other objections: we know, it is true, that the forests of certain departments of France were anciently much more extensive than at the present day; but we know also that in many portions of that country the soil has been bared of its forests, and then, in consequence of the depopulation of great provinces, left to reclothe itself spontaneously with trees, many times during the historic period; and our acquaintance with the forest topography of ancient Gaul or of mediæval France is neither sufficiently extensive nor sufficiently minute to permit us to say with certainty that the sources of this or that particular river were more or less sheltered by wood at any given time, ancient or mediæval, than at present. I say the sources of the rivers, because the floods of great rivers are occasioned by heavy rains and snows which fall in the more elevated regions around the primal springs, and not by precipitation in the main valleys or on the plains bordering on the lower course.

"The destructive effects of inundations, considered simply as a mechanical power by which life is endangered, crops destroyed, and the artificial constructions of man overthrown, are very terrible. Thus far, however, the flood is a temporary and by no means irreparable evil, for if its ravages end here, the prolific powers of nature and the industry of man soon restore what had been lost, and the face of the earth no longer shows traces of the deluge that had overwhelmed it. Inundations have even their compensations. The structures they destroy are replaced by better and more secure erections, and if they sweep off a crop of corn, they not unfrequently leave behind them, as they subside, a fertilizing deposit which enriches the exhausted field for a succession of seasons. If, then, the too rapid flow of the surface-waters occasioned no other evil than to produce, once in ten years upon the average, an inundation which should destroy the harvest of the low grounds along the rivers, the damage would be too inconsiderable, and of too transitory a character, to warrant the inconveniences and the expense involved in the measures which the most competent judges in many parts of Europe believe the respective governments ought to take to obviate it.

"But the great, the irreparable, the appalling mischiefs which have already resulted, and which threaten to ensue on a still more extensive scale hereafter, from too rapid superficial drainage, are of a properly geographical, we may almost say geological, character, and consist principally in erosion, displacement, and transportation of the superficial strata, vegetable and mineral—of the integuments, so to speak, with which nature has clothed the skeleton frame-work of the globe. It is difficult to convey by description an idea of the desolation of the regions most exposed to the ravages of torrent and of flood; and the thousands who, in those days of swift travel, are whirled by steam near or even through the theatres of these calamities, have but rare and imperfect opportunities of observing the destructive causes in action. Still more rarely can they compare the past with the actual condition of the provinces in question, and trace the progress of their conversion from forest-crowned hills, luxuriant pasture grounds, and abundant cornfields and vineyards well watered by springs and fertilizing rivulets, to bald mountain ridges, rocky declivities, and steep earth-banks furrowed by deep ravines, with beds now dry, now filled by torrents of fluid mud and gravel hurrying down to spread themselves over the plain, and dooming to everlasting barrenness the once productive fields. In surveying such scenes, it is difficult to resist the impression that nature pronounced a primal curse of perpetual sterility and desolation upon these sublime but fearful wastes, difficult to believe that they were once, and but for the folly of man might still be, blessed with all the natural advantages which Providence has bestowed upon the most favoured climes. But the historical evidence is conclusive as to the destructive changes occasioned by the agency of man upon the flanks of the Alps, the Apennines, the Pyrenees, and other mountain ranges in Central and Southern Europe, and the progress of physical deterioration has been so rapid that, in some localities, a single generation has witnessed the beginning and the end of the melancholy revolution."

He cites statements made by Surell and by Blanqui which have been already quoted. He says, in connection with their statements relative to Dévoluy, Barcelonette, and Embrun,—" It deserves to be specially noticed that the district here referred to, though now among the most hopelessly waste in France, was very productive even down to so late a period as the commencement of the French Revolution. Arthur Young, writing in 1789, says,—' About Barcelonette, and in the highest parts of the mountains, the hill-pastures feed a million of sheep, besides large herds of other cattle;' and he adds,—' With such a soil, and in such a climate, we are not to suppose a country barren because it is mountainous. The valleys I have visited are in general beautiful.' He ascribes the same character to the provinces of Dauphiny, Provence, and Auvergne, and though he visited, with the eye of an attentive and practised observer, many of the scenes since blasted with the wild desolation described by Blanqui, the Durance and a part of the course of the Loire are the only streams he mentions as inflicting serious injury by their floods. The ravages of the torrents had, indeed, as we have seen, commenced earlier in some other localities, but we are authorized to infer that they were, in Young's time, too limited in range, and relatively too insignificant, to require notice in a general view of the provinces where they have now ruined so large a proportion of the soil."

After giving a picture of the devastations wrought by the Ardéche, which I shall afterwards have occasion to quote, he goes on to say,—" As I have before remarked, I have taken my illustrations of the action of torrents and mountain streams principally from French authorities, because the facts recorded by them are chiefly of recent occurrence, and as they have been collected with much care and described with great fulness of detail, the information furnished by them is not only more trustworthy, but both more complete and more accessible than that which can be gathered from any other source. It is not to be supposed, however, that the countries adjacent to France have escaped the consequences of a like improvidence. The southern flanks of the Alps, and, in a less degree, the northern slope of these mountains and the whole chain of the Pyrenees, afford equally striking examples of the evils resulting from the wanton sacrifice of nature's safeguards. But I can afford space for few details, and as an illustration of the extent of these evils in Italy, I shall barely observe that it was calculated ten years ago that four-tenths of the area of the Ligurian provinces had been washed away or rendered incapable of cultivation in consequence of the felling of the woods.

"Highly coloured as these pictures seem, they are not exaggerated, although the hasty tourist through Southern France, Switzerland, the Tyrol, and Northern Italy, finding little in his high-road experiences to justify them, might suppose them so. The lines of communication by locomotive-train and diligence lead generally over safer ground, and it is only when they ascend the Alpine passes and traverse the mountain chains, that scenes somewhat resembling those just described fall under the eye of the ordinary traveller. But the extension of the sphere of devastation, by the degradation of the mountains and the transportation of the débris, is producing analogous effects upon the lower ridges of the Alps and the plains which skirt them; and even now one needs but an hour's departure from some great thoroughfares to reach sites where the genius of destruction revels as wildly as in some of the most frightful of the abysses which Blanqui has painted."

"According to Arthur Young," who travelled in France, Italy, and Spain, in 1789, says Marsh, "on the lower Po, where the surface of the river at high water has been elevated considerably above the level of the adjacent fields by diking, the peasants in his time frequently endeavoured to secure their grounds against threatened devastation through the bursting of the dikes, by crossing the river when the danger became imminent and opening a cut in the opposite bank, thus saving their own property by flooding their neighbours'. He adds, that at high-water the navigation of the river was absolutely interdicted, except to mail and passenger boats, and that the guards fired upon all others; the object of the prohibition being to prevent the peasants from resorting to this measure of self-defence."

Streffleur quotes from Duile the following observations: "The channel of the Tyrolese brooks is often raised much above the valleys through which they flow. The bed of the Fersina is elevated high above the city of Trent, which lies near it. The Villerbach flows at a much more elevated level than that of the market-place of Neumarkt and Vill, and threatens to overwhelm both of them with its waters. The Talfer at Botzen is at least even with the roofs of the adjacent town, if not above them. The tower-steeples of the villages of Schlanders, Kortsch, and Laas, are lower than the surface of the Gadribach. The Saldurbach, at Schluderns, menaces the far lower village with destruction, and the chief town, Schwaz, is in similar danger from the Lahnbach."

PART III.

LEGISLATIVE AND EXECUTIVE MEASURES TAKEN BY THE GOVERNMENT OF FRANCE, IN CONNECTION WITH REBOISEMENT AND GAZONNEMENT, AS REMEDIAL APPLICATIONS AGAINST DESTRUCTIVE TORRENTS.

THE term *reboisement* is one of modern origin, because that to which it is applied is only of modern date, and I know not an English term of similar import. By a periphrasis the thing may be described, but the conventional term is more convenient than the constant use of a periphrasis would be, and more explicit than a literal rendering of the term would be, or any English synonym with which I am acquainted.

The term is applicable, strictly speaking, to re-planting with trees a place or a district previously clothed or adorned with forests. It is held by some students of forest science that there is a tendency in many species and genera of arborescent vegetable productions to encroach upon and take possession of all unoccupied land, and in the struggle of life to dispossess other plants—if these have previously taken possession of the land,—and if these cannot submit to their domination. The names of places innumerable in various parts of Europe—Britain and the Continent alike—are terms applicable, strictly speaking, only to various forms of wooded land, and supply a presumptive proof that these were once forest homes.

Marsh says,—"We may rank among historical evidences on this point, if not technically among historical records, old geographical names and terminations, etymologically indicating forest or grove, which are so common in many parts of the eastern continent now entirely stripped of woods—such as, in southern Europe, Breuil, Broglio, Brolio, Brolo; in northern, Brühl, and the endings dean, den, don, ham, holt, horst, hurst, lund, shaw, shot, skog, skov, wald, weald, wold, wood.

"The island of Madiéra, whose noble forests were devastated by fire not long after its colonization by European settlers, takes its name from the Portugese name for wood."

And history, properly so called, confirms the conclusion that the whole of Central Europe at least may be considered as having been one vast forest, such as now extends over the northern governments of Russia and the northern territories of America. And there are numerous indications, both historical and physical, that the whole of the High Alps had been at one time richly wooded. Hence originated the application of the term in question to the projected sylvicultural operations there, and its subsequent application to all similar operations wheresoever prosecuted.

The term *gazonnement* I have also retained, being unable to render it by any English synonym which would be equally explicit and equally comprehensive.

The English term turf is generally, though not necessarily, associated exclusively with grass, or with turf which is largely composed of grass. *Gazonnement* is used in regard to a turf formed largely, and in many cases

P

exclusively, of what may be designated herbs, in contradistinction to grasses; and the term is more convenient and less pedantic than any I could devise.

Though it is only of late that prominence has been given to *reboisement* and *gazonnement* in the legislation of France, the evil they are employed to arrest and remedy early commanded the attention of her legislators.

In 1669 was issued an Ordinance by Colbert, regulating woods and waters in which *déboisements*, or the destruction of woods, is forbidden to communities. There is evidence that a great part of the Alps had by that time been completely *déboissée*, or cleared of forests.

This and similar *déboisements* the forest economists and students of forest science in France sought to remedy by an extensive system of sylviculture,—replanting trees where forests had been destroyed, and planting trees where never tree had grown before.

An edict, issued by Humbert Dauphin in the 14th century, forbids clearings in the Briançonnais, assigning, among other reasons for doing so, the resistance presented by the woods to avalanches and other evils.

The archives of the Benedictine Monastery of Boscodon, preserved in the church *Notre-Dame-d'Embrun*, embody a record of a great many contentions, or legal proceedings, relative to forest depredations. It is the most common subject of these archives during a period extending over five centuries, and one which provoked numerous formal excommunications. From these archives it may be seen, by a host of facts, that the forests had then come to be a rare and precious thing—the result of long-continued fellings,—and in the fate of the monastery already referred to we see the consequence of these.

Latterly, previous to the employment of *reboisement* and *gazonnement* as means of extinguishing torrents, dikes or embankments were what were chiefly employed as means of arresting the ravages and devastations of these. When a river-bank, throughout a considerable extent, was destroyed by a torrent, the proprietors affected thereby met and constituted a *syndicate* or council. An application was made to the prefect; he appointed an engineer of roads and bridges to examine the locality, and if necessary to prepare a specification of the works required for the defence of the river. The work, when approved, was decreed. The engineer superintended the execution of it, and sanctioned delivery. The expense was then apportioned amongst those who were interested, conformably to a scheme prepared by the syndics.

The whole procedure is prescribed by a special decree, which subjects torrents to a defined *régime*, and places them under the immediate superintendence of the Administration. The following is a translation of this decree:—

"Decree of the *4th thermidor, an XIII.*, relative to torrents of the department of the High Alps:—

"Art. 1. In the communes of the High Alps, which are exposed to the eruptions and inundations of rivers and torrents, the mayors, after having submitted the matter for consideration to the municipal councils, shall make application in the usual form to the prefect of the department for authority to execute repairs, or other necessary works. In urgent cases they may summon the municipal councils for this purpose without a special permission.

"Art. 2. The prefect shall appoint an *ingénieur des ponts et chaussées* to examine the spots exposed, to prepare a plan of the places, and to prepare specifications and estimates, which shall be communicated to the municipal councils; and after they have made their remarks, the prefect shall give the authority if required.

"Art. 3. If the works to be executed affect only private parties, the prefect shall nominate a commission of five individuals from among the principal proprietors interested, who shall choose from amongst themselves a syndic, to deliberate on the utility or the inconveniences of the works demanded.

"Art. 4. The prefect shall then commission an engineer to prepare plans and estimates, which shall be communicated to the commission, as is prescribed for the municipal councils in Art. 2.

"Art. 5. In cases where the works to be executed would effect many communes who would not act in concert, the demand of the municipal council making the application shall be communicated to the other municipal councils, and the prefect shall then proceed, with regard to all the councils, conformably to Art. 2.

"Art. 6. When the neglect—be it of one or more private parties, be it of one or more communes—to make dikes, curages—*i.e.*, clearing away of stones and deposits in the channel, or artificial structures, along a torrent or an unnavigable river, shall expose the territory abutting upon it in a way prejudicial to the public weal, the prefect, on complaints which may have been made to him, shall order a report of an *ingénieur des ponts et chaussées*; this report shall be communicated to the parties interested, with injunctions to give their answers in writing within eight days, and the council of the prefecture shall decide on the disputes which may result.

"Art. 7. If a dike interest a commune in general, and some private parties oppose the construction of it, the municipal council shall be consulted, and the opposition shall be submitted to the council of the prefecture.

"Art. 8. In all the cases specified, when the time allowed shall have expired, if all the parties interested shall have given their consent, or if there be no protests, the adjudication of the works, according as they have been determined and resolved upon, shall be made in the usual forms before such functionary as the prefect shall appoint in the presence of the parties interested, or those there duly summoned by posted bills and the usual ordinary publication of such announcements.

"Art. 9. The amount thus adjudicated shall be apportioned according to the extent of interest attaching to their property, according to a scheme of division which the prefect shall make legally obligatory, accord to the law of the *14th floréal, an XI.*; and the council of the prefecture shall decide protests relative to this partition of the expense.

"Art. 10. The adjudicators shall be paid the expense of their adjudication, in virtue of an order delivered by the prefect, on certification of the works having been taken over and delivered by the engineer charged with the management of the works. The parties liable shall be forced to pay in the form prescribed by the law of the *14th floréal, an XI*.

"Art. 11. No proprietor can be taxed for contributions to such works in the course of any one year, beyond a fourth part of his net revenue, after deduction of every other tax."

Of this decree Surell says, that it did great service to the department of the High Alps, because it subjected to a fixed rule all the works which otherwise would have been executed as chance might determine, perhaps with mutual detriment and damage to one and another of the works executed. And he adds, that, if it has not yielded all the fruit which might have been expected, it may be well to take into account the hostile spirit which generally animates the proprietors of the opposite banks of a river, and which, unhappily, often prevents their union and co-operation in the construction of such an embankment as would be the only means of rendering the defence perfectly harmless and productive of the greatest amount of benefit possible.

Valuable information in regard to the subject of this decree may be found in *Notice des principales lois décrets ordonnances, &c., relatif aux rivières, torrents, &c., par* Morisot, *Chef de bureau à la prefecture des Basses-Alpes, 1821.* There exists also in the papers preserved in the prefecture of the High Alps an excellent *règlement*, which developes fully the decree of the *4th thermidor*, which was drawn out in 1802 by M. Gauthier, councillor of the prefecture.

By an Act of 16th September 1806, obtained on demand of M. Ladoucette, who had been prefect of the High Alps, this decree was extended to the Drome and the Lower Alps.

In a law bearing the date of 16th September 1807, there were embodied several enactments somewhat at variance, if not directly opposed, to the requirements of that decree. And the question was subsequently raised, whether this law were not virtually an abrogation or rescinding of the decree. To those who are desirous only of learning what may be learned relative to practical measures, sanctioned and tested, and approved or abandoned by the Government in dealing with this matter, this question is of importance mainly as indicative of the importance attached to the subject by the Legislature and Administration of the country. The importance of this legislation, under this view of the case, arises from the probability which there is that it will be long before the more efficient remedies proposed by Favre, Dugied, and Surell, will be extensively adopted in newly-peopled territories, and from the probability that meanwhile the adoption of the less efficient measures which occupy only a secondary position in the comprehensive projects submitted by them may be advocated as temporary, if not as final, measures to be adopted ; and it may be advantageous to know what has been done in similar circumstances by others, and with what results.

According to Art. 33 of the law of 16th September 1807, it is enacted,—" When it is proposed to construct sea-dikes against rivers, streams, and torrents—navigable and not navigable alike—the necessity for this being done shall be determined by the Government, and the expense borne by the property thus protected in the ratio of the interest in the work, excepting in cases in which the Government shall deem this of public utility, and grant all necessary assistance from the public treasury. . . .

"Art. 34. The forms of procedure hereby established, and the intervention of a Commission, shall be applicable to the carrying out of the preceding Article." . . .

It is in these forms of procedure alone that there is aught opposed to the decree previously enacted. By this law there were established two commissions, the *Syndicate*, and another designated a Special Commission. By the decree there was established only the *Syndicate*, and the powers of this were not exactly equivalent to that of the two commissions now established. The law of 1807 gives to the Special Commission not only the right to prepare or to verify and sanction a roll of the valuation of lands interested in the works, but also the power of regulating ex-propriations, or transfers to the government, of lands requiring to be used in the execution of the enterprise, where this cannot otherwise be effected. In regard to the inquest *de commodo et incommodo*, or the necessity of the measure and the disadvantages which the execution of it might entail on any whose interests might thereby suffer,—it had been the custom that every proposal should be lodged in the Mayor's Office, with summons to all concerned to put themselves in communication with that office. Objections were recorded, discussed by the Syndical Commission, addressed to the prefect, and then submitted to the Council of the Prefecture. By the law of 1807 it was enacted that objections and protests should be sent by the prefect before the Special Commission, which should decide finally on these.

The construction of a dike, according to the decree, required no other sanction than that of the prefect, who considered the projects of the engineers, decided whether the construction should or should not be carried out, and gave a final decision on all disputed points. The formalities prescribed by the law of 1807 are more complicated, and require the intervention of the superior administration.

In regard to the course usually followed in preparing the roll of contributions towards meeting the expense of constructing such a dike, required of the several parties interesed, M. Surell supplies the following information:

"The work is begun by arranging all the properties interested in a certain number of classes, determined by the greater or lesser probability of their suffering from inundation. There are thus classified together all properties which have nearly the same chance of being invaded by the torrent; and to each class is given a number, designed to be representative of this probability alone : one class, considered twice as liable to invasion as another, is marked by a number double that of the latter This done, the area of each property is multiplied by the number assigned to the class to which it belongs, as indicative of the chance to which it is exposed. And the total expense is assigned proportionally to the product."

Thus is the roll of liabilities prepared; but it is pointed out by Surell that it is neither the most exact, nor the most equitable plan which might be adopted. There may be two properties equally liable, and of equal extent, of very different pecuniary value. The soil of the one may be better than that of the other; it may have been more improved, it may be of easier access, or it may have upon it dwelling-houses or mansions, and other amenities, which would enhance its price if it were sold. Should this be destroyed, the proprietor would suffer a proportionally greater loss; having thus a greater interest in the maintenance of the dike, he ought to be required to pay a greater contribution towards the execution of the work ; and it follows, that the rule commonly adopted by the syndic, according to which both pay alike, is neither rational nor equitable. The payment ought to be calculated in the same way as the law of probabilities is applied at the gaming-table. And for this purpose there should be determined—(1) The probability of inundation in regard to each property ; and (2) The value of each property menaced. And the product of these, multiplied the one by the other, will give the proper proportion of the expense to be borne by each property.

The determination of the first of these factors is somewhat difficult, but it is not impracticable. In normal cases of inundations, of a temporary character—the waters returning again to their usual bed—there may be ascertained what marks exist of inundations which have occurred within the most protracted period during which they can be enumerated—say 80 years. These indications supply a point of departure, in determining with exactitude levels following the inclination of the river bed, so arranged as to include in one line all the corresponding marks of one flood. The whole ground, from the river to the most distant part reached by an inundation, would thus be divided into a certain number of zones, subject each to a different chance of inundation. All the lands included in one zone would constitute a class subject to the same chance. What this is has next to be determined; and that may be done thus:—If the portions included in the zone nearest to the river have been flooded upon an average three times every year, the chance of inundation may be represented by the fraction $\frac{150}{150}$. While the zone most remote, if flooded only once in the course of the fifty years, would be represented by the fraction $\frac{1}{150}$. The value of the property in each zone being then determined, representing the value by $v\ v'\ v''$, and the chances by $p\ p'\ p''$, the equivalent of the extent to which the different classes were interested would be expressed by the products $p\ v,\ p'\ v',\ p''\ v''$, &c.; and to determine the quota of each party interested, it is only necessary to multiply the value of the property by the chance of the class to which it belongs, and the product by a constant co-efficient, determined in such a way that the sum of the shares of all interested will equal the whole expense. This co-efficient may be determined by the following equation, in which the total expenses, represented by S, co-efficient $= \dfrac{S}{p\ v\ \times\ p'\ v'\ \times\ p''\ v''}$ &c.

It is an intricate question, and leaves much to be determined by the syndic; but the classes once formed, what follows is rigorously just.

It has been stated, that it has been questioned whether the law of 1807, in superseding, abrogated the decree of *4th Thermidor, an VIII.*

The question was raised in the Chamber of Deputies, on the 12th April 1837, by M. Jaubert, acting in the name of a commission appointed to examine a proposed law relative to the joint action of proprietors in works undertaken on rivers of greater and lesser size. This they considered it did, but others thought differently. That question has not now the interest it then excited. In the one may be seen a development of the other. Both related exclusively to the construction of dikes as means of protecting the land against the devastations of rivers and torrents.

In 1797 appeared the work by Favre, advocating the creation of plantations as a means of more efficiently securing the object desired. The date of M. Dugied's work, advocating the same measure, I have not ascertained. In 1841 was published, printed by order of the Minister of Public Works, the work by Surell, shewing the primary and almost absolute importance of plantation, while the topical application of dikes may be necessary as a secondary and subsidiary means of preventing devastations. And the legislation of the present is of national application; these laws were of more limited local application.

In entering upon the consideration of this later legislation, it may be mentioned that from the first it was maintained by Surell that as things then were in France nothing satisfactory could be done without Government interference; that the problem to be solved was the prevention of the formation of new torrents, and the arrest of the ravages which were being made by torrents already formed; that the battle-field must be in the basin drained by the torrent, and that a system of extinction must be followed; that in view of the public interest it was vain to trust to the prudence of communes, to the publication of information and warnings, or to moral suasion in any form; that both the number of cattle depastured and the extent of the pastures must be restricted, and the introduction of cattle and sheep, other than those of the commune, prohibited; that agricultural operations which do not promote the carrying away of the soil should be allowed without restriction, but that such as have that effect should only be tolerated on slopes not exceeding a prescribed inclination—compensation being given, if necessary, for loss which might be sustained in consequence of this restriction, but enforcing it with rigour: the circumstances of the locality being exceptional, exceptional legislation might be requisite. Further, the forests having an exceptional importance—being required not only to meet daily recurring wants of the population, but to preserve the very soil—it was necessary that their conservation should be secured, and measures taken to effect their extension. And to these preliminary measures had to be added the more direct measures detailed or suggested in his treatise, of which a *résumé* has been given.

There might be private rights in the way of the execution of these works, and the conflicting claims of the public interest and of private property must be reconciled. This might be done, according to circumstances, by the Government taking possession, with compensation to the proprietor, as is done in carrying out other works carried out for the public good, or by requiring of the proprietors that they themselves should plant the ground with woods, and giving to them every just and reasonable assistance in the execution of the work.

The poverty of many of the proprietors might make it impossible for them to meet the expense; and the restrictions imposed upon them by some of the preliminary measures required would entail upon them a considerable sacrifice. The measure, moreover, was one affecting the public interest as really as do many of the public works—such as embankments, roads, bridges, and the improvement of mountain passes,—while the outlay by the State might be counterbalanced in a great measure by a diminution of outlay on these, through the prevention of injuries now done to these by torrents; and, as shown by M. Dugied, the forests would in course of time become a source of revenue, and, according to others, they would tend to improve the climate. Such are some of the considerations adduced by M. Surell as reasons for the Government taking up the work.

Having done this, he gave details of what measures he considered would be requisite in carrying out the work: the initiative to be taken by the Government; the objections to be anticipated, and the means of meeting these, whether they might take the form of an allegation that it was impossible to plant the mountains with woods, or of an allegation that if this were done it would not suffice to cause the torrents to disappear, or of objections to different regulations which it would be necessary to carry out in connection with the execution of the work, some relating to rights of property, some to rights of pasturage, some to one thing, some to another.

And, having done this, he proceeded to draw a fascinating picture of the wide-spread felicity which was to follow the execution of the project.

He had previously given the saddening picture of Dévoluy, which I have cited in the Introduction. And he proceeds to show what had been done in France when it was perceived that the fruits of the country were being destroyed; he details the evils which followed in the train of that destruction, the alleviations of these secured by the inhabitants of the plains, but which were unattainable by the inhabitants of the mountains, and the privations to which they had been in consequence reduced.

"There may be seen," says he, "here in one small valley (that of Lagrave) the inhabitants reduced to the necessity, in order to heat their houses and cook their provisions, to burn cow-dung formed into bricks and hardened in the sunshine. This disreputable fuel saturates with its smoke their huts, their clothes, the air which they breathe, and even the food which they eat—the whole atmosphere of the country is filled with it. Now if they have recourse to such a fuel, it is not that the country is absolutely devoid of fuel; it is, on the contrary, very richly supplied, as there are many beds of anthracite under active exploitation. But one may easily imagine that this mineral, being very heavy, if it be necessary to transport it on the backs of mules or of men to great elevations, across rocks and perilous slopes, the fatigue and consequent price of transport will raise the cost of it to such a point that the great bulk of the poor people must renounce the use of it. And the consequence is, these mines, which would be so valuable in a plain, here benefit only such of the inhabitants as live in the immediate vicinity of them, and they remain almost unused by all living beyond a radius of some leagues from the spot."

He contends that, in order that the mountains may be habitable, they must be wooded; and that the total annihilation of forests will necessitate the emigration of the population. But the difference between the destruction of forests on the mountains and on the plains, says he, stops not here. "If a forest disappear on the plain it is to give place to agriculture, it is the substitution of one product of the soil for another, and the substitution often leaves nothing to be regretted. But if, on the other hand, you fell an old forest which covers the flanks of a mountain, immediately everything is upset and overturned. The storms and the ravines cut up the slopes, the vegetable soil is soon washed away, and with it all fertility and verdure. No more fields! No more cultivation! Delivered defenceless to the attacks of the waters, eaten down to its very entrails by the torrents, and sinking at last under its own weight, the mountain, as if crushed and spread out, is soon rolling its material into the plain, and this it buries under its debris and involves in its own ruin. It is true, it happens here as in the plains, that wood is every day felled to free soil for the plough, and those who root out the trees only do so for the profit which follows. But we must not confound the ephemeral and illusory profits which are obtained by them with the lasting advantage and real benefit which follow such operations in the plains.

"The first years following immediately the rooting up of trees on a mountain produce excellent crops, because of the quantity of humus left behind them in the soil by the trees. But this valuable earth, the less stable in proportion as it is productive, does not remain long on the slopes; at the end of a short time it is dispersed, the sterile subsoil makes its appearance, and the unreasonable proprietor loses his property from having wished to constrain it to produce more than its nature would permit. We see too often the old story of the goose which laid golden eggs practically exhibited in the mountains, notwithstanding the instructions a thousand times repeated by experience. A recent case in point, says he, is supplied by the rooting out of woods on the mountains of Champsaur.

This rooting out of woods on declivities is always followed with disastrous consequences; and the destruction of forests, practised almost always without inconvenience on the plains, becomes, on the contrary, in the mountains the most disastrous of disturbing operations. It breaks up the equilibrium of the land, and brings back the disorder of ancient chaos. After having wrenched from the inhabitants the usufruct of the forests, it carries off the soil which nourished them, thus pursuing man with hunger, if he submit unresistingly to the privation of wood.

And in eloquent and stirring appeal, called forth by what he foresaw, he urged the *reboisement* of the mountains, whatever the expense might be.

The appeal appeared to have been made in vain—if in this world, in which no atom of matter appears to be destroyed, and no form of physical force to be lost, any counsel, good and true, can be given in vain. Years, at least, passed away—as nearly fifty years had passed away after Fabre had spoken something similar—and nothing was done. But at length, in process of time, there was a resurrection of the two witnesses, their testimony was again called for; and the fulness of time being come, their testimony was listened to, and their counsels were adopted.

In November 1840, the year before the publication of M. Surell's work, there occurred a destructive inundation of the basin of the Rhone. Occurring at that late season of the year, all the crops had been gathered in, but the damage, notwithstanding this, was estimated at 72,000,000 of francs, or well-nigh £3,000,000 sterling.

Several smaller floods of the Rhone subsequently occcurred in 1846, at a somewhat earlier period of the year, and occasioned a loss of 45,000,000 of francs, or £1,800,000.

"If these floods," writes Dumond, "instead of happening in October, between harvest and seed-time, had occurred before the crops were secured, the damage would have been reckoned by hundreds of millions."

These inundations in 1840 and 1846 made the question of forests an order of the day; remedial measures, which were demanded and were opposed on all hands, became the subject of careful consideration and study; and the Government was about to promote a general law for the regulation of all rooting out of woods, and to reform the forest code, when the revolution of 1848 broke out. The effect of this upon the forests was soon felt, but in another way. Within thirteen days after installation the Provisional Government authorised, by decree of 9th March, the sale of a considerable portion of the crown forests, and all the forests of the civil list.

On 30th June 1848, the State ceded to the Bank of France 75,000 hectares of forest, as security for a loan of 150,000,000 of francs.

On 4th December 1848 the National Assembly discussed the forest budget; in vain did the Minister and the Director-General defend their Administration; in vain did the tribune re-echo the famous words of Colbert —*La France périra faute de bois*; retrenchment and economy were the order of the day, and the different forest services saw themselves threatened with dissolution. Particulars are given in the *Annales forestiers*, of December 1848 and January 1849.

This, however, proved but a passing storm. On the establishment of the empire the Forest Administration, promptly re-constituted, shared after 1852 the great impulse which was given to public works.

There are decrees dated 17th and 27th March, and laws of the 12th April 1853, of 5th May 1855, of 28th July 1860, and of 8th June 1864, which have authorised alienations or extensive fellings of the State forests. These may be considered comparatively unimportant operations, and the proceeds of them were to be employed in works of reproduction. But when, in 1865, the Government proposed the alienation of forest domains to the extent of 100,000,000 of francs, to be applied to the commencement and prosecution of public works, public opinion was roused, and, alarmed by the proposal, publicists of every shade, politicians, savants, littérateurs, &c., combined their efforts and raised a crusade against the *projet de loi*, which made it necessary to withdraw it.

Meanwhile another inundation, or the cotemporaneous occurrence of a number of inundations, had given a new direction to men's thoughts on the subject.

In the month of May 1856 violent and almost uninterrupted rains fell throughout France, and most of the river-basins of the country were inundated to an extraordinary extent. In the valleys of the Loire and its affluents about a million of acres, including many towns and villages, were laid under water, and the amount of the pecuniary damage was almost incalculable.

The flood was not less destructive in the valley of the Rhone, and an invasion by a hostile army, it was said, could hardly have been more disastrous to the inhabitants of the plains than was this terrible deluge.

"In the fifteen years between these two great floods," says Marsh, "the population and the rural improvements of the river valleys had much increased. Common roads, bridges, and railways had been multiplied and extended; telegraph lines had been constructed,—all of which shared in the general ruin, and hence greater and more diversified interests were affected by the catastrophe of 1856 than by any former like calamity. The great flood of 1840 had excited the attention and roused the sympathies of the French people, and the subject was invested with new interest by the still more formidable character of the inundations of 1856. It was felt that these scourges had ceased to be a matter of merely local concern, for, although they bore most heavily on those whose homes and fields were situated within the immediate reach of the swelling waters, yet they frequently destroyed harvests valuable enough to be a matter of national interest, endangered the personal security of the population of important political centres, interrupted communication for days and even weeks together on great lines of traffic and travel, thus severing, as it were, all South-Western France from the rest of the empire, and finally threatening to produce great and permanent geographical changes. The well-being of the whole commonwealth was seen to be involved in preventing the recurrence and in limiting the range of such devastations."

"The inundations of 1846, and more especially those of 1856," wrote Cézanne, "compelled attention to be given to the conservation of the forests. In proportion to the greatness of the prosperity which prevailed, and the profound feeling of security which had lulled so many to sleep, the more severe and the more unexpected seemed the disasters thus occasioned."

A great complaint arose ; this was followed by a keen controversy ; the Head of the Government took part in this ; a letter from him to the Minister of Public Works, under date 19th July 1856, published over the whole of France, gave a *résumé* of the popular movement ; and, founded on information elicited, there was issued in his name, under date of 5th January 1860, a programme of procedure, which was followed by a *projet de loi*, which was submitted to the Emperor with the following report by M. Magne, Minister of Finance :—

"PARIS, *2nd Feb. 1860.*

SIRE,—" The attention of your Majesty has been given at different times to the dangers which result from the deforesting of the mountains. At the time of the inundations of 1859, you were led to point out that deforesting was one of the causes of the evils which had then afflicted the country ; in the programme traced in the letter of your Majesty is included, in the innumeration of the great administrative measures destined to develope the public prosperity, the clearance of forests on plains, and the reforesting of the mountains.

"A law passed in the last Legislative Session has given new facilities for the clearance of forests situated on the plain. This law, long waited for, is one of the recent benefits conferred by the Imperial Government ; it realizes its liberal views in what relates to woods belonging to private proprietors. There remains to be proposed, as a necessary complement to this, a law for the reforesting of the mountains.

"No legislative measure of any importance on this subject has been adopted by the Governments which have preceded that of your Majesty. The old edicts, and the ordinances anterior to 1789, contain only exceptional measures to arrest the progress of deforesting. The *code forestier* was conceived in the same spirit ; there is found there a series of arrangements designed to restrain the abuse of depasturing, but only one arrangement was introduced to promote reforesting ; it is Art. 225, which exempts from taxation, for a period of twenty years, woods sown and planted on the summit and on the declivity of the mountains. The law adopted last year, in regard to the clearance of woods belonging to private parties, prolonged this exemption of taxation to 30 years.

"But, notwithstanding the new extension given by the Government of your Majesty to this favourable arrangement, one knows not how we can await the very important results which must follow. The sowing of trees and plantations, especially those on the mountains, profit the future more than the present, and general interests more than the interests of individuals: hence the necessity for efficacious measures, and for the direct intervention of the State. This intervention has for a long time been urgently called for. Since 1843, sixty-three general councils have urged the necessity of measures being taken for the reforesting of the mountains. A report and a *projet de loi* were prepared by the director-general of forests in 1845. This *projet de loi*, remitted for examination to a commission composed of forest administrators and distinguished *savants*, was amended in many parts, and submitted to the Chamber of Deputies in the session 1847. The report which was presented by the commission admitted the importance which attached to the question, but also the uncertainty which prevailed in opinions relative to the measures which should be adopted ; and nothing came of this *projet de loi.*

"But the greater part of the general councils have not ceased to call, year by year, for legislative measures, designed to favour the reforesting of the mountains. Many have even voted subventions with this view. There may be cited, more especially, the general councils of the *Puy-de-dome*, of the *Lozère*, of the *Bouches-du-rhône*, of *l'Ariége*, and of *l'Ain*; and lastly, a certain number of communes have imposed on themselves sacrifices, and have taken the initiative in works of reforesting; but these efforts, which attest the urgency of the need, are not in keeping with the magnitude of the evil, and they must remain, moreover, inoperative in securing the co-operation of the State. It is this co-operation which your Majesty has sought to secure to the population of the mountains.

"The region in which reforesting is becoming most urgently necessary comprises a certain number of departments, furrowed by many chains of mountains, of which the principal, and the most deforested, are the Alps, the Pyrenees, the Cevennes, and the mountains of Auvergne. It is in these chains of mountains which the principal affluents of our rivers, and the rivers themselves, of which the basins are most exposed to inundations—the Rhone, the Isére, the Loire, the Durance, the Garonne, &c.—take their rise.

"Statistics have been prepared at different times to determine the extent of lands susceptible of reforesting in the mountain regions of France. These, carried out more fully and completed of late years by the forest administrators, have been verified by special reports which had been required of conservators in 1859. The results have been tabulated, and show that lands susceptible of reforesting, in the departments the most threatened by the denudation of their declivities, may be estimated proximately to be in extent 1,133,000 hectares. These lands belong to the State, to communes, and to private proprietors.

"No legislative arrangement appears to be necessary in regard to lands belonging to the State; it suffices to secure the reforesting of these that special credits be introduced into the budget of the administration of the forests. Your Majesty's Government has already taken the initiative in this matter, and since 1855 a sum of 500,000 francs has been appropriated annually to works of replenishing in the State forests. This appropriation has allowed of a great reduction of the void spaces existing in the forests, and of works being executed during the last five years on lands situated on the mountains or on the declivities, and thus has led to the reforesting of fourteen thousand hectares. By continuing this appropriation of 500,000 francs for a certain number of years, it is believed that the forty thousand hectares of lands belonging to the State in these departments of the mountains may be completely replanted with woods.

"But it is not the same with lands belonging to communes, to public bodies, and to private proprietors; the replantings executed by them on these lands are the result of a few isolated efforts—trials left to themselves, without direction and without encouragement. The State ought to interfere, to give to these works the impulse demanded by the general interest; and a law is required to point out the importance of this joint action, and to determine the conditions of it.

"For the greater part of the lands situated on mountains, the intervention of the State can only consist in subventions granted to private proprietors, to communes, and to public bodies. These subventions might consist, in those which relate to private proprietors, in supplying to them **plants and seeds** before the execution of the works, and in the subsequent distribution of premiums; in those which relate to communes and to public bodies, in subventions in money which might be granted before the execution of the works, but the grants should be proportionate to the wants, to the resources, and to the sacrifices made by the several departments and by the several communes.

" The *projets de loi*, which have been proposed for the replanting of the mountains with woods, in 1845 and 1847, have recognized the necessity of authorising, in a public interest so very great, the distribution of subventions and of premiums, as well as the supply of plants and of seeds. This first part of the *projet de loi* need not then raise the question of principle. The Administration will only require to take the necessary measures to see that the subventions be distributed with discernment. In point of fact, it is not necessary that the whole of the lands susceptible of replanting should be covered with woods; in many places a covering of the land with turf may suffice to ensure the maintenance of the land on the mountain, and, where the planting with wood is a recognized advantage, the subventions ought not to have as their result to substitute the action of the State for the initiation of the work by the individual.

" Communes which may be disposed to demand too high subventions, regard being had to the sacrifices which they impose upon themselves, should only receive from the State co-operation subject to certain conditions, which may perhaps appear somewhat severe, such as a proportional participation in the forests created on the communal lands. It would not be just, indeed, that certain privileged communes should be able to draw to themselves all the benefit of the subvention. The benefit, in order to its being shared by a great many, should not be applied to each otherwise than in a certain proportion. If this proportion be exceeded, the pecuniary co-operation of the State should assume another character, and the subvention become an advance to be paid back, at least in part, to the public treasury, through a cession of a portion of the lands, the principal value of which will have arisen from the replanting of them with woods.

" At the same time, it is impossible not to foresee that, notwithstanding the subventions offered, and notwithstanding the advances which the State may be willing to make, there may be communes, or private proprietors, utterly unable to execute the replanting, and yet on certain determinate places, replanting may be demanded, not only by manifest public interest, but, so to speak, by an imperious necessity.

" There are, on the mountains, places which are more especially threatened by the violence of rivers, by the impetuosity of torrents, and by the fall of avalanches or of rocks. Such, for example, are certain lands on steep declivities, situated on the sides, or at the *debouche* of torrents; such are villages exposed without shelter to catastrophes which are in some measure periodical. The reformation of wooded masses, designed to arrest the ravages of waters, and to divert from the places imperilled disastrous effects of great natural disturbances, is in the highest degree a work of public interest. In such masses of woods as are desired every thing combines to offer resistance to the scourges which desolate the mountains : the roots of trees keep the ground in its place and consolidate the soil, the branches form a shelter against the storm and wind, the leaves fertilize the light bed of vegetable earth covering the rock.

" The *reboisement* presents, then, in these exceptional cases, and at certain determinate places, a character of public utility such that the necessary works to reconstitute masses of woods should be rendered obligatory, and, if need be, be executed at the expense of the State.

"Imperial decrees, issued after the observance of forms of procedure which shall give satisfactory guarantees to all interests, should specify the boundaries of these lands. The Council of State would then have to ascertain whether within these exceptional boundaries ex-propriation for the sake of public utility could not be applied to lands belonging to private proprietors, and whether the temporary occupation of lands belonging to communes ought not to take place conformably to the principles laid down by the law of 1857, relative to the plantation of the communes of the Gironde and of the Landes.

"But the provisions of this part of the *projet de loi* should be applied with such reserve as not to lead to hasty changes in the general habits of the population of the mountains. It should be applied, in the first instance, to places in regard to which it is already seen and acknowledged that the replanting of them with trees would be a benefit. If in certain communes it be the case that the population are without cause disturbed by every attempt at replanting, considering this as a hindrance to the enjoyment of the right of pasturage, there are, on the other hand, others struck by the imminence of the dangers by which they are threatened, or pressed by the scarcity of wood in regions in which the snow lies on the ground eight or ten months of the year; and, considering the replanting of woods as a measure of protection and safety, they urgently solicit it, as is notably the case in the departments of the Haute-Loire and of Puy-de-Dome. In certain mountain countries, then, the co-operation of the population is now certain to be given to works of replanting. This co-operation guarantees success in it, and the importance of the results to be obtained will little by little enlighten the communes which are less favourably disposed towards the advantages of the measures prescribed by the Government.

"The Administration, however, should not forget that pasturage is one of the necessary conditions of life to the dwellers on the mountains. The interest of the shepherd population ought then to be treated with the greatest care. But this same interest is closely allied to that of the operations of replanting, for the abuse of depasturing is not less hurtful to the conservation of the pasturages than it is to the conservation of the forests. In the day when the forests shall disappear from the mountains it may be predicted with some measure of certainty that the day is not distant when the pasturages shall disappear in the train of the forests.

"In the department of the Basses-Alpes, for instance, where the abuse of the pasturage, and the incursion of stranger flocks, known by the name of *transhumant* flocks, have occasioned disastrous consequences, the pasturage resources have rapidly diminished with the destruction of woods on the declivities, and the latest statistics have attested the impoverishment of the land and the emigration of the population. All the prefects of this department for forty years past have reported the progressively increasing seriousness of the state of things there. Besides, do not the forests themselves supply in the mountains what is required in pasturage? If, during a period of some years, the shepherd would respect the forest sowings, the plantations, and the young fellings, till the wood has become capable of self-defence, the animals might then enter it, and there find abundant nourishment. And does not the pasturage present more valuable resources in the forests of the mountain than it does on the denuded slopes, where vegetation tends to disappear and to give place to a sterile soil?

"The replanting there is not less necessary in the interest of the shepherd of the mountain, than in the interest of the agriculturists of the valley, who are threatened by inundations; and the legitimacy of the exceptional measures in certain determinate cases is justified by public interests of the very highest order. It would be possible, moreover, to moderate the dreaded effect of these measures: there might be granted, for instance, to private proprietors, after the *reboisement* of their lands, the power of re-entering on the proprietorship of these lands on repayment to the State of the indemnity of expropriation, and the expense of the works. The replanting with woods being effected, the public interest is secured, and the proprietor might be permitted to exercise a kind of action of recovery within a determinate period. On the other hand, the State would thus recover a portion of the advances made, and might apply the amounts received to new works of *reboisement*.

"So, also, might communes be permitted to recover possession of their wooded lands, on reimbursement to the State of the advances made; but more than this, they might be allowed, without making any reimbursement, to resume possession of one half of these lands, on ceding the other half of them absolutely to the State.

"These varied combinations will be appreciated at their true value by the Council of State, which will know how to reconcile the requirements of the public interest with the guarantees and arrangements due to private proprietors, and to communes.

"It only remains to intimate to your Majesty the financial measures which it appears ought to be adopted, in order to the carrying out of the *projet de loi*.

"A sum of ten millions should be appropriated to subventions, and to works of replanting with woods on the mountains. The necessary resources for meeting this expense should be obtained through the sale of woods belonging to the State to a corresponding amount of ten millions.

"The alienation of these woods should take place successively, throughout a period of ten years, in such a way as to proportionate each year the resources obtained by the sale of woods on the plain to the allocations granted to the budget for replanting the mountains with woods. The Forest Administration should be charged with the double operation, and the Minister of Finance should be charged to see that the advances made from the treasury be covered by the payments received in the year.

"For the success of the operation of replanting, as well as for the successful operation of the alienation of woods, it is not needful to urge on precipitately either the works to be executed, on the one hand, or, on the other hand, the sales to be effected. An allocation of a million per annum, devoted to replanting, is sufficient for the distribution of subventions and important premiums, and for the undertaking of somewhat considerable sowings and plantations within the exceptional boundaries specified by the Imperial Decrees. The corresponding annual alienations of woods to the value of a million can occasion no perturbations in the sales of landed property, or of fellings of timber. Hitherto, the success of alienations of woods has always been neutralised by the mass of operations going on at the time to meet the urgent requirements of a period of crisis. The alienation which will take place in carrying into execution the present law will be made in circumstances much more favourable, and there may be anticipated good results.

"The woods of which the projected law proposes the alienation are those comprised in Table I. of the law of the 5th March 1855; so the alienation of them has already been authorised by the *Corps Législatif*, as the sales authorised towards the raising of fifteen millions did not amount to more than about a sum of six millions.

The woods designated in that table are, moreover, in the conditions determined by the programme of your Majesty. They are, in general, in lots of moderate contents situated in fertile plains, or woods, the clearing away of which has been already authorised. From the point of view of forest economy, alienations limited to woods so circumstanced presents no inconvenience, while the reconstruction of extensive masses of woods on the mountains is seen to be of the first importance.

"Such, Sire, is the general purport of the *projet de loi*, which I propose to your Majesty to submit to the consideration and examination of the Council of State. The eminent men who compose that Council will know how to improve the arrangements of that law. I have endeavoured, in the preparation of it, to enter into the spirit of the great foresight which determined your Majesty to grant the concurrence of the State to the important work of the *reboisement* of the mountains.—I am, with the most profound respect, Sire, your Majesty's most humble, most obedient servant and faithful subject, P. MAGNE."

In accordance with usage, the *projet de loi*, or draft of the law proposed, was submitted to a Committee, along with an *Exposé des motifs*, or statement of reasons for its enactment, to which they were required to give consideration. And, in accordance with a report made by them, the law was enacted in the following terms :—

"Law, of the 28th July 1860, on the *Reboisement* of the Mountains.

"Art. 1. Subventions may be granted to communes, to public bodies, and to private individuals, for the replanting with woods of lands situated on the summits or on the declivities of mountains.

"Art. 2. These subventions may be made in grants of seeds, or of seedlings, or in premiums of money.

"These to be granted on account of the utility of the works, in view of the general interest; and in the case of communes and public bodies, regard is to be had to their resources, to the sacrifices they have made, and to their need, and also to the sums granted by general councils for reforesting.

"Art. 3. Premiums in money, awarded to private individuals, cannot be delivered until after the execution of the works.

"Art. 4. In any case in which the public interest requires that the work of reforesting be made obligatory, in consequence of the state of the ground, and the dangers resulting from this to lower-lying lands, this is done thus:

"Art. 5. An Imperial decree, issued through the Council of State, declares the public utility of the works, determines the boundary of the lands in which it is deemed necessary to carry out the work of reforesting, and limits the time within which the work must be done.

"This decree is preceded by (1) an open inquest or inquiry in each of the communes interested in the works; (2) a deliberative discussion of the subject by the municipal councils of these communes, together with those of the more important of those which are circumjacent; (3) the opinion of a special commission, composed of the prefect of the department, or his delegate, a member of the general council, a member of the council of the arrondisement or district, and two of the landed proprietors of the communes concerned; (1) the opinion of the council of the Arrondissement, and that of the General Council.

"The *procès-verbal*, or attested minute specifying the lands, the diagram, or chart of the lands, and the specification of the works proposed, prepared by the Forest Administration with the co-operation and approval of an engineer of roads and bridges or mines, are to remain deposited in the office of the mayor during the inquest or inquiry, the duration of which is limited to a month, dating from the time of the prefectoral resolution, which prescribes the opening of the inquest and the convocation of the municipal council.

"Art. 6. The Imperial decree is to be published and posted up in the communes concerned.

"The prefect, moreover, is to serve on the communes, on the public establishments, and on the private individuals concerned, extracts of the Imperial decree, containing severally what therein relates to the lands belonging to them.

"The notification made to them is to state the limit of time within which the works of *reboisement* should be completed, and, if there be occasion for it, the subvention granted by the Administration, or the advances which they are prepared to make.

"Art. 7. If the lands comprised within the boundary determined by the Imperial decree belong to private individuals, these are required to declare if they intend themselves to effect the *reboisement*, and in that case they are held bound to execute the works within the period fixed by the decree.

"In case of refusal to do so, or of failure of execution of the engagement undertaken, it is competent to proceed to expropriation on the ground of public utility, observing the formalities prescribed by the Title II. and those following of the law of 3rd May 1841.

"The proprietor expropriated in the execution of this Article has a right to obtain reintegration in his property after the reforesting, on repayment of the expense of the expropriation, and of the works, principal and interest; or he may relieve himself of repayment of the expense of *reboisement* by ceding half of the property.

"Art. 8. If the communes or public establishments refuse to execute the works on lands belonging to them, or if it be impossible for them to execute these in whole or in part, the State may do so, either by amicably obtaining possession of the part of the lands which they do not wish, or are unable, to reforest, or by undertaking the whole of the works at its own expense. In the latter case it retains the administration and the use of the reforested lands until the advances made have been reimbursed, principal and interest. But while this is the case, the commune will enjoy the right of pasturage on the reforested lands as soon as the woods shall have been sufficiently protected from injury.

"Art. 9. The communes and public establishments can, in every case, exonerate themselves from repayment to the State by giving up their right of property in half of the re-wooded lands. This cession of right of property must be made, on pain of forfeiture of privilege, within a period of ten years from the notification of the completion of the works.

"Art. 10. The sowing or replanting in each commune cannot be made on more than on one-twentieth of the extent of the lands annually, unless a resolution of the municipal council authorise the works being carried on on a more extensive scale.

"Art. 11. Forest warders of the State may be appointed to the surveillance of the sowings and plantations within the boundaries fixed by the Imperial decrees. Offences within these boundaries, proved by these guards, are to be prosecuted as offences committed in the woods subjected to forest *régime*; and the execution of the sentence is to be enforced conformably to the Articles 209, 211, 212, and to the §§ 1 and 2 of Art. 210 of the *Code forestière.*

"Art. 12. The first paragraph of Art. 224 of the *Code forestière* is not applicable to *reboisements* effected with subvention or premium granted by the State in execution of the present law.

"The proprietors of lands replanted with woods, with premium or subvention of the State, cannot depasture these without a special authority from the Forest Administration, until such time as the woods shall have been recognised by the said Administration as sufficiently protected.

"Art. 13. A regulation by the Public Administration shall determine— (1) the measures to be taken for the determination of the boundary indicated in Article 5 of the present law ; (2) the rules to be observed for the execution and conservation of works of *reboisement*; (3) the mode of determining the advances to be made by the State, the proper measures for securing the reimbursement of these, principal and interest, and the rules to be followed for giving up proprietorship of lands which Article 9 authorises communes to cede to the State.

"Art. 14. A sum of 10,000,000 [francs] is appropriated to the payment of expenditure authorised by the present law, to the extent of 1,000,000 per annum.

"The Minister of Finance is authorised to alienate, with power of uprooting if necessary, woods belonging to the State, to the value of 5,000,000 of francs.

"These woods cannot be taken except from amongst those entered in table B. annexed to the present law. The alienations may be made successively within a period which shall not exceed ten years, reckoning from 1st Jan. 1861.

"The Minister of Finance is in like manner authorised to sell the above-mentioned woods to communes on an approved valuation, and on conditions determined by a regulation of the public Administration.

"The 5,000,000 of francs, necessary to complete provision for the expenditure authorised by the present law, shall be provided by means of extraordinary fellings of wood, and, if necessary, from the ordinary resources of the budget."

On the same day was enacted a land improvement law, entitled *Loi sur la mise en valeur des communaux*, providing for the utilization of commons by means of State aid in the drainage, or other measures required to fit for agricultural or sylvicultural operations, uncultivated lands and marshes belonging to communes, or sections of communes, the utilization of which might be deemed beneficial.

Under date of 17th August 1860, M. Vicaire, director-general of the Forest Administration, addressed to the Forest Conservators a circular relative to *reboisement*, of which the following is a translation :—

"The question of the replanting of mountains with woods, so important in view of inundations, is about to receive a practical solution. The Imperial Government, which does not shrink from any expense required to give legitimate satisfaction to the wishes of the country, is of opinion that the time has come to bring to a close the discussions to which this grave question has given rise in scientific societies and in general councils.

"His Excellency the Minister of Finance, faithfully following the Imperial programme of 5th January 1860, has shown, in a remarkable report of the 3rd February following, the necessity of replanting the mountains with woods, and the measures to be adopted to effect this. Shortly thereafter, the Council of State, adopting the views of the minister, presented a *projet de loi*, which has been adopted, with marked approval, both by the Corps Legislative and by the Senate, after careful consideration. The ardour with which the Legislature of the State has set about realising the generous thought of the Emperor testifies to the greatness of the enterprise, and to the great interest taken in it by the country.

"To the Forest Administration is assigned the honour of carrying this into execution, and it will not, I feel assured, come short of its mission.

"The mode of execution being to be made the subject of a regulation issued by the public Administration, it becomes of importance that there should be obtained forthwith the data needed for its preparation. I therefore invite you to consider carefully the provisions which it may appear to you it should contain.

"As is the case with all new laws, it may be, the law for the replanting of the mountains with woods will, on its first application, give rise to some difficulties. May I ask of you to consider well those which you may consider it likely to give rise to in your district, and to point out to me the best means of removing them.

"It cannot be expected that until after the publication of the regulation by the public Administration you should be able to write me fully on this matter. At present, therfore, I limit myself to soliciting your attention to provisions of the law, the execution of which is independent of that regulation, and to measures which should precede the application of the regulations which may be issued.

"Your first endeavour should be to determine the localities in which the work of *reboisement* will be most useful; and it should not be allowed to escape your attention that, according to the prescriptions of the law, Government aid should be given exclusively to the replanting of lands situated on the summits or the declivities of mountains.

"The Administration cannot extend the resources placed at its command to all the lands which may fulfil these conditions; you should, therefore, endeavour to ascertain to which a preference should be given. In what relates to the works entrusted to your consideration you should avoid making choice, in the commencement of the enterprise, of lands the replanting of which would present an excess of difficulties, that you may not bring the enterprise into contempt, which might result in cooling the zeal of communes and of private landholders.

"You should select as much as may be possible, according to the climate, the nature of the soil, and the exposure of the lands, the kinds of trees of which the successful growth would be most certain, and the propagation of which would be most useful, and prescribe only the culture which would be most suitable for them.

"It may be it will only be after many trials, which may be like groping in the dark, that you will be able to make your final selection; and I cannot too strongly recommend to you to multiply your experimental trials and to make them with all necessary care.

"It is desirable, for instance, that you should carry on your operations on a great many different spots.

"The work of replanting woods can only be carried out successfully if it secure the sympathy of the people of the locality, and nothing likely to secure this should be neglected.

"It is then necessary that you should make yourself well acquainted with the wants of the communes, and lay yourself out to reconcile with these, as far as possible, the measures to be adopted; and if present profit cannot be combined with the interests of the future, to sacrifice these; it is nevertheless necessary to give due consideration to this, and to reckon it a matter of no small importance.

"According to the idea of the Legislature, the encouragements given should be given as much as possible in the form of grants of seed and of plants.

"The Administration is already engaged in carrying out drainage and forming *sécheries* [for the drying of seeds], wherever it is practicable to do so with advantage.

"The case of establishing nurseries pertains more especially to you. You will be supplied with all the funds needed to make these in sufficient numbers and under the most favourable conditions.

"You cannot give too much attention to the extension, diffusion, and development of these valuable works, whether they be carried out on account of communes or be executed on account of the State.

"Whenever lands of a certain extent are to be replanted with woods, it will always be found useful to locate one or more nurseries near to these, so as to avoid the risk of failure consequent on too prolonged transport of the plants.

"Article 2 of the law bears, that in the distribution of subventions to communes and public bodies, regard is to be had to their resources, to their sacrifices, and to their wants, and also to the amount granted by the General Councils for *reboisement*.

"The application of this provision will require, on your part, much care, discernment, and tact.

"The distribution of subventions will give occasion for the forest officials entering into frequent communication with the representatives of communes and of public bodies; I need scarcely remind you that all your communications should be characterised by the greatest cordiality and amity.

"The General Councils will be led to consider whether, in carrying out the views of Government, they ought not to devote a portion of their resources to the work of *reboisement*. Instructions, which may be necessary to their giving a deliverance on this subject, you should supply without loss of time to the prefects, and, if necessary, take the initiative in communications of this kind.

"The law divides itself into two distinct parts, relating severally to encouragement and to coercion; we have only to do with the first of these here.

"The happy results which may be obtained by encouragement may render the application of coercive measures the more rare. This is a consideration which should lead you to see that nothing be neglected which at first sight gives assurance of success.

" The whole country will watch with the greatest solicitude the results of the new law. It reckons on your zeal; we must not disappoint the expectation.

" I shall annually report to the Minister the works executed each year, in accordance with the law; it will be a pleasure to me, in doing so, to make special mention to His Excellency of your zeal, and of that of the agents who shall have best seconded you in this work, which is so important."

The following decree, embodying the statute of the Public Administration for the enforcement of the law of 28th July 1860, was issued 27th April 1861.

"CHAP. I.—OF REBOISEMENT FACULTATIFS, OR SANCTIONED AND AIDED OPERATIONS.

" *Arts 1, 2, 3, and 12 of the law of 28th July 1860.*

" Art. 1. Proprietors of lands situated on the summits or the declivities of mountains who desire to avail themselves of the subventions granted by the State, in terms of Arts. 1 and 2 of the law of 28th July 1860, should address their demand to the Forest Conservator.

" If it be a commune or a public body, the demand should be addressed to the prefect, who will transmit it to the conservator, with a letter of advice.

" Art. 2. Lands belonging to communes or public bodies, on which works of *reboisement* are undertaken by help of subventions granted by the State, are subjected to the *régime forestière.*

" The works on these, as well as those of conservation, or of full maintenance, are to be executed under the control and surveillance of the forest officials.

" If the lands belong to several communes, and it be necessary to the successful prosecution of the *reboisement* that the work be carried on simultaneously in all of these, there is to be created, conformable to Arts. 70, 71, and 72 of the law of 18th July 1837, a syndic council for the purpose of effecting an execution of the works.

"In case of the non-execution, or of the bad execution, of the works certified by the forest officials, the prefect passes a resolution enjoining the restitution to the State of the subventions which have been allowed.

" Art. 3. Premiums in money, obtained by private proprietors after the execution of the works, are to be paid on presentation of a minute of acceptance of works, drawn up by the local forest official in the form of minutes of acceptance of works of improvement in the State forests, and on the advice of the inspector and the conservator.

"Subventions in seeds, or in plants, delivered to private proprietors before the execution of the works, are to be estimated at their money value. The valuation is to be notified to the proprietor, and accepted by him. The amount of this is to be repaid to the State in case of the non-execution of the works, the misappropriation of the seeds or plants, or of the bad execution of the work certified, as has been prescribed in Art. 2 of the present regulations.

" Art. 4. The allocation of subventions exceeding in amount 500 francs is to be made by the Minister of Finance; the allocation of subventions under 500 francs in amount is to be made by the Director-General of Forests.

"Art. 5. When works of replanting have been executed on lands belonging to private proprietors by help of subventions, the proprietors, before admitting cattle into the plantation, must address an application for permission to do so to the conservator, who is to cause the state of the young woods to be ascertained by the forest official, and to determine accordingly, under power of appeal to the Minister of Finance.

"If the proprietor fail to conform himself to the decision given, the whole or part of the subventions granted may be charged against the proprietor.

"CHAP. II.—OF REBOISEMENTS OBLIGATOIRES, OR ENJOINED REPLANTINGS.

"*The determination of the boundaries within which it is necessary to execute the reboisement.*

"Act. 6. When the Forest Administration considers that it is proper to proceed to determine the boundaries of the lands on which it is necessary to execute works of *reboisement*, the Director-General of Forests is to give notice to the prefect of the forest agents designated for the preparation of the minute of specification of the lands, the diagram, or plan of the places, and the proposed project of the works to be executed.

"The prefect is to designate the engineer of roads and bridges, or of mines, whose consent to the operation is required.

"Art. 7. The minute of specifications is to be accompanied by a descriptive memoir, indicating the object of the enterprise, and the benefits expected to attend it.

"The diagram, or plan of the place, is to be prepared in accordance with the land-register of the district. It is to indicate, in regard to each plot, the number specifying it in that register; the superficial contents; the name of the proprietor; and if it belong to a commune, or to a public body, the sum total of the superficial contents of lands belonging to the commune, or to the public body.

"The project of works to be executed is to indicate the lands which it is designed to replant; it fixes the time within which the works should be executed; and it contains (1) an approximate estimate of the expense, and a project of the partition of this expense among the different proprietors; (2) an indication of the subventions which might be offered to each proprietor; (3) a valuation of the actual revenue from each lot, of the value of the ground, and of the value of the crop; (4) any other statistical information which might be useful, if known.

"Art. 8. The documents spoken of in the preceding Article are to be addressed by the Forest Administration to the prefect, who is to proceed to institute in each commune the inquest prescribed by Art. 5 of the law of 28th July 1860.

"The project of operations is to remain deposited in the mayor's office for a month; at the expiry of this time a commissioner, designated by the prefect, is to receive, at the mayor's office, during three successive days, declarations from inhabitants regarding the public utility of the projected works. These days are reckoned from the advertisement, given by means of publication, and posted notices. The authority for such advertisement, and the publication of the order of the prefect which appoints the opening of the inquest, must be a certificate from the mayor.

"After having closed and signed the register of the declarations, the commissioner is to transmit this immediately to the prefect, with advice, and the other documents of instruction which have served as a ground for the inquest.

"Art. 9. The Municipal Council of each commune concerned, summoned for the purpose by injunction from the prefect, is to examine the documents in question, and, after a delay of a month, to give its opinion by a resolution adopted in conjunction with the prescribed addition of others, equal in number to that of the officiating members of the municipal council. This resolution is to declare, if such be the case, whether the municipal council authorises the works of replanting to be carried out to a still greater extent than that specified by Art. 10 of law of 28th July 1860. The minute of this resolution is to be added to the documents connected with the inquest.

"Art. 10. The commission instituted by the second paragraph of Art. 5 of the law of 28th July 1860 is formed by the prefect in each of the departments traversed by the line of works.

"This commission meets at the place indicated by the prefectoral resolution, and on the fifteenth day from the date of that decree. It examines the documents giving the requisite instructions, and the declarations delivered to the registrar of the inquest; and, after having deliberated on these in company with any persons whom they may consider it would be well to consult, and with the information which they consider necessary, they give their opinion both on the utility of the undertaking, and on the various questions submitted by the Administration.

"These different proceedings, of which a minute is to be prepared, must be completed within another period of one month.

"Art. 11. The prefect, after having taken the opinions of the Council of the Arrondissement and of the General Council, is to forward all the documents relative to the case, together with his own opinion, to the Minister of Finance, who, after having previously consulted with the Minister of Agriculture, of Commerce, and of Public Works, and the Minister of the Interior, if there be occasion for it, is to submit to us his report.

"A deliverance will then be given by us on the question of the public utility of the works, our Council of State having heard the case.

"Art. 12. A duplicate of the decree which declares the public utility of the works is to be transmitted by the Director-General of Forests to the prefect, who is charged with the fulfilment of the formalities prescribed by Art. 6 of the law of 28th July 1860.

"At the same time the Forest Administration is to notify to the prefect, in regard to each plot in the register, the works to be accomplished, the conditions under which they are to be executed, and the time within which this must be done, the offers of subvention made by the Administration, and the advances of money to which they are prepared to consent.

"CHAP. III.—OF THE EXECUTION AND MAINTENANCE OF THE WORKS.

"CHAP. 1.—*Lands belonging to private proprietors comprised within the boundaries specified by the decretal declarative of public utility.*

"Art. 13. Within a period of one month reckoned from the notification which is made to him of the decreet declarative of the public utility, the private proprietor of the lands comprised within the boundary shall declare whether he intends to execute the works himself, or to give up the execution of them to the Forest Administration.

"This declaration is to be made in duplicate, and transmitted to the sub-prefecture of the locality in which the places are situated, or in which they are registered. These duplicate declarations are revised by the sub-prefect, who is to return one to the party by whom the declaration is made, and to transmit the other immediately to the prefect.

"If the private proprietor wishes to execute the works himself, his declaration is to contain, in addition, proof of his possessing means of doing so.

"Art. 14. Failing the deliverance of this declaration, the private proprietor is to be held to have refused to undertake the works at his own expense.

"Art. 15. The works executed by the private proprietor, with or without subvention, are to be subject to the surveillance of the Forest Administration.

"Art. 16. The Forest Administration is to proceed to the execution of the works to be carried on on the lands of the expropriated proprietors.

"The completion of the works is to be notified by the Forest Administration to the expropriated proprietor. The notification is to contain, moreover,—(1) a detailed account of the amount, principal and interest, of the cost of the works executed from the period of expropriation; (2) an estimate of the annual expenditure deemed necessary for the conservation and maintenance of them.

"Art. 17. When, in accordance with Act. 7 of the law of 28th July 1860 the expropriated proprietor wishes to avail himself of the right to obtain reintegration, he is to make the declaration to the sub-prefecture within the five years following the notification made to him, in terms of the preceding article, and to notify in this deed whether he intends to obtain his reintegration by reimbursing to the State the advances made, or by giving up to the State a half of the property.

"These declarations are to be registered, and of this a certificate is to be given.

"Art. 18. If the proprietor makes choice of reimbursing the advances made by the State, he is to produce, in support of his declarations, the necessary proof that he is in circumstances to reimburse the indemnity of expropriation, and the expense of the works, both in their first establishment and their maintenance—principal and interest.

"This declaration and documents in support of it are to be addressed within a month to the Minister of Finance, who is to decree and determine the forms, and the time within which the proprietor shall be reintegrated.

"Art. 19. If the proprietor offers to give up to the State one-half of the property, proceedings are to be taken by a forest agent, and by the proprietor, or his delegate, to divide the land into two lots of equal value.

"In case of dispute in regard to the formation of these lots, it is to be determined by a third party, a skilled umpire, named by the president of the tribunal.

"The appropriation of the lots is to be determined by drawing of lots, if the parties cannot come to an amicable arrangement.

"If a part of the works has been executed by the proprietor, this is to be taken into account in making the division, by a proportional deduction being made from the lot which falls to the possession of the State.

"CHAP. 2.—*Lands belonging to communes or to public bodies comprised within the boundary specified by the decretals declarative of public utility.*

"SECT. 1ST.—*The execution of works to be carried on on the lands belonging to communes or public bodies.*

"Art. 20. Within a month from the date of the decreet declarative of public utility, communes and public bodies, proprietors of lands, comprised within the boundary, are to notify to the prefect, by an explanatory declaration, whether their intention be to execute the works with their own resources, in whole or in part, on the prescribed conditions; or to leave to the State the care of charging itself with the works, at its own expense, subject to reimbursement; or, in fine, amicably to cede to the State, in whole or in part, lands belonging to them comprised within the boundary.

"Failing the communes or public bodies notifying their intentions within the period stated, the State is to undertake the works at its own charge, conformably to the provisions of Art. 8 of the law of 28th July 1860.

"Art. 21. Lands belonging to communes or to public bodies, comprised within the boundary specified by the decreet declarative of public utility, are to be subject absolutely to the *régime forestière*.

"Art. 22. When the commune or the public body shall have notified its intention to execute the works, the Municipal Council, or the Administrative Commission, is to grant each year the funds judged necessary for the execution of new works, and for the maintenance of works accomplished.

"Art. 23. The execution of the works is to take place under the surveillance of the forest agents.

"In case of non-execution, or of bad execution, certified by the conservator, a decision by the Minister of Finance is to ordain, if the measure be proper, that the State shall undertake the work at its own charge, in terms of Art. 8. of the law of 28th July 1860.

"When the lands belong to several communes, and the successful prosecution of *reboisement* requires the works to be carried on together, there is to be created, if all the Municipal Councils charge themselves with the undertaking, a syndical commission for the prosecution of the execution of the works, conformably to Arts. 70, 71, and 72 of the law of July 18, 1837.

"SECT. 2ND.—*Determination of the advances made by the State to communes or to public bodies, and measures proper to secure the reimbursement of these.*

"Art. 24. When the communes or public establishments decide to leave the works to the charge of the State, the Forest Administration is to cause them to be executed in accordance with the forms used in the matter of works of amelioration in the forests of the State.

"The statements of expenses are to be prepared conformably to the rules of office accounts in the Forest Administration.

"In the same forms are to be prepared the statement of annual expenses of maintenance.

"Art. 25. If the works concern several communes, the partition of the expense is to be made according to the form required by Art. 72 of the law of 28th July 1837.

"Every year there is to be delivered to each of the parties interested a statement of the expenses incurred on account of the party by the Administration.

"After the completion of the works, an account-general of the expenditure is to be ordered by the Minister of Finance, and a copy is to be delivered to each of the parties interested.

"The sums forming the amount of this account, and constituting principal, are to bear a charge of simple interest at 5 per cent. from the completion of the works.

"Art. 26. The works effected by the State are to be maintained by the care of the Forest Administration. The advances of the State for this object, ordered each year by the Minister of Finance, are to bear interest at 5 per cent. per annum.

"A copy of the account is to be delivered to the parties concerned, along with the statement of the expenses previously incurred.

"Art. 27. Demands of revision or rectification of the annual accounts of the expenses of the establishment, or of the maintenance of the works, should, on pain of forfeiture of right to these, be brought before the Councils of Prefecture within six months of the notification of said accounts. After that time the accounts become fixed.

"Art. 28. The accounts of these products, and that of the expenses, are to be made out and approved each year by the Minister of Finance, and a copy is to be certified to the parties concerned. Within six months after this notification the parties concerned can, as in the case of the account of works, make the demand indicated in the preceding article.

"The value of these products is to be deducted from the interest due to the State, or otherwise, in the next place, from the principal constituted by the expenses incurred in the establishment and the maintenance of the works.

"Art. 29. When the State is entirely reimbursed, the advances made by it—be it by products gathered by it, or be it by payments made by the parties concerned—these are forthwith to be put again in possession of the lands administered for them by the State, under such reservations as result from their being subjected to the *régime forestière*.

SECT. 3RD.—*Rules to be followed in giving up of lands which Art. 9 of the aw of 28th July 1860 authorises communes to cede to the State.*

"Art. 30. If the commune, or the public body, relieve themselves of all repayment by ceding a half of the lands replanted, the Municipal Council of the Administrative Commissions is to adopt a resolution to this effect, which is to be notified to the prefect within the period indicated by the second paragraph of Art. 9 of the law of 20th July 1860.

"Art. 31. Proceedings are forthwith to be taken by a skilled person nominated by the prefect, and a Forest Agent designated by the Forest Administration, to divide the same into two lots of equal value.

"The appropriation of the lots is to take place by drawing of lots, if the parties concerned cannot come to an amicable arrangement of this. This proceeding is to take place in the presence of the sub-prefect of the Arrondissement.

"If a part of the works has been executed by the commune or the public body, this is to be taken into account in the division, and a proportionate reduction is to be made in the share which falls to the lot of the State.

"CHAP. 3.—*General directions.*

"Art. 32. Before commencing the works within the extent of the boundaries fixed by the Imperial decrees, there is to be made, at the expense of the State, a determination of the boundaries, and, if need be, a marking off of the said boundaries on the ground."

In order that the superior local officers of the Forest Administration might be fully acquainted with what it was desired should be done in the **carrying out of the decree, the Director-General of the Forest Administration,**

M. Vicaire, subsequently issued the following explanations, in a circular addressed to forest conservators, under date of 1st June 1861 :—

".PART I.—REBOISEMENTS FACULTATIFS, OR SANCTIONED AND AIDED OPERATIONS.

" The 1st Article traces out the course to be followed by private proprietors, the communes, or the public establishments, in the applications for aid that they may have to make. By the terms of Art. 2 of the law of 28th July 1860, the aid in the reforesting of the lands situated on the summits or the slopes of mountains is granted on account of the utility of the work as regards the general good, having respect, in reference to the communes and the public establishments, to their resources, their sacrifices, and their need, as well as to the sums allotted by the General Councils for reforesting. The communes, whose territory is situated in the regions where the *reboisement* of the mountains is in the highest degree important to the public good are in general very poor, and often they have no other resources than what they derive from pasturage. Every reduction in the extent of the lands free to all excites among the inhabitants of these regions great apprehension. Great efforts in the initiative could not, therefore, be expected from them, and it is better in such cases to be lavish with encouragement. The Government will contribute very largely to the expense of the work, whenever the communes placed in these circumstances show their goodwill to the work.

" When, on the other hand, the lands belong to proprietors more advantageously situated in regard to pecuniary resources, it is better to be less ready in giving aid, and to apportion this more strictly to the efforts and sacrifices of the proprietors.

" The law for the *reboisement* of mountains is essentially a law of general interest, and it is in this point of view that it is necessary to regard it in considering the demands for aid.

" With a view to securing proper order and regularity in the consideration of such demands, they ought to be given in before the 15th of July of the year previous to that in which the aid is required. Those which arrive subsequently to that time will be carried over to the next year, excepting, however, cases in which you may decide that it is better to proceed without delay to the decision. The requests will be summed up in the form of ordinary *reconnaissances*, and must reach the Government before the 1st of September, with your observations and information, at the same time with the accounts of the demands for seeds which you annually furnish for resowing void places in the Government forests. It is not necessary to say that the rule in regard to this can only be followed out when the demands for aid shall have been established in an orderly manner. The demands which reach you this year, or at the commencement of the next, will be attended to with the least possible delay.

" It is necessary to take every precaution to insure the proper application of the aid. To this effect the demands should be made out upon formulas, conformed to the models 1 and 2 hereto annexed. Notice should be given of the granting of the aid, in the form of models Nos. 3, 4, 5, and 6. As you will see, these different formulas have been arranged so as to make known to the parties for whom the aid is destined the obligations to which they are subjected, and at the same time to give to the Government the right of exercising, in case of need, its right to reimbursement.

"The preceding directions are only applicable to the authorised replanting of woods for which demands for aid may in future be sent in. The works already undertaken, by the help of credit placed at your disposal upon the fund appropriated to the expenses of the replanting of the mountains with woods, will be continued according to the plan on which they were commenced.

"Article 2 relates to the work to be done in the lands belonging to the communes, or to the public establishments. These lands being by clear title subject to the forest department, it will not be necessary to get for them special applications; it will suffice that you address to the Administration, at a fit opportunity, the necessary instructions, in order that they may be inscribed upon the roll of the communal woods or public establishments subject to the Forest Administration.

"When there may be occasion to form a municipal commission, you will make known to the Government the measures taken to this effect. Aid will only be granted when the said commission shall have been regularly constituted.

"You will take care that the control and supervision of works by the forest agents shall be efficiently maintained.

"The aids for the execution of works of reforesting on lands belonging to private proprietors are principally granted in kind. But aid in money can be granted to private proprietors under the head of money advanced. The *exposèe des motifs*, or explanation of the grounds of the law of 28th July 1860, says, on this subject, that the private proprietors themselves shall be admitted to the benefit of aid in money, at least in certain exceptional cases, where the actual expense would be too great compared with the profits, necessarly remote, and in which the work should present a character of public utility sufficiently obvious in order to its appearing just on the part of the State to support outlays, of which the public would in part reap the fruit.

"When a private proprietor shall have framed a request, the same form will be followed as in the demand for aid in kind. The agents will visit the localities, certify the state of the property, appraise the expenses of the work, as well as their utility in regard to the public interest, and estimate, principally as regards the latter point, the amount of aid to be allowed.

"The third Article traces the course to be pursued in the payment of this.

"As regards the payments in kind, the model formula No 3. has been so prepared that the deed of notification forms, between the Adminstration and the receiver of the aid, a contract of a kind supplying a reciprocal guarantee on behalf of the subsidised proprietor, and on behalf of the general interest.

"Articles 4 and 5 have no need of explanation.

"When the staff of agents and *employés* of the ordinary service shall be acknowledged to be insufficient to insure the execution of measures relative to the replanting of woods, further measures must be taken for this object. Already, at a certain number of points, posts of brigadiers and warders have been created for this special object. I recommend to you to see that the persons appointed to these posts shall render all the services rightly required of them.

"Part II.—Reboisements Obligatoires, or Enjoined Operations.

"*Specification of the boundaries of the lands on which it is necessary that reboisment be effected.*

"Article 6 gives to the Administration of Forests the charge of marking out the lands on which it is necessary to execute the work of *reboisement*.

"Of all the measures which are prescribed for the execution of the law of the 28th July 1860, there are none to refer to which is more important than the marking out of these. I call your special attention to this point. The object of the law is the protection of the soil against the ravages of inundations, and the falling away of the slopes. In order that the work of *reboisement* may have any efficacy as regards hydraulic results, and the retention of the soil, it is indispensable that they should not be limited to scattered points. The overflow of the water-courses during storms or heavy rains is caused, as you know, by the sudden flowing in of the waters into the beds of rivers and torrents. These water-courses are formed by the union of the streams, more or less considerable, which rise in the bosom of the mountains. If the surface of the inclines where the streams rise were properly clothed with vegetation, the water, restrained on all sides in its progress, would flow without violence into the bed of the river, which would only overflow in those rare and exceptional circumstances, in which occur great meteorological phenomena, against which all obstacles are powerless.

"By an analogous operation the presence of vegetation on the surface of an incline prevents its falling away, by dividing the lesser courses of water and preventing their augmentation. According to these considerations, which I cannot avoid mentioning here, but which your experience of mountainous regions must enable you to appreciate, you will understand that the consideration of the lands on which it is necessary to undertake such works, ought, with a view to securing the important result desired, to be directed to both of these aspects of the effects anticipated.

"In every case the operation must be determined by circumstances relating it may be to the water-courses, or relating it may be to the mountain declivity; and it will be for you to judge and decide at what point it is most urgently required that a commencement should be made.

"Article 6 bears that 'the Director General of Forests shall make known to the prefect the forest agents designated to prepare the report on the character and condition of the lands, the chart of the localities, and the specification of the works proposed.'

"The operations must be carried on in general by special agents, who shall enjoy the same advantages as the agents composing the Commissioners of the Cantonment, or of Management of Forests.

"The forest agent shall have for associate the engineer of roads and bridges, or of the mines, designated by the prefect. I do not require to recommend to the agents of the Forest Administration to maintain in their relations with the agents of the Administration of roads and bridges, or of mines, the spirit of cordial co-operation which ought to animate all the functionaries of the State in their common efforts for the advancement of the public interest.

"Art. 7. The report of the inspector, the descriptive memoir, the plan of the localities, and the scheme of projected works, form conjointly and exclusively the basis of operations. In terms of article 5th of the law of 28th July 1860, this work, first submitted for examination for a month, during which period all parties interested may become acquainted with it, and then presented for the discussion of the Municipal Councils, is to be examined in succession by a special commission, by the Council of the Arrondissement, and by the General Council. It is not until after the Ministry of Agriculture, of Commerce, and of Public Works, and the Ministry of the Interior, if such there be, have been consulted that the Ministry of Finances is to submit it to the Emperor the order to be issued, the Council of State having been heard on the question of the public utility of the works. It is of importance then that the work be prepared with due care, and contain sufficient indications, that in passing through this long process of examination no considerable element of defect may appear.

"I cannot prescribe any determined form to be followed in these documents, the instructions which would be necessary would vary with the special circumstances of each country.

"It would be, besides, premature to lay down at the outset of an operation regulations so important, which experience had given no opportunity of ratifying. I leave to the agents to consider, under your direction and in concert with the engineers, what may be the most convenient manner of presenting the different elements of the work. I consider, however, that I ought to address to you on this subject some general observations.

"I have stated to you above the considerations in accordance with which ought to be carried out the inspection of the lands which it is necessary to cover with vegetation, to accomplish the object of the law on the *reboisement* of the mountains. Amongst these lands, doubtless, many, through their state of complete denudation, cannot be converted immediately into what can properly be called woods. The agents ought to make known by what preparation, be it by putting them totally or partially under enclosure for a time more or less prolonged, or be it that by the natural or artificial production of vegetation of any inferior kind, they may be rendered fit for the reception of seeds or of plantations. This indispensable preparation comes directly within the range of the works of *reboisement*.

"There are also lands which by reason of their situation appear naturally destined for use as pasturage. The conversion of these lands into woods would be of no utility, as preserved in the condition of pasturage they render to the inhabitants the least expensive and best services possible in what therein concerns the general interest, and the maintenance of these pasturages in good condition suffices in many cases to retain the water, and the land. The scheme of operation ought then to divide the lands into three classes, namely,—(1) Those in which we may proceed immediately to direct works of *reboisement;* (2) those in which these works ought to be preceded by a natural or artificial preparation ; (3) in fine, those which ought to be left free for the growth of pasturage, subject to appropriate regulations. The two first classes alone are subject to the application of the law of 28th July 1860, on *reboisement.*

"It will suffice to indicate in regard to lands of the third category, the regulations to which it would be well to subject the exercise of pasturage.

"In relation to the subvention which might be offered to each proprietor for the execution of the works within the prescribed limits required by public utility, you have only to take into account the resources of the parties interested, their requirements, the sacrifices which they are disposed to make, and the amount of the sums allocated by the General Council.

"Article 10, of 28th July 1860, bears, 'that the sowings or plantations cannot be made annually in each commune over more than the twentieth part at most of the area of these lands, unless a decision of the Municipal Council authorise works over a more considerable extent.'

"When an area surveyed comprises more than the twentieth of the lands belonging to a commune, if this commune refuse to allow the execution of works over a more extended area, it will be well to select and indicate the portion of the lands on which it will be of most utility to execute these works.

"Article 8 indicates the manner in which ought to be conducted the examination prescribed by article 5 of the law of 28th July 1860. Measures to this effect are to be taken by the prefect, when he is supplied by the Administration of Forest with the papers enumerated in article 7. You will address these papers to the Administration when the work for any one complete undertaking shall be finished, whatever may be the importance of the work, be it for a water-course from a river, from a secondary affluent, or even from a torrent, or, what is better still, be it for a mountain declivity.

"Articles 9, 10, 11, and 12 require no explanation. I shall only call your attention to the institution of the commission, of which article 10 defines the powers. According to the terms of section 2 of article 5 of the law of 28th July 1860, it is required that one member of that commission shall be a forest agent. You will understand the necessity of not designating to the prefect for that important mission any but an agent capable of worthily representing the Administration, if you do not judge it proper to reserve it for yourself, because of the interest attaching to the projected works.

"PART III.—OF THE EXECUTION AND OF THE MAINTENANCE OF THE WORKS.

"CHAP. 1.—*Lands belonging to private proprietors, comprised within the boundaries specified by the decree declarative of public utility.*

"The Administration will forward to the prefects, with the documents mentioned in section 2nd of article 12, formulas in accordance with the models Nos. 9 and 10, hereto annexed, in order that they may be enabled to furnish particulars, at the same time calling their attention, in accordance with section 2nd of article 6 of the law of 28th July 1860, to the extract of the Imperial decree, containing the indications relative to the lands belonging to them.

"Article 13 gives to private proprietors the option of undertaking the immediate execution of the works under their own superintendence, and at their own expense, with the subventions granted by the State, if there be any, or of giving up to the State the execution of the said works.

"In the first case, advice must be given to the party interested of the allocation of the subvention in the formulas 8 and 9 hereto annexed.

"According to the terms of paragraph 4th of article 13, the private proprietor who wishes himself to execute the works ought to give proof of his possession of the means of doing so. The public interest being sufficiently protected by the power given to the State by article 7 of the law of 28th July 1860, to recover by expropriation of the property, in case of non-fulfilment of the engagement made, there is no occasion to be offensively exacting in requiring the production of such proof. A declaration from the mayor of the commune, agreeably to the indications borne by the form No. 7, it appears to me, may be considered in most cases a sufficient guarantee.

"When the private proprietor has declared his intention to give up the execution of the works to the State, it is done by the Administration in the same form as that for the communal lands and those of public establishments, in which the State proceeds directly to the works; and the re-entry on possession of the said private proprietor takes place according to the same mode, and on the same conditions.

"The operation is considered an amicable one between the State and the party interested, and is not to take the legal form of expropriation, excepting in cases of disagreement.

"Articles 14 and 15 require no explanation.

"Articles 16, 17, 18, and 19 detail the course to be followed in case of expropriation, in terms of sec. 3, 4, and 5 of article 7 of the law of 28th July 1860. It is but very rarely that there will be occasion to have recourse to this measure. The exposition of the motives of the law on *reboisement* bears, in regard to this subject, 'That this exceptional remedy of the expropriation of private property shall be a rare and exceptional appliance.' It has not been without great reluctance and repugnance that the legislative body has consented to introduce into our code a new case of expropriation.

"When the Administration shall meet with a refusal, or with an insufficiency of resources of a private proprietor for the execution of the works, and all attempts at persuasion, and all the offers of subvention, have come to nought, against a declared opposition, or an absolute inability, it will become necessary to have recourse to expropriation, But every time that this shall occur you shall refer to the Administration, which will address to you timeously the instructions of which you have need. It will consequently be of no advantage to indicate here general rules in relation to this.

"CHAP. 3.—*Lands belonging to communes or public bodies, comprised within the boundaries specified by the decretals declarative of public utility.*

"SECTION 1.—The execution of works to be carried on on such lands.

"The Administration will address to the prefects the documents mentioned in 2nd section of article 12 of the decree of 27th April last, formulas conformable to the forms 11; 14, and 15 hereto annexed, in order that they may be able to transmit them to the Municipal Councils, or to the Administrative Commissions, directing their attention, in execution of the arrangements of par. 2nd of article 6 of the law of 28th July 1860, to the extract of the Imperial decree containing the indications relative to the lands which belong to them.

"Art. 20 indicates three different courses which may be followed in carrying out the work on communal lands, or the lands of public establishments within the limits. The commune, or public establishment, which does not wish to submit the whole of its lands to the same *régime*, ought to make as many special declarations as this land contains of portions destined to have a different course adopted in the execution of the work upon it.

"In case of allocation of a subvention, advice is to be given to the party interested, according to the forms Nos. 12 and 13.

"The amicable cession to the State of communal lands, or the lands of public establishments, in terms of the article 8 of the law of 28th July 1860, will present in many cases great advantages. On one hand the State will thus find facilities resulting from the suppression of pasturages, and from exclusive direction, without disputes in regard to the works ; on the other hand, poor communes will thus have the means of deriving advantage from lands which procure for them at present only insignificant resources, and of which the *reboisement* would entail expenses which perhaps they would **never be able to reimburse.**

"In such cases you should use all your influence with the Municipal Councils, and call in that of the prefect to induce them to treat amicably with the State.

"Communal lands, or the lands of public establishments within the limits, being subjected absolutely to the forest *régime*, in terms of Article 21, it will suffice that you address to the Administration, at the proper time, the information necessary to have those lands inscribed among the communal woods, or the woods of public establishments subjected to the forest *régime*.

"Articles 22 and 23 need no explanation.

"SECTION 2.—Specifications of the advances made by the State to the communes, or to the public bodies, and measures proper to secure the reimbursement of these:

"Works at the expense of the State on lands belonging to communes and to public establishments differ in nothing from the works carried on by the Forest Administration on the grounds belonging to the State. The agents directly superintend and maintain these works, without any intervention of the Municipal Councils, or of the Administrative Commissions.

"It is only required to keep a separate account of the expenses relative to each commune, or to each public establishment, up to the time when the State being completely reimbursed the advances made to the commune, or to the public establishment, they re-enter on possession of the lands subject to the forest *régime*.

"There will be sent to you timeously, if there be need, more detailed instructions for the keeping of this account, and for the annual notification to the parties interested of the expenses incurred on their behalf by the Administration.

"SECTION 3.—Rules to be followed in the giving up of lands which Art. 9 of the law of 28th July 1860 authorises communes to cede to the State:

"In terms of Art. 9 of the law of 28th July 1860, the communes and the public establisments may, in any case, relieve themselves of repayment to the State by ceding the proprietorship of half of the lands re-wooded.

"Articles 30 and 31 of the decree of 27th April 1861 prescribe the course to be followed in such a case.

"The execution of these dispositions ought not to take place but in a future pretty distant. I reserve to myself to give to you, in good time, the instructions which you may then require.

"CHAP. 4—*General Directions.*

"Art. 32. It is necessary that the bounds of the lands comprised within the limits fixed by the Imperial decrees should be determined in such a manner as to prevent all subsequent dispute. This end may doubtless be attained in both cases without its being necessary to proceed very stringently to work.

"In order to avoid long delays, and the expense inseparable from such, it will almost always suffice to proceed to a conjoint reconoissance in a manner almost analogous to that of the partial determination of boundaries.

"The deeds concerning the communes and the public establishments are neither subject to the formality of official seal nor to that of registration.

"In what relates to private proprietors, the demand for subvention, like as do all petitions, requires the official seal. The petitioners are authorised, by express exception, on paying ready money, to get the official seal applied to those formulas, either before or after the examination of the demand, provided in every case that this be before the despatch of the document to the conservator. In regard to deeds, entitled 'Advice of allocation of a subvention,' copies are exempt from seal and registration. Despatches which shall be delivered otherwise than to the public functionaries, for the service of the Administration, and with notice of this destination, must be written on stamped paper at 1 franc 25 cents per sheet.

"The instructions which I have just given you for the execution of the principal arrangements of the law on the *reboisement* of the mountains may possibly prove incomplete. If the preceding explanations do not appear to you sufficient, I shall supplement them by special instructions.

"The season being already very much advanced it is of importance to set to work immediately. The watchmen belonging to the State or to communes, on duty in the districts where the works are carried on, shall be put at the disposal of the agents whenever you judge this to be necessary. These overseers, by reason of their knowledge of the localities, will be for the material portion of the works very useful auxiliaries.

"When the number of the overseers shall be insufficient, you can propose to create for the object to which it refers the special employment of watchmen, or of brigadiers, who shall be ultimately appointed to the oversight of the sowings and plantations, in terms of article 11 of the law of 28th July 1860.

"I have already had occasion to point out to you the great importance which the Administration attaches to the work of *reboisement*. I shall not insist on this matter.

"The agents and the overseers who take part in the works in an active and useful way shall acquire special titles at the good pleasure of the Administration. Every time also that the work of *reboisement*, whatever be its nature, shall occasion to them extraordinary fatigue or expense, there shall be allowed to them either in the form of special indemnity, or in the form of an annual gratuity, a remuneration proportionate to the burdens borne and to the services rendered.

"In order to keep the Administration acquainted with the progress of the works, you should cause to be sent to them, in the months of July and January, statements in the forms Nos. 16 and 17 hereto annexed, in which shall be given information relative to the works done during the preceding half-year."

It has not been deemed necessary to append the schedules referred to.

It may have been observed that care was taken not to run counter to the prejudices and feelings of those who were likely to be benefited more immediately by the operations proposed; and by a ministerial decision of the 21st November 1861, there were instituted annual conferences of the agents employed in the superintendence and execution of the works of *reboisement*. The following is a *résumé* of the first of these conferences which was held on the 9th, 10th, and 11th of December of that year, at Valence, for the region of the Alps; at Aurillac, for the region of the mountains of Central France; and at Tarbes, for the region of the Pyrenees.

There are stated the questions discussed, and the annotations of the Administration, the whole being arranged under different headings.

Résumé, etc. :—

"DISPOSITIONS MANIFESTED BY THE POPULATION OF THE DIFFERENT DISTRICTS.

"The inhabitants of the mountains, chiefly preoccupied with the interest of pasturage, do not welcome in general, but with a certain apprehension, any measures relative to *reboisement*. Nevertheless, the personal proceedings of the agents, with the concurrence of the prefectoral authority, have already overcome much of the resistance of the Municipal Councils. In many departments, amongst which may be cited the Cantal, l'Ariège, Vaucluse, a good many of the communes have voted subventions for the replenishment of denuded mountain lands belonging to them. In the Arrondissement of Saint-Girons seventeen communes, according to the specifications of the inspector, have given up in 1860 and in 1861 either the twentieth part of the price of the fellings sold, or the proceeds of damages, or amends pronounced by the civil courts in their favour, to be employed in works of *reboisement*. There has been occasion to remark that on many points the mass of the population is favourable to the operation, and that resistance is offered only by some more or less influential members of the local Administrations having a personal interest in securing that the pasture lands be not diminished.

"There is reason also to acknowledge that the rapidity of the success of the works has had the good effect of bringing the communes to enter into the scheme of *reboisements*. This result has been notably the case in the Puy-de-Dôme, where important works of *reboisement* have been completed for some years, and where the Administration meets now but rarely with opposition, and this opposition is overcome without difficulty.

"As for private parties, they hesitate generally to undertake works of *reboisement*, the fruits of which they can only reap after long delay. They dread the expense of the works, and the difficulties of surveillance, and they are kept back by their ignorance of what means to employ to accomplish conveniently the replenishments. A great many of them, more especially in the Loire, have manifested a desire to see the direction of works of *reboisement* on their properties entrusted to the agents of the Forest Administration, and the example in this matter is found to be contagious. The fact has been established in the Ardèche, where some private parties, having made demands for subventions on the invitation of the forest agents, have been speedily followed by many proprietors. The number of demands of this kind in the department in question has risen to no less than 365 in 1861.

> "REMARKS.—The report given of the state of mind in the mountainous regions, relative to *reboisement*, indicates the means to be employed to enlist the sympathies of the population in the operations. To multiply the personal proceedings,—to make a good selection of ground for first experiments, in order to arrest the eye, and to convince the indifferent and the incredulous,—to call in the conjoint action of the prefectoral authority at all times when resistance, resulting from personal interest, is shown in the Municipal Councils—such are the general means which may be employed by the agents. The Administration on its part will support their proceedings, and will be liberal in encouragement whenever the general interest may appear to demand the powerful concurrence of the State.
>
> "To act on such minds too much cannot be done to diffuse information of the advantages realized by *reboisement*. The commune of Bourg,

Lastic, in the Puy-de-Dôme, possessed a piece of ground of 64 hectares, covered with heaths, which they could not dispose of in 1834 at the price of 7,000 francs. At this time a sowing of the ground with Scotch firs was undertaken, at the expense of the commune, with the assistance of the departmental treasury. The expense was not great. To-day the ground is valued at 70,000 francs, and the commune begins to obtain from it products which in a few years will be very considerable. The commune of Durtol, in the same department, possessed a wood of 47 hectares, planted with Scotch fir some fifteen or seventeen years before, in which they have lately carried out a thinning which has brought into the communal chest a sum of nearly 16,000 francs. Such cases are of a character to remove hesitation.

"As regards private proprietors, the applications for subventions, which have been made successively in the Ardèche, are an indication of what will occur, most likely, everywhere where the bite has been given. The Administration will agree, moreover, to cause the works of *reboisement* to be directed and superintended by its agents or by special overseers, whenever a certain number of private proprietors, resident in the same district, shall express a desire for this, and the measure shall appear necessary to the success of the works, and to their development.

"REBOISEMENTS FACULTATIFS, OR SANCTIONED REBOISEMENTS.

"The opinion was expressed that no applications for subventions should be entertained which are made by private proprietors for the planting of small widely separated pieces of ground, and which would require the Administration to expend money unprofitably, without the possibility of superintending and controlling such widely scattered replenishings.

"REMARKS.—Certain rules, most assuredly, ought to be observed in the allocation of subventions. The *reboisement* of a territory which is not attached to any similar operation completed or to be undertaken, in most cases, will be of no advantage to the general interest, and will not be of such a nature as to be encouraged by the State. It will therefore be well, in case of requests for subventions, to find out in what way the projected *reboisement* is related to the public interest, and to keep this relation in view, when allocations of money, seed, or plants are in question. Of course, at the beginning of the enterprise, operations aided by the State will be a good deal apart; it cannot be expected that all the proprietors in a given area will resolve to effect these *repeuplements* contiguously. But it is necessary to prevent the distances being so great as to make the control of the subventions and the superintendence of the works too difficult.

"It was proposed that rewards should be given to communes or to private proprietors who shall be the first to enclose their lands.

"REMARKS.—The law regarding mountain *reboisement* limits its operation to works of *reboisement* strictly so-called. No portion of the funds devoted to this work can be employed as premiums to proprietors who may take the initiative in the enclosure of all, or of a part, of their estates. But this can be always done as regards the communes, by appealing to the law concerning bringing in the waste communal lands.

"The proposal can be made at the proper time to the Superior Commission charged with presiding over the combined operation of the aforesaid law, and the law of *reboisement*.

"Questions relating to forest improvement have remained till now, and especially in the south, too much confined to a narrow circle. It is very important to make them known in every possible way. A periodical publication has just appeared under the title of *Revue agricole et forestiere de la Provence*. Everything relating to forests, and especially the question of the *reboisement des montagnes*, are to be therein treated of, with the necessary developments. An appeal has been made for help from those who wish to popularise forest science.

"REMARKS.—Government cannot hesitate to encourage the enlightenment of the popular mind respecting questions connected with the prosperity of the forests. A subvention of 500 francs has been granted to the *Revue agricole et forestiere de la Provence* from the funds for mountain *reboisement*. It is desirable that the employés should lend their help to this work of enlightenment.

"REBOISEMENTS OBLIGATOIRES, OR ENJOINED REBOISEMENTS.

"Important *reboisements* have been effected in certain departments, more especially in the *Puy-de-Dôme*, and in the *Haute-Loire*, with the help of the enactments in the last paragraph of Article 90 of the forest code. Those employed have enquired if they may not continue to proceed in the same way wherever it is possible. Government will thus possess an additional means of carrying on mountain *reboisements*.

"REMARKS.—The law of 28th July 1860 has not abrogated any of the enactments of the forest code, and there is nothing to hinder Article 90 from being applied wherever this means of *reboisement* can be advantageously employed.

"The agents employed have usually agreed upon the best way of finding out where compulsory *reboisements* ought to be effected. For example, suppose a river, resembling a torrent like the Durance, the flow of which it is necessary to restrain: the first thing done is to study the whole basin, beginning at the source of the stream, attentively following its course, either on the spot or on a map furnishing sufficient details of the principal and second tributaries; and after this preliminary study, operations are projected at different points in the basin in the order of urgency. They have proceeded in this way in the Basses-Alpes, in the Hautes-Alpes, in La Drôme, where all the operations, either projected or in the course of execution, aim at regulating the flow of the Durance and its tributaries, such as the Ubage, the Bléone, the Asse, the Buech; in La Drôme, L'Aigue, L'Ouvers, Le Bez.

"REMARKS.—If it be necessary to concentrate operations in *reboisement* where they are only sanctioned, this proceeding is much more important when they are declared to be of public utility. Isolated observations should not be made, but on the contrary all should be connected with a plan of operations converging to the same end. It is very essential to demonstrate by facts the advantage of these operations. It is necessary, where the examination of a given area is determined, that it should be pursued through all parts of the area where the rush of water is to be restrained, so that when the enterprise is completed the demonstration of the effect of the *reboisement* should be perfect and conclusive.

"A question has been raised as to the relative importance of *reboisement* and *gazonnement* for the consolidation of the soil and the creation of obstacles to the sudden overflow of streams. Several engineers, especially in the Alps, appear disposed to think that *gazonnement* is often the most suitable means of attaining the proposed end. Other *experts* are of opinion that if in certain cases *gazonnement* may appear enough, *reboisement* will more slowly but more completely and durably effect a result.

"REMARKS.—There seems to be attributed to *gazonnement*, especially in the Hautes-Alpes, in L'Isére, and in the La Drôme, a power almost as great as that of *reboisement* for restraining torrents. This is a little exaggeration. The Administration does not deny the utility of restoring the turf, but works of this kind should be undertaken on the vast bare surfaces which extend above the region of forest vegetation. Executed simultaneously with *repeuplements*, they give powerful aid in hindering the rush of torrents into the valleys; in order to seek this result by a double means, the Administration has promoted the formation of a higher commission for the simultaneous execution of the two laws on *reboisement* and reclaiming of waste communal lands. But everywhere where *repeuplement* is practicable this latter seems to promise to be the most efficacious means. The employés are mistaken if they think themselves obliged everywhere to propose immediate *reboisements* with valuable trees. When the soil is nearly exhausted, and requires to be renewed before being fit for the production of forest trees, it should be planted with bushes or hardy shrubs, such as exist here and there on the barest parts of the mountains. This work is included in the category of *repeuplements*, properly so called, and constitutes a real *reboisement*. The circular No. 806 contains, on this point, pages 7 and 8, all the necessary hints. According to the enactments of this circular, the examination of ground for compulsory *reboisement* should include grounds to be *reboiséd* either with permanent trees or with preparatory plantations, and grounds on which it is necessary to carry out works of *gazonnement*.

"The replenishings with woods may be effected through the operation of the law for bringing in the waste communal lands. These undertakings should furnish all necessary hints; and those of them which include operations belonging to both categories will be handed over to the high commission appointed by the decree of 7th November 1861.

"In La Haute-Loire, the employés entrusted with the survey of districts for compulsory *reboisement* have declared that they are often at a loss, on account of the peculiarities of the soil, consisting of waste pasture, partly wooded, but forming no greater obstacle to the torrents than if it were entirely bare. They have asked the conference to decide whether districts of this kind, which do not cover less than 65,000 hectares in the department, may be included in the *périmètres obligatoires*. The employés, assembled in conference at Aurillac, did not hesitate to answer in the affirmative, at the same time referring the question to the Administration.

"REMARKS.—The principal object of the law of 28th July 1860 is the creation of barriers to the sudden descent of torrents into the valleys. There is no doubt that districts sparsely covered with trees, having no hydraulic effect, should be included in the extent to be *reboiséd* when there is an opportunity for fixing or determining a *périmètre obligatoire* or area of enjoined *reboisement*.

"The form to be given to enterprises of compulsory *reboisement* has been the subject of a detailed examination. It has been acknowledged that up to this time these enterprises differ very slightly, and that experience will supply the most useful indications for the simplification and modification of these projects.

"REMARKS.—It does not seem that the proper time has come for prescribing a determinate form to enterprises of compulsory *reboisement*. The number of those examined by the Administration, up to this date, is not large enough to enable one to decide on the best form for these undertakings. On the other hand, no great difference has been observed in the plans presented by the officials of the different districts. The only remark that there is any need to make is that some officials have assigned too long a time—10 or 20 years—for the completion of the work. The Administration has pointed out that such a delay is incompatible with the rapidity which, from every point of view, is seen to be very desirable. It has just repeated that, when *reboisement* with long-lived trees is not immediately possible, the ground can be stocked with shrubs of an inferior order: an operation which can almost always be effected at once, and which is really included in the category of *reboisements*, properly so-called.

"It should be added to the instructions given—(1) That when a proprietor possesses several pieces of ground in the périmètre, these pieces may be grouped together if they fall under the application of similar measures; (2) That it is not necessary to point out the subvention to be allotted to each piece, but that those pieces may be grouped together for which the same proportionate subvention is proposed, and the importance of each group may be known by the proportional per cent. of the expense; and (3) That pieces may be grouped together, the value of which has been fixed by their yielding the same amount yearly.

"SUBVENTIONS.

"Various observations have been made upon the allocations of Government subventions for works of *reboisement*. One employé has expressed the opinion that it will be difficult, according to circumstances and according to locality, to grant subventions of variable importance, and to absolve in certain cases, the communes from all expense, on account of considering, as a direct participation in the expense of the *reboisement*, the allocation of a subvention from the treasury of the Department. This employé has, in fine, requested that a maximum should be fixed, for example, say 80 per cent. of the expense for the communes, and 60 per cent. for private individuals, a maximum which must not in any case be exceeded in the offers of subventions from the State or from the Department.

"REMARKS.—The subvention is in its very nature variable. It depends on the importance of the *repeuplement*, or restocking, with regard to the public interest, the attitude of the public mind in the district with regard to mountain *reboisement*, the more or less easy position of the proprietors, and on various other analogous circumstances which it does not appear necesssary to detail. From thence it follows that the Administration should specially reserve the power of taking into consideration on each demand, the amount of the subvention to be granted. A maximum cannot therefore be fixed. As regards the communes, the Administration intends, where required, to consider the subventions voted by the General Councils of the departments as a direct participation in bearing the expense of the works.

" Several employés have given an opinion that subventions ought to be offered in preference to proprietors whose land is included in the *périmètres*, so as especially to encourage *reboisements* of acknowledged public utility.

"REMARKS.—The law grants subventions in cases of sanctioned *reboisement*, and in cases of compulsory *reboisement;* the Administration will proportion in both these cases the amount of the subventions to the expected result of the enterprise, regard being had principally to the public interest.

"METHOD OF CARRYING ON OPERATIONS.

"*Nurseries.*—After different opinions had been expressed in regard to the extent which should be given to nurseries, it was agreed that this should depend on the yield of the nursery and the extent of territory to be *reboiséd*.

" There were various opinions expressed upon the point, whether it would be better to form great central nurseries which would cost less and be more easily superintended, or to form a great number of small nurseries scattered over the district to be *reboiséd*, which would have the advantage of placing plants more within the reach of the districts to be re-wooded.

"REMARKS.—The chief effect of establishing large central nurseries in close proximity to the great populous centres, is to attract public attention, and by degrees to invite the proprietors of waste mountain land to *reboisement*, by the facilities which are offered them for procuring all that is needful for the operation. Nurseries of this kind can also be better and more cheaply taken care of. At the same time nothing is absolutely fixed on this point, and there is no reason to prefer one system to the other.

" There were also diverse opinions expressed in regard to how nursery ground should be selected. Some thought that nurseries ought to be formed on the best soil of the district to be *reboiséd*, so as to produce healthy plants. Others were of opinion that nurseries should be formed where there were average conditions of climate, fertility, and altitude, so as to produce plants which would run no risk of dying from a too rapid change when transplanted.

" From the same view, an opinion was expressed that in general it was not good to manure the soil ; but that in cases where manure appeared necessary to pulverise the soil or to repair its losses, vegetable compost should be used, and more especially that which was collected in the woods.

"REMARKS.—If the nursery can be placed where the soil is good and at a moderate distance from the districts to be *reboiséd*, it will evidently be of advantage to the State to become its proprietor. There are nearly always dangers in fixing the position of a nursery, if care be not taken to stipulate in the leases the guarantees necessary to protect the interests of the State. There is reason to believe that in most cases the purchase is of greater importance than the situation, since the State can always, when necessary, sell the land which has been improved by culture, when it becomes useless as a nursery.

"It is agreed that the ground should be thoroughly pulverised and dug to at least the depth of 30 centimètres, or 12 inches. All were not of one opinion as to the quantity of seed necessary for stocking a nursery. As to pine seeds, such as are principally made use of in mountain *reboisement*, the calculation towards which most opinions seemed to converge, was from 8 to 10 kilogrammes of seed per *are*.

"Sowing the entire nursery, and extracting the plants from a third of the extent at the end of two years, with an immediate re-sowing of the ground, and so on for the two other thirds, appeared to some an economical plan, yielding satisfactory results. By this system the plants would be used without being previously planted out.

"Others thought that with regard to nurseries there should be less thought of the expense than of the benefit to be expected; and that it was, above all things, necessary, especially at the beginning of a great enterprise, to employ all possible means to ensure its success; and that, with this in view, the ground should be divided into strips, which should alternately be sown and left unoccupied; that the young trees should be planted out carefully, to allow of a proper development of the root; and finally, that the sowings should be graduated in such a way as to obtain a difference of age favourable for transplantation.

> "REMARKS.—The idea underlying this suggestion is a sound one. Attention should be given primarily to the efficiency of the nursery, and the question of saving expense be considered secondary to this.

"For stocking the nurseries, it has appeared right to employ, as much as possible, seed grown in the locality, or in the immediate neighbourhood. It has been thought good to employ shelter of every kind,—branches, stretched out cloth, straw quilted between canvas or cord, fern, and screens of arbor vitae. Some were of opinion that the plants should be watered, but with much caution; and it was thought that, although in certain localities indispensable, there was in most cases the inconvenience of accustoming the plants to a moisture which would not always be maintained, and of thus making them more sensitive to the action of heat.

"Other operations, such as *binage* and hoeing, were considered by every one to be indispensable.

"All were also unanimously of opinion that the nursery should be enclosed, and that nurseries of any extent should be provided with a hut as a shelter and tool-house.

> "REMARKS.—There have been recommended, as sufficient and economical fences, either simple ditches, wide enough to present obstacles to the incursions of animals, or parallel lines of wire fencing, fixed at regular distances to wooden posts.

"As a useful precaution in transporting, it was recommended to cover the roots with a mixture of clay and cow-dung.

"Opinions were very various as to the season for sowing, mode of culture, and several details as to keeping in proper order.

> "REMARKS.—Experience alone can provide useful hints as to what is most suitable to each locality.

"CHOICE OF TREES.

"The employés have not been able as yet to submit well prepared returns as to the kind of trees to be used in mountain *reboisement*. Up to this date, the trees principally used have been the *épicéa*, or Norway fir, the

Scotch fir, the black Austrian pine, the pine of Aleppo, the Corsican pine, and the ailanthus, which have generally succeeded,—the larch, which has failed in certain places because the ground was too damp and the elevation too low,—and the acacia, which has failed when planted at too great an elevation, but has succeeded lower down. The Atlas cedar has been used in several districts.

"Deciduous trees, such as the white oak, the green oak, the Liege oak, the chestnut, the willow, the white poplar, and the birch, have been successfully planted in several places; shrubs, such as *l'amêl*, *anchier*, the sumach, the hazel, &c., have already afforded good results in preparing the soil for a stock of valuable trees.

"REMARKS.—It is well to attend to the indications supplied by Nature in each locality, especially where there is any question of replanting with shrubs or inferior vegetation. There is nothing to hinder a trial of new essences, or kinds of trees, when this is made with requisite caution. Thus, the ailanthus, recently tried in several places, has everywhere yielded good results. The same can be said of the Austrian pine, which almost universally succeeds in calcareous soil, and at the most varied altitudes. It will be only after a number of experiments that it will be possible to classify with any amount of precision the kinds of trees, by regions and by zones of altitude.

"In Germany, a mixture of Norway firs and larches is generally considered a good one.

"A mixture of oak and Scotch fir is also recommended at points where the former has a chance of succeeding.

"One cause of the failure in sowing larches is having placed the seed at too great a depth. Larch seed should be covered very lightly with earth.

"MODE OF EXECUTION OF WORKS OF REBOISEMENT.

"After preparing the ground—in doing which, especially on the slopes, great care should be taken not to disturb the soil too much—it is necessary to proceed with the work of *repeuplement*, or restocking with trees. Opinions are divided as to whether sowing should be preferred to planting, or *vice versa*.

"Many are inclined to think that sowing should be employed, as more economical in temperate districts, where success is sure, but that planting is to be preferred at greater altitudes.

"There is, moreover, a mode of sowing, known as *semis à la niege*, which has been several times employed successfully, and which will facilitate the stocking of large surfaces at the small outlay of from 25 to 30 francs per hectare.

"Sowings of larch seed on the snow have several times succeeded in the Hautes Alpes, and in the Basses Alpes, and it is proposed to make similar experiments in these districts with other seeds.

"When the ordinary mode of sowing is employed, it is advantageous to sow early, that is to say, at the beginning of spring, so as to avoid the too sudden effects of the summer heat.

"Sowing by means of *potets placettes* has appeared most suitable for clothing uneven surfaces, or friable soil. The quantity of seed to be used is calculated, on an average, at 3 hectolitres of mast per hectare, at 6 or 8 kilogrammes of Scotch fir, or of similar seed, and at 6 kilogrammes of larch. This quantity ought to be doubled when the sowing is done in strips. These quantities are, besides, essentially variable, according to circumstances and locality.

"REMARKS.—The Administration thinks it proper to recommend the sowing of seed upon the snows. Although its success has not yet been tested in a sufficient number of places, there is reason sufficient to employ it with different kinds of seeds, and in different places, where it may be likely to succeed. It is not necessary to enlarge on the advantages of so simple and economical a mode of *repeuplement*.

"On volcanic soil, covered with scanty heath, good results have been obtained by sowing broadcast, without any further preparation than a simple *ecobuage* when the long thistle heath hinders the seed from reaching the ground.

"The necessity for early sowing cannot be too much insisted on,—in March, for example, when they can profit by a few fine days, often very soon followed by snow and rain. Germination then takes place under favourable conditions, and the young plant is able to resist the great heat, which would have killed it if the sowing had been deferred until the last snow had melted.

"It is desirable to form artificial shelter wherever it is possible. When planting is the mode chosen, the season chosen is not the same in every district. Opinions are not agreed on this point. In Provence it has appeared that almost invariably planting in autumn is to be preferred, because of the early season of the droughts, and on account of the scarcity of workmen who are resuming their agricultural employments.

"The age at which plants should be used is very variable. In the high regions of the Drôme and the Isère, it has been remarked that plants should be strong in proportion to the elevation of the district. It is good at such points only to plant trees which are four or five years old at the least.

"The quantity of plants per hectare is necessarily variable, only strictly local indications can be given on the point.

"The expense of the stocking per hectare has not yet received a sufficiently approximate calculation. Experience alone can furnish the data necessary.

"Several employés have considered the question, whether the mode of working by contract might not be advantageous and economical for the preparation of the soil; without being quite decided, they are inclined to think that this mode of proceeding may be useful.

"REMARKS.—It does not appear that there are as yet sufficient grounds for throwing open the operations to private speculation. It is only after they have been for a long time carried on economically that it will be advisable to substitute contract for Government management.

"The scarcity of workmen in certain districts, and especially in the Alps, has engaged the attention of the agents, who have expressed a wish that the Administration would interfere and obtain from the Minister of War the paid assistance of military workmen when circumstances permit.

"REMARKS.—The Administration will most willingly negotiate this matter with the Minister of War when it becomes necessary. But to do this it must be able to specify and define the proposal, and let him know the situation and extent of the operations, their duration, the time when they will take place, the number of workmen required, the point from which they should be sent, the pay which they will receive, &c.

"Several employés are of opinion that the special staff of the *reboisements* should be in keeping with the increasing development of the operations, and that the employés composing this staff be entrusted with the execution of the enterprises which they have suggested, with the assistance of the local employés, during the disposable time left to these latter employés by the requirements of their ordinary duties.

"REMARKS.—The Administration proposes to entrust special agents with the work of *reboisement*, not only in what regards the preparation for the undertaking, but also the execution. This service will also be placed in due time in a position to grapple with new exigencies as they may present themselves.

"At the same time, the Administration does not intend that the employés of the ordinary service shall consider themselves relieved from all participation in the operation in question.

"Negotiations with proprietors of waste mountain land, for the purpose of engaging them in *reboisement*,—the giving due notice in regard to the demands for subventions for *reboisements facultatifs*,—the supervision of execution of operations of *repeuplement*,—the giving of assistance in operations of enjoined *reboisement* when they take place,—will be a part of the functions and duties of the officials attached to the ordinary service. The Administration has pleasure in believing that all the officials will assist the enterprise with all necessary zeal and devotion.

"Such are the principal questions which have occupied the employés assembled in conference on the 9th, 10th, and 11th December, at Valence, Aurillac, and Tarbes.

"These agents have, moreover, given a concise account of the operations already completed, and of those which are projected. The repetition here of this account would be uninteresting. Concerning the completed works, the Administration will find more circumstantial details in the statements Nos. 16 and 17, which should be produced, in accordance with the circular No. 806. As to the projected operations, special notes will be supplied by the conservators, each regarding what concerns his own circuit.

"*(Signed)*—H. VICAIRE, Director-General of Forest Administration.

"Paris, 10th January 1862."

The following is a *résumé*, or abstract, of the official report of operations carried on in 1861 submitted by the Administration:—

"A. *Reboisements facultatifs, or sanctioned operations.*

"If the comparative unimportance of the greater part of the works, and if the wide dispersion of these do not permit them to be included to-day in the general system of defence against torrents, they tend at least efficaciously to accomplish the object which the Legislature had in view.

"These *reboisements*, though partial, are in effect creating woods, which, though now isolated, by successively effected combinations will prepare for the future, masses of important forests. On the other hand, the rendering productive lands which have remained until this time unproductive constitutes a true agricultural progress.

"There have been received, in 1861, 695 demands for subventions, almost all of which have been approved. And the communes have competed energetically with private proprietors.

"The extent of communal lands rewooded by aid of subventions is 2653 hectares. The regions of the Alps, and the Pyrenees, and the central plateau, have been the sites of the greater part of these *reboisements*. The lands replanted with woods by private proprietors comprise 584 hectares.

"Besides, there have been executed works of *reboisement* on 1402 hectares of State lands on the mountains.

"In all, 4639 hectares. The expenses have been 372,000 francs, or 80 francs per hectare; the proportion of this paid by the State has been 200,000 francs.

"B. *Reboisements obligatoires, or enjoined reboisements.*

"Of 1,100,000 hectares of lands capable of being rewooded, the resources put at the disposal of the Forest Administration did not admit of actual *reboisement* over an area of more than 80,000 hectares. It has been necessary first of all to determine what were the localities where the works were most urgently required, and it is towards the origin or source of the water-courses that the explorations have been directed.

"There were taken into consideration 129 projects of *reboisement*, embracing 107,474 hectares of land situated in the Alps, the Pyrenees, and the mountains of central France. These lands are not all designed to be replanted with trees; a portion will require to be preserved as pasturage, subject to regulation of trespass; the remainder may be successively replanted in definite portions annually, either immediately or after the preparation of the soil by the erection of fences, and by the natural or artificial production of vegetation of an inferior order."

A second conference of agents employed was held in 1862, on 8th September, and following days, at Clermont-Ferrand, for the regions of the mountains of Central France; at the same time, at Carpentras, for the region of the Alps; and on 15th September, and following days, at Foix, for the region of the Pyrenees. Of these conferences, the following is a résumé, with annotations by the Administration :—

"I. REBOISEMENTS FACULTATIFS, SANCTIONED REBOISEMENTS.

"*First Question.*

"Up to the present time numerous applications for aid in carrying out sanctioned works of *reboisement* have been made by communes, by public bodies, and by private proprietors. The Administration has reason to believe that the parties for whom this aid is desired, and more especially the private proprietors, do not always possess the information necessary to enable them to make the most of such aid.

"The Administration has reason to believe that it can in general rely with confidence only on works of restocking woods effected by itself, or under its direction. Again, looking at the subject from another point of view, the *reboisement* carried out on scattered patches, and often very imperfectly, will not effect the object in view excepting in so far as they shall by combination form, sooner or later, a sufficient and effective protection to the soil.

"In accordance with these views, it is not unreasonable to enquire whether it would not be desirable to specify zones, beyond which aid shall not be given for *reboisements facultatifs*, excepting in special cases of considerable extent, presenting an indisputable character of public utility, and holding out in every way a probability of success. Within these zones the work would be carried on under the superintendence of the overseers actually employed in the vicinity, or of special overseers, the number of which it will be necessary to determine.

"*Opinions and Propositions of the Agents.*

"The members of the conference at Clermont were not altogether agreed as to the circumstances in which the specifying of such zones should be carried out. Some were of opinion that the measure, though unnecessary in regard to communal lands—for the reforesting of which the Administration is furnished with sufficient authority—would produce good effects in its application to private lands. Others have alleged, against the proposed measure, the difficulty of its application. And the conference, without pronouncing finally upon the question, has expressed the opinion that the creation of such zones might be considered a very useful measure; but it was remarked that the measure should not have the effect of binding agents, or fettering their personal action, so as to prevent this aid being given to demands relating to lands situated out of the zones, when the reforesting of these lands should present a marked character of public utility.

"The agents of the conference at Foix expressed the opinion that it is useless, and would be inconvenient, to establish zones for communal reforesting.

"The members of the conference at Carpentras acknowledged that it is indispensable, in order to the good use of subventions, that sanctioned works of reforesting should be carried out by the Administration, and that consequently it would be desirable to fix the zones. They think, however, that the grounds are not yet sufficiently explored, nor the spirit of the population sufficiently known, for this demarcation to be possible at present. The agents were desirous also that, except in exceptional cases in sanctioned works of reforesting, a minimum extent—say, for example, of 10 hectares—should be specified for every one operation, whether on land belonging to one proprietor or to more.

"*Remarks and Instructions of the Administration.*

"The Administration does not think proper to insist on the *reboisements facultatifs.*

"The work of *reboisement* on the mountains is so recent that it is deemed desirable still to allow every latitude in framing the demands for aid.

"Every demand brought forward will, as heretofore, be the object of a special examination, without there being fixed a minimum of operations.

"Only, the demands being for the most part called forth by the direct intervention of the agents with the proprietors, the measures ought to be directed as much as possible in such a way that the works effected, or to be effected in the same region, should concur to a common object. With these views it would not be possible to estimate at present the number of persons whom it would be necessary to appoint over the works. The creation of new employments of the guard for reforesting will go on. The creation of new employments for those engaged as guards in connection with *reboisement* will go on as heretofore, and, until fresh orders, according to the present scale and the requirements of the service.

"*Second Question.*
"Hitherto the applications for aid have been transmitted to the Administration as fast as their submission by the proprietors of the reforested lands. The number of these applications having now become considerable, it appears necessary now to group them so as to send up several at one time.
"It is desirable to ascertain at what time the transmission of these should take place, so as to secure at the proper season the execution of the works.
"*Opinions and Proposals of the Agents.*
"The agents at the conference at Clermont advised that the applications for aid in reforesting by communities should be produced before the 15th July for the autumnal labours of the same year and the succeeding spring, and that these demands should reach the Administration by September 1st.
"At the conference at Foix the agents deemed that the demands relating to the works in spring might be drawn up before October 1st, and sent at that date to the Administration, and that for the reforesting in autumn they should be sent in by May 1st.
"The conference was of opinion that no date should be fixed for sending the requests to the agent, but that the latter should be authorised to put off till the following year the preparation of every demand which might not reach him two months before the general despatch relating to the season for executing the works.
"According to the opinion of the conference at Carpentras, it is better to fix two dates for sending off the reports—*i.e.* 1st October, for the works to be carried on in the spring of the following year, and June 1st for those of the autumn of the same year.
"*Remarks and Instructions of the Administration.*
"It seems to be well to fix upon two periods for forwarding collected applications for aid on sanctioned *reboisements* to the Administration, accompanied by a report on the same—that is to say, June 1st for the autumn works; October 1st for the works of the ensuing spring. Any demand which shall not reach the superintendent of the department a month before each of these fixed times may be carried over by the chief of the department to the following season, or following year; and this chief will be the judge in the case in which a demand for aid should be separately sent in to the Administration out of the prescribed periods.

"*Third Question.*
"The Administration has caused to be prepared a form of report for instruction on applications for aid. It is desirable to ascertain whether this formula embraces all the points on which information is required.
"*Opinions, &c., of the Agents.*
"The three conferences have proposed to adopt the formula proposed by the Administration, with some slight modifications.
"*Remarks and Instructions of the Administration.*
"The model of the report has been sent to the conservators by the Administration in sufficient number of copies for the requirements of the service.
"When an agent shall have to send at one time a number of applica-

tions, it will be unnecessary to add to each a special report. The attestation of the requests can be verified by a collective statement in writing, conformably to the directions of the formula.

"II. REBOISEMENTS OBLIGATOIRES, OR ENJOINED REBOISEMENTS.

"*Fourth question.*

"In the plans for enjoined reforesting, the estimate of the expense of the works, as well as the division of this expense among the parties interested, the fixing of the subsidy, and the estimation of the revenue and of the value of the lots, can be given approximately. If further simplifications should be desirable, the agents are desired to prepare a statement of these, after having discussed the subject in conference.

"*Opinions, &c., of the Agents.*

"The conference at Clermont remarks that, in a certain number of projects already presented, it has been thought best to unite in one group the different lots belonging to the same proprietor. This measure, which is tedious and laborious, does not appear to the agents to be of much use.

"The same conference remarks that the Administration has returned several plans of enjoined reforesting, on account of stipulations for too long a delay in the execution of the works. The agents think that a considerable latitude should be left in this respect, as well as for the other details of execution.

"*Remarks, &c., of the Administration.*

"The union of different lots belonging to the same proprietor, useful in certain cases, is not prescribed absolutely. The Administration leaves to the agents to decide whether or not there be reason for doing so.

"The Administration considers that it is desirable to push on energetically the execution of the works, and that, in this point of view, it is inconvenient to stipulate in the plans for the long delays which interested parties may desire the Administration to sanction.

"It is certainly necessary to allow all possible discretion in execution; but it is necessary also to avoid raising hindrances of a kind to paralyse the effort of the Administration, and to hinder it in giving, or trying to give, a fresh impulse to the progress of the works, as circumstances may require.

"The agents submit to the Administration the question—Whether the directions, relative to the designation of two proprietors as members of the special commission, instituted by the fifth Article of the law of the 28th July 1860, should be understood of two proprietors of each commune comprised within the area of *reboisement*, or only of two proprietors for all the communes? The agents of the conference at Foix have not pointed out any further simplification required.

"The composition of the commission lies with the prefect; it belongs, then, to this magistrate to interpret, according to his judgment, the directions of the law in this respect.

"The Administration deems that in appointing two proprietors to take part in the commission, the Legislature had in view to introduce into this commission members possessing knowledge of localities and their requirements, and not of persons directly interested in the operation only.

"With this view two proprietors would suffice not only for one périmètre, or area of *reboisement*, but also for several périmètres in the same district.

" In the conference at Carpentras, the agents have expressed the opinion that it is better to leave as much latitude as possible to the verbal report, and to the written account of the examination of the land, and that the specifications of the works proposed for execution should be on printed forms, supplied by the Administration.

" *Remarks, &c., of the Administration.*

" The form of the specification of works proposed for execution being susceptible of variation according to circumstances, it does not appear desirable to provide a separate form for this any more than for the report of survey and descriptive memoir.

" Some agents have expressed an opinion that it would help the full understanding of the report if there were indicated on the charts the lay, or inclination of the land, the general water-shed, and the flow of the waters.

" *Remarks, &c., of the Administration.*

" These suggestions are valuable, and the Administration strongly recommends to the agents to carry them out in the preparation of plans for projected works.

" *Fifth Question.*

" It should be ascertained whether it would not be well to prepare for every périmètre, or specified area of *reboisement*, a detached statement, in which shall be anticipated all the cases mentioned in chaps. i. and ii. (secs. 1 and 2) of part iii. of the decree, regulating the reforesting, of April 27 1861. In this statement should be classed in order all the lots included in the périmètre, with indications of the course followed with each of them, and of the expense of the same. In it especially should be included all the necessary elements for establishing the annual accounts mentioned in Arts. 26, 27, and 28 of the above-mentioned decree.

" *Opinions, &c., of the Agents.*

" The agents of the conference at Clermont expressed the opinion that it is impossible, in the case of a périmètre composed of a great number of separate lots of small extent, to keep a separate account for each lot of the quantity of seed and of the manual labour employed ; that it is only necessary to keep one account of the whole of the lots requiring the same expenditure in execution of the work reforested the same year ; and that it would be easy to estimate from this account, at the end of the year, the cost per hectare of each parcel, as well as the expense to each proprietor.

" Some agents proposed the keeping of two statements for each périmètre— the one to contain statements of all the lots included in the périmètre, and in which account would be kept of the changes effected on each lot by each proprietor ; the other, which should be destined for the total of the lots reforested the same year, supplying all the information relative to this reforesting.

" The conference expressed definitively the opinion, that on account of the variety of conditions occurring in different districts, the Administration cannot well prescribe the use of the same form of account in all ; and that at present uniformity should only be required in half-yearly estimates, to be transmitted to the Administration in accordance with the circular No. 806.

"The agents of the conference at Foix proposed the formation of a statement comprising seven principal divisions, each corresponding to 'one of the seven cases provided for in chapters I. and II. of title III. of the decree of 1861'; each one of these principal divisions to be divided into columns, corresponding yearly to the different details of the work.

"The conference at Carpentras submitted a form of statement which it considered would meet the case.

"*Remarks, &c., of the Administration.*

"The different forms of statements presented by the conferences have not appeared sufficiently definite to produce a form adequately simple and clear to meet all the instructions proposed. This is naturally explained by the fact that, as the works of obligatory reforesting have scarcely begun to be executed, the progress of these works has not yet presented the opportunity of these being studied. In these circumstances, the Administration does not feel called on to prescribe a uniform rule, which probably would require to undergo numerous modifications in the course of its working.

"The Administration thinks it proper to pospone the settlement of this question till it shall be more enlightened in regard to details to which the question relates. At present it confines itself to recommending to the agents to supply all the information likely to be useful, whether this be done in the form of a statement, the substance of which is optional, or under the form of instructions, prepared with care and method, in the file of papers supplied for each périmètre.

"As to what relates to the determining of expenses, as it would be impossible, as the agents of the conference of Aurillac have deemed it to be, to keep an account in detail of the expenses belonging to each lot, the expenses will be calculated annually, when there is occasion so to do, from the account of the expense established by hectare for the whole of the lands comprised in the périmètre.

"*Sixth Question.*

"It will be desirable to consider, if it would not be well to establish, for each périmètre, a statement in which the periodical phases of the operation of reforesting shall be recorded.

"*Opinions, &c., of the Agents.*

"The agents presented different forms of statements intended to meet the requirements of the question.

"*Remarks, &c., of the Administration.*

"The same remarks and instructions as have been given in relation to the preceding question, are applicable.

"III. EXECUTION OF THE WORKS.

"*Seventh Question.*

"In what cases is it proper to proceed to the restocking of woods by planting? and in what by sowing?

"*Opinions of Agents.*

"According to the agents at the conference at Clermont, the sowing, being more economical than plantations, making more certain the retaining of the land and opposing greater obstacles to the flow of waters, ought to be preferred to plantations in view of the object of the law of 1860; but when the works are at great altitudes on steep declivities, not well adapted for retaining the seed, and in certain soils, such as the chalks of La Bresse, or the calcarious schists of La Lozère, it will be necessary to have recourse to plantation.

"At Foix, the agents express the opinion, that it is better to proceed exclusively by means of plantation in the elevated parts, and only to employ sowing in connection with planting in the places of medium height, and low parts, where frosts are less to be feared.

"At Carpentras, the agents were of opinion that plantation is preferable to sowing, looking, and looking only, to culture; but that sowing, being more economical, it is better to employ it when it appears to offer sufficient chances of success.

"*Remarks, &c., of the Administration.*

"Without its being possible to point out exactly the cases in which the one or the other of the two modes of procedure should be followed, it seems expedient to admit, as a rule, that plantations, being subject to fewer destructive agencies than seed-beds, it is better to plant under rigorous and peculiar conditions of climate, locality, or soil. The essential point is to ensure the success of the reforesting of the locality. The question of economy ought undoubtedly to be one of great consideration, but whenever success appears to be certain by one mode and to be doubtful by another, there should be no hesitation in employing the former.

"*Eighth Question.*

"Discuss the kinds of trees selected; the mode of plantation, singly, or in clumps, &c.; the number of plants per hectare; the season best for the execution of the work; the expense, per hectare, of restocking woods.

"*Opinions, &c., of the Agents.*

"The agents attending the conference at Clermont have experimented successfully with the larch in reforesting bare lands.

"The Norway pine and the pine of the country have given results which are pretty satisfactory, and they appear to be such trees as should be employed in regions of medium altitude.

"The oak, planted but only to a limited extent in the Puy-de-dôme and in the Haute-Loire, has succeeded well.

"The ash, whether planted in large clumps, or intermixed with resinous trees, promises to succeed well in the Haute-Loire.

"*Remarks, &c., of the Administration.*

"The indications reported by the agents are based on experiments actually made, and the Administration has nothing to add to the contrary. It can only recommend to the agents carefully to note all the facts observed in the different regions, with a view to obtaining, when requisite, instruction from these. It is by continual experimenting that the Administration will gradually come to give the operation a more and more satisfactory direction.

Hitherto planting single trees has alone been attempted, and this has succeeded very well. It is only from next year that the nurseries of Arpajon, of the Puy, and of the Mende, will present sufficient resources to permit of the experiment of planting thickets being made. In any case this latter mode could not be at great altitudes, plants of three, four, and five years growth alone succeeding under such conditions.

"The number of plants ranges from 7,000 to 11,000, according to the conditions of soil, exposure, inclination, &c. The most favourable season is spring, in climates where the winters are very severe. In the middle, or low lying districts, autumn seems preferable, on account of its permitting the young plant time to get strength to resist the great heats of summer. The spring seems to suit better for the plantations of resinous trees. There is reason to believe that plantations of these, when the sap begins to move, succeed more certainly. Broad-leaved plants seem to accommodate themselves better to the autumn planting.

"The expense of carrying on the work of plantation amounts, for the hand labour, to 70 francs, by the hectare, in the Loire, to 57 francs in the Haute-Loire, to 48 francs in the Cantal, to 38 francs in Puy-de-Dôme.

"Seedlings brought from the depths of forests, and planted in various localities experimentally, with a view to determining the economical importance of such a procedure, have not given satisfactory results.

"*Remarks, &c., of the Administration.*

"The Administration is aware that plants from the source mentioned have no great value. But in order to avoid the expense of purchase, it was necessary to try to derive some advantages from the resources offered by the forests, until such time as the nurseries shall yield plants.

"The conference at Foix was of opinion that it is best to employ the indigenous products of the Pyrenees—such as the Mugho, or dwarf pine, the Scotch fir, the birch, the silver fir, the ash, the beech, the oak, the evergreen oak, the great maple, and the chesnut; and to continue the experiments which had been made with plantations of the Norway pine, the Austrian pine, the acacia, the silver fir, the ailanthus, the larch, the pine of Aleppo, and the mountain pine. These kinds to be distributed according to the altitude, to the local conditions, and to the results of experience.

"The planting in separate holes plants transplanted from nurseries seems to offer the best chances of success. The planting in clumps is, however, preferable, when disposing of very young plants taken from a plantation near the lands to be reforested.

"The number of plants on each hectare may vary from 10,000 to 2,500, this last number being applicable more especially to saplings, and to the chesnuts, if it be desired to obtain from them poles of good growth at an early age. The season of spring being almost unknown on the mountains, where great heat succeeds, almost always without interval, to the cold of winter, the autumn is in all cases the most convenient season for planting. The price of hand-labour varies from 50 to 100 francs per hectare. The purchase of plants has occasioned an expense of not less than 10, 15, and 25 francs per thousand plants.

"*Remarks, &c., of the Administration.*

"The minimum of 2,500 plants per hectare appears very small. The reforesting of the mountains having especially for its object to cover the soil, independently of the addition of future produce, it is better to avoid planting the trees separately, at great distances apart.

"The nurseries belonging to the Administration promise soon to supply plants at less expense than that at which at present they can be obtained

"The agents at the conference at Carpentras stated that they had employed, on L'Isère, and the High and the Low Alps, the white oak up to 1000 mètres of altitude, the acacia up to 900 mètres, in all exposures. The ailanthus had as yet given too little experience for deducing from its use any certain remarks. The Scotch fir, the Norway fir, the Mugho, and the larch, have been employed with success in different situations. In the departments of Vaucluse, of the Gard, of the Bouches-du-Rhone, of the Var, of the Maritime Alps, and of L'Herault, there was reason to think that the trees which should be used principally were the white oak, the green oak, the acacia, the maritime pine, the Norway fir, and the larch—at all altitudes, and in situations pointed out by experience. The planting in holes, taking the precaution to disturb the soil very little, and to procure for the young trees natural shelter—such as bushes, rocks, and the stones which are found on the land—seemed to be the most suitable system of planting.

"The mode of planting in clumps, which is very costly, should only be used for resinous trees, and in situations where to secure success is difficult. But this proceeding will always be most advantageous when it is not necessary to regard the question of economy. The best plants are generally transplanted plants of two or three years.

"*Remarks, &c., of the Administration.*

"The last mentioned method has been made the subject of experiment with success. It is not well to attach too much importance to the expense which it occasions : in the first place, the plants being very small, their price is not great; in the second place, the preparation of the soil is very easy; finally, as this process is almost always successful, it must be employed without fear in difficult situations, apart in some measure from the question of expense.

"The number of plants per hectare to be employed varies from 10,000 to 16,000 for separate plants, according to the conditions of exposure and soil, and the kind of tree, etc. For planting in clumps, the number would be from 30,000 at the rate of 3 plants per hole, and 10,000 holes per hectare. The planting in autumn is generally preferable, as giving time for the plants to be in a state to resist the spring frosts and the early heats.

"The cost of manual labour varies from 40 to 100 francs. The cost of plants varies too much to allow of an estimate approximately correct being made.

"*Ninth Question.*

"*Sowing.*—Discuss the choice of kinds of trees, the fitness of each mode of sowing, (sowing in rows, in holes, in the open bed, etc.,) the quantity of seed to be used per hectare, the fit season for carrying on the works, the expense of the work per hectare, etc.

"*Opinions, &c., of the Agents.*

"The agents of the conference at Clermont reckoned that in the central region, wherever the climate is mild, and the altitude a medium one, (800 metres and under,) the oak and chesnut should be employed in preference to every other tree, and if the soil is of poor quality, the resinous trees, the acacia, and ailanthus.

"In the regions where the climate is more severe, and the altitude greater, recourse should be had to the Scotch fir, the Austrian pine, the Corsican pine, the mountain pine, the Norway fir, and the larch. The Atlas cedar, the larch, and the Siberian cedar can be used for the greatest heights.

"No tree, except perhaps the fir-tree and the beech, ought to be rejected in so far as the sowings are made *in loco.*

"**The least costly and most simple method of sowing, practised for a long** time in the Puy-de-Dôme, is sowing at hap-hazard on short heath, or after *écobuage* if the heath be too high. But this system is not practicable everywhere. The method of sowing most usually employed is sowing in rows, or in holes, according to circumstances. In both cases much disturbance of the soil is to be avoided.

"The quantities of seeds necessary are, for the oak and chesnut, 6 to 10 double décalitres; for resinous seeds of small size, 10 to 12 kilogrammes, on ordinary land, and a third more if the conditions be unfavourable; for the Austrian pine, 12 to 15 kilogrammes; for the maritime pine, 20 to 25.

"The most favourable time for sowing is the autumn for broad-leaved trees, and spring for the resinous.

"The spring sowing should be as early as possible in February or March. The cost of hand-labour is, for sowing in *bandes*, from 30 to 35 francs, per hectare; and for sowing in *potets*, from 25 to 30. Reforesting in resinous seeds costs on an average, in central regions, 70 or 80 francs per hectare.

At Foix, the members of the conference were of opinion, that the choice of trees depending essentially on the nature of the land, and on its exposure and altitude, nothing decisive can be pronounced on this head. At the same time it may be concluded, that in elevated regions there will be used with success, the Norway fir, the larch, and the black Austrian pine; in the regions of medium altitude, the pine, the Norway fir, the beech, and the pine intermixed with the beech; in lower regions, broad-leaved trees in general, the chesnut, the green oak, the ash, and the ailanthus.

"The method of sowing in *potets* seems to be the most advantageous. The quantity of seed to be used is from 10 to 15 kilogrammes per hectare.

"Spring in general is the best season for sowing, especially for resinous seeds. The expense of sowing can be approximately, and in a general way, reported at 100 francs per hectare—*i. e.*, 60 francs for manual labour, 36 for the purchase of the seeds, 4 for unforseen expenses.

"At the conference at Carpentras, the agents estimated, that for sowings the trees to be preferred are generally the same as those pointed out for planting, with the addition of the Corsican pine, the cedar, the pine of Aleppo, and the shrubs intended for the preparation of certain soils, or for the prevention of erosion of hill sides, such as the box-tree, *l'argoussier l'amélanchier*, the barberry, the juniper tree, etc. In L'Ardèche, the sowings of the Norway fir do not offer sufficient chances of success.

"Sowing by *bandes* is preferable whenever it can be employed, but it has the inconvenient disadvantage of loosening the soil too much on the inclines. The method of sowing in *potets* will be more generally employed. Complete or full sowing is the only method possible on rocks, on ground difficult of access, stony parts, and volcanic scoriæ. The quantity of seed to be employed per hectare is from 7 to 10 kilogrammes for resinous trees, and 3 to 6 hectolitres for the oak. Opinion was much divided on the choice of season. The result appeared, however, to be generally that for resinous trees, and in friable earth, spring ought to be preferred; whilst autumn appears to suit better for the oak.

"The expense of manual labour may be estimated at 60 francs per hectare for sowings *par bandes*, at 35 francs for sowings *par potets*.

"The price of seeds being approximately, on an average, 3 francs per kilogrammes, the expense will be from 70 to 100 francs per hectare.

"*Remarks, &c., of the Administration.*

"The quantities of seed mentioned by the agents at the conference at Clermont will require to be increased, in so far as larch is concerned, the seed of which in general only succeeds in the proportion of 40 to 50.

"Sowing in *potets*, or drills, seems generally recognised to be the most advantageous.

"There is a mode of sowing called *semi a la niege*, which consists in sowing seeds broadcast on the snow, which melting away deposits these on the soil, and causes them in some measure to sink partially into the ground.

"*Opinions, &c., of the Agents.*

"No trial of sowing *a la niege*, as it is called, has yet been made in the central region; and all experiments in the Pyrenees have failed. In the Alps it was that this mode of sowing was first intoduced, some fifteen or eighteen years ago. The experiment was made in the department of the Basses-Alpes, on a calcareous soil, for a long time unused and covered with grass, and with a northerly exposure; it succeeded perfectly. The experiment was renewed in 1862, in the same department, on 200 hectares, and in the Hautes-Alpes on 40 hectares, with fir, larch, cedar, Norway fir, and Scotch fir. The fir did not succeed; the larch succeeded only in part on grass lands, and with a northerly exposure; the cedar succeeded well; as to the Norway fir and Scotch fir the result has not been established.

"There were used from 6 to 8 kilogrammes of seeds per hectare; the manual labour cost only 2 francs.

"An attempt made in the Drôme, at 700 mètres of altitude, on limy soil, and in a northeast exposure, with the maritime pine, succeeded to a medium extent.

"The sowing should be made on soft snow, and in a settled temperature, in order to avoid the floodings caused by the southerly winds and warm rains.

"*Remarks, &c., by the Administration.*

"The so-called *semi a la niege* is very economical, and for this reason one might be tempted to employ this method for the reforesting of large surfaces. But experience in this matter gives reason to conclude that the results, always uncertain, are generally unsatisfactory. It does not appear that there is any reason for classing this kind of sowing in the category of regular modes of reforesting. But it may be considered as an expedient capable of being employed with success in certain cases. The attempts made up to this time are, however, too few for a certain deduction to be drawn on this point. It might be usful to try further experiments when the conditions shall appear more favourable. Manual labour being at a very low price, there would be no difficulty in increasing the quantity of seed sown, which appears to have been too small in the attempts made in the Basses-Alpes.

"IV. PREPARATORY WORKS IN REFORESTING.

"*Tenth Question.*

"*Nurseries.*—A moderate number of nurseries, of which some are of great importance, have been created by the forest agents, with all the care and intelligence necessary.

"It is desirable to discuss the processes of extracting and packing the plants, as well as the precautions to be taken at their despatch and receipt, in order to insure their growth; to study the method of sowing adopted in the nurseries (in *bandes* or in *potets*),—the quantity of seed used per hectare,—the means used for protection,—the expenses of the works. The system of repeated transplanting may be discussed. As soon as the beds produce plants fit to be used, it will be important to have kept, by the official specially charged with the nursery, a register, in which shall be inscribed the number of disposable plants, and the numbers taken away and sent off; and the conference is to consider the plan that should be adopted in keeping this register, of which an extract shall be periodically addressed to the Administration, that it may know the number of plants ready for use.

"*Opinions, &c., of the Agents,*

"At the conference at Clermont very circumstantial details were presented—taking, for example, the nursery of Arpajon, the creation of which has been attended to with great care, and the state of which is very satisfactory. It may not be uninteresting to reproduce details which answer to the questions put by the Administration, which may serve as useful indications of what may advantageously be done.

"Before being sown the bed should be prepared. The preparation consists, after having cleared and cleaned the ground, in mixing the natural earth with heath mould for leaf trees, and in adding to the soil some kind of manure. The ground is then carefully broken up.

"*Remarks, &c., of the Administration.*

"If the ground be encumbered with weeds it may be well to raise on it a crop of potatoes to secure the destruction of the weeds before appropriating it to the growth of forest seeds. And too much digging, or displacement of the soil, should be avoided.

"The ground may then be divided into beds, a mètre, or 40 inches, in breadth, raised above the level of the ground, and separated by footpaths; and the beds about 8 or 10 mètres long must then be surrounded by sheltering screens or fences of the Chinese arbor vitæ. While these shelters are growing to a convenient height, their place is supplied by artificial shelters, either formed of straw, or of osier, or hazel lattice work placed nearly vertically, or linen stretched over boards. The sowing is done in the first 15 days of April, or later, if possible, in moist weather. It does not seem necessary to cover the seeds with earth, it is enough to pass the roller over the bed after scattering the seed, and it is covered with moss reduced to small pieces and watered. The quantity of seed to be used per *are* is 12 kilogrammes for pines with small seeds; 15 to 18 for larch trees, the Norway fir, the black pine of Austria; 26 to 30 for the fir; for the oak 1 hectolitre; for the chesnut 6 double decalitres. The seeds gathered in the country have given much better results than those obtained from purchased seeds. The beds must usually be watered every day until the plants have gained some strength. After the first year the plants can be used. They are then, according to the expression used by the nursery-men, in the condition of *pourettes.* They cost little, 1 to 2 francs the thousand, are easily dug up, and are removed at little expense. But the chances of such young plants taking root being necessarily limited, it is only prudent to use them in moderate conditions of soil and altitude.

"*Remarks, &c.*

"The lifting of *pourettes* in the way described is employed with advantage in planting in tufts. The earth raised is divided into clods containing each a certain number of plants, and these plants are conveyed in the clod to the place in which they are to be planted;

fragments, containing two, three, or four plants, to be put into the place together, are broken off, and at least one of these almost always grows.

"To obtain hardier plants more likely to take root under severer conditions, it is necessary to wait nearly three years, and to have them transplanted. The design of this operation is to place the young plants in circumstances favourable for the development of the fibrous root. It is employed for plants of a year old, and should be done in spring, in order not to expose the young subjects to the risk of being raised above the ground by the effects of frost.

"It has been attempted to avoid the expense of this difficult and costly operation. As regards the oak, one agent has mentioned a process which it may not be uninteresting to bring under the notice of the agents. This process consists in artificially causing the acorns to germinate during the winter, to cut off the radicle and to sow in the seed-bed the acorn thus mutilated. It has been remarked, that the extinguishing of the radicle led to the formation of lateral roots, and to suppress the growth of the descending taproot.

"*Remarks, &c.*

"Transplanting does not appear to be always necessary. In the nurseries it is practised at different periods of the plants growth. If, when the plants are required, the best and most fibrous-rooted alone are made choice of, the removal of these will have the effect of relieving the others, and so favouring their development; in this way plants of different stages of development may be successively removed, and this kind of periodical thinning has for its result, to permit the plants of inferior growth to acquire sufficient strength. This removal is facilitated by the arrangement of the plants being sown in furrows on the flat beds. When it is necessary to thin the plants, there is dug along the furrow a hollow into which the plants are turned; the proper choice is then very easily made, and the plants remaining in the furrow are easily re-arranged. Finally, the operation of transplanting can sometimes be replaced by the cutting off of the root in the ground, by the use of the spade used at Hagenau (coupe-pivot), which ends in a diamond-shaped edge.

"The cutting of the root has for its effect to favour the development of a fibrous root This economical and beneficial operation, however, can only be practised in the earths into which the edge of the spade easily penetrates. It has not been attempted in the case of resinous trees; it would not be without interest to make some attempts in this direction.

"The sowings in the nurseries are exposed to ravages by rats, field mice, mole crickets, moles, birds, &c. It does not appear to be necessary to enter into the details of the methods employed to combat these various enemies. The methods followed, moreover, have succeeded only imperfectly, and it will be necessary to devise others more efficacious. To prepare the plants for sending away, a dry day must be chosen, the digging must be done with the spade, 100 plants are united in one clod, the roots of which are immersed in a bath of well-tempered clayey earth, and they are covered with dry moss; they are then placed in layers in a box with open bars, the spars of which are covered with dry straw. A rapid conveyance is chosen, in order not to leave the plants for more than five or at most seven days in the boxes. On arrival, the plants are immediately unbound and sorted.

"*Remarks, &c.*

"It is by the spongeoles or extremities of the fibres that the roots draw from the earth the nourishment of the plant. It is therefore in the highest degree necessary to protect these delicate organs. For this purpose the bath of tempered clay is a very useful precaution. Before putting the plants into the earth, it will be well to leave the roots nearly 24 hours in urine.

"This operation has the effect of singularly reviving the vegetative power of the plant.

"To show the importance of the services that the nurseries are expected to render, the conference at Clermont cited the results of the nursery of Arpajon, formed scarcely two years before. It appeared from the accounts, kept with care in this nursery, that it would contain 32,489,000 plants of various kinds, of the value at a commercial valuation of 159,622 francs.

"The details given render necessary a similar circumstantial account of the observations made by the agents at the conferences at Foix and Carpentras on the subject of nurseries.

"The principal points of the discussion, with those which have called forth differing opinions, will alone be requisite.

"At Foix, the agents considered that the operation of transplanting is too expensive, and requires too great an extent of land, to be followed. A method of taking up plants analogous to that which has been above described, in the opinions and instructions of the Administrations, seems to be almost sufficient to take the place of transplanting.

"Watering appears necessary to be practised with moderation, on account of the expense which it causes. At Foix and at Carpentras, the observations relative to the digging up and packing of plants, as well as the precautions to be taken at their despatch and receipt, do not differ from those presented at the conference at Clermont, and reproduced above. At Carpentras the sowing in furrows has seemed, in all points of view, that deserving to be preferred for nurseries. The quantity of seed necessary to be used has been estimated at 10 or 15 kilogrammes per *are* for the resinous plants, at 1 or 2 hectolitres for the oak, and at 10 or 15 kilogrammes for the acacia and ailanthus.

"The agents have unanimously expressed the opinion that it is advisable to diminish the sheltering fences as soon as the plants acquire strength, and that it is necessary to make them always sufficiently low to enable the light to reach easily the plants. The transplanting, which appears to the agents at the conference at Carpentras indispensable for the oak, is considered less necessary for pines and the Norway fir.

"Watering, if not indispensable, is at least useful to the resinous trees, and it must, when once commenced, be assiduously pursued.

"The agents of the three conferences have presented plans of a register for the record of the plants ready for sending out, and those sent out.

"*Remarks, &c.*

"The form to be adopted temporarily for this register is the following, which must be tried upon formulas prepared in writing in each conservatory till the time when a definitive model, made from experiments, shall be adopted :—

NURSERY OF ————.

PLANTS FIT TO BE DISPOSED OF.

Scotch Fir.			Black Pine of Austria.			Norway Fir.						Observations.
1 year.	2 years.	3 years.	1 year.	2 years.	3 years.	1 year.	2 years.	3 years.	1 year.	2 years.	3 years.	

PLANTS SENT OFF.

Scotch Fir.			Black Pine of Austria.			Norway Fir.						Observations.
1 year.	2 years.	3 years.	1 year.	2 years.	3 years.	1 year.	2 years.	3 years.	1 year.	2 years.	3 years.	
												The date of despatch and destination are marked in this column.

" This register will be kept by the official in charge of the nursery under the inspection of the chief of the district.

" At the end of every month the work will be repeated by making an entry in each column of the plants then fit to be disposed of, and of those despatched.

" An extract of this register is to be sent to the Administration at the periods to be hereafter specified.

" *Eleventh Question.*

" *Sécheries.*—The Administration having ascertained that seeds prepared in the Government drying booths are superior in quality to seeds obtained by purchase, it is desirable to consider whether it be not desirable to erect additional drying-booths *(sécheries)*, or seed depôts.

" *Opinions, &c., of the Agents.*

" The agents at the conference at Clermont have observed that the seeds furnished by the Goverment *sécheries*, or drying-booths, of Murat (Cantal), or gathered in the country, are incomparably superior to the seed purchased. This *sécherie* furnishes only 4000 kilogrammes of seeds yearly, which is a quantity much below the requirements of the departments of La Lozère, of Puy-de-Dôme, of Cantal, of the Haute-Loire, and of the Loire; the quantity required annually being estimated at 10,000 kilogrammes, it is desirable to establish a new *sécherie*. This the conference proposes should be located at Puy. Subsequently another might be constructed at Marvejols.

" A member of the conference at Clermont showed that the Corsican pine might be advantageously employed in *reboisement*,—that the price of the seed of this species is from 7 to 12 francs per kilogramme,—and that, without doubt, it would be possible to establish in Corsica one or two *sécheries*, by which the seed of the Corsican pine would be furnished at the cost of 4 francs at most.

" The agents at the conference at Foix state that the *sécherie* of La Lagonne is capable of furnishing annually 4000 kilogrammes of seeds. They think that it would be desirable to establish in the inspection district of Simoux, within reach of the nurseries, a *sécherie* for fir-tree seeds.

" At Carpentras the agents expressed in general terms the opinion that there would be great advantage in erecting *sécheries*, or seed depôts, wherever the existence of extensive masses of forest admit of this being done.

" *Remark of the Administration.*

" Notes have been taken of the different opinions expressed by the agents on the question of establishing new Government *sécheries*.

" *Twelfth Question.*

" Examine, and say whether it would not be desirable to gather seeds in the Government forests under the charge of the local officials, and to have them put into places of deposit, from which at a fit time they might be sent off to the places to be reforested.

" This measure promises to the agents of the three conferences to be productive of advantageous results. The agents of the conference at Carpentras express the wish that the gathering should also be made in the communal forests.

" V. Measures for Securing Order.

" *Thirteenth Question.*

" Discuss the measures taken to insure the thorough execution of the works, and to justify the use made of the credit accounts opened by the Government.

" *Opinions, &c., of the Agents.*

" At Clermont, the conference expressed the opinion that the first condition of the success of the works of *reboisement* is, that the direction and oversight of these works should be entrusted exclusively to the agents, and that it is necessary that a guard should be constantly stationed at the wood-yard of the works. This obligation necessarily causing heavy expenses to the officials, it is desirable to extend the arrangements of circular No. 708 to each official compelled to sleep away from home.

"*Remarks, &c., of the Administration.*

"The Administration is of opinion that the works being executed under the direction of agents gives the only security for success. It does not seem possible always to exact the presence of an official on the spot. This ought, however, to be secured whenever it is possible. The Government has the intention to remunerate in a suitable degree the officials who accomplish onerous duties, and who render important service. In this respect no absolute rule can be fixed, the rewards must vary with circumstances.

"Special propositions on this head may be sent in; and, in order to prevent these coming at all times of the year, (which occasions serious loss of time to the Administration) it is desirable, henceforward, to collect them into two despatches, added to the forms ordered, Nos. 16 and 17 (Circular 806).

"In justification of the credits opened, the agents stated that in the delay of a month, which occurred in the settlement of accounts, the agents have produced tables with margins of the accounts of the day's work of the labourers, as a return for the sums put to their account; for the supplies, they have presented receipted bills, all according to the rules of debit and credit.

"The conference at Clermont submitted to the Administration the following question: When the aid granted, according to the estimate of the works, exceeds that estimate, should the extra sum be granted to the applicant, and if the expense is less than the allotted aid, should a credit for that extra sum be demanded?

"REPLY, &c.—The estimate of the expenses of the works can only be an approximate one. Consequently, when the aid is greater than the actual expense, the course to be pursued is to extend the works over a space proportionate to the excess; or, if that extension is impracticable, to leave unemployed the said excess in money, which will thus be disposable for other works. If the aid granted be less than the expense would be, the works should be reduced by an extent corresponding to the difference of means.

"The agents at the conference at Foix proposed that the good or bad execution of the works should be established by a minute declarative of the facts, and to extend the application of this measure to the sanctioned *reboisements* of communal lands, when undertaken by individuals with the grant in aid.

"*Remarks, &c.*

"Nothing would be gained by the establishment of the good execution of the works by such a minute; and in regard to the bad execution of the work, it is already prescribed by the regulation to supply grounds for exacting repayment, in whole or in part, of the grant made.

"The agents at the conference at Carpentras, in giving account of the means employed to insure the good execution of the works, and the payment of expenses, stated, in regard to delivery of orders, that where this is done it is by small coupons, which has facilitated payment and rectification of accounts, and has rendered unnecessary the making of advances, and repayments of these, which it is always desirable to avoid in accounts.

"*Remark.*

"To this there is nothing to object, but the fear that it may lead to too great complication of accounts.

"*Fourteenth Question.*

"The allotting of aid having, up to this time, been made as fast as the production of the demands, the Administration have been obliged to leave to each conservator the care of procuring by purchase the seed required, and the necessary plants. It would seem to be more simple, more convenient, more regular, and doubtless more economical, that the Government should centralize the orders for, and the despatch of, these seeds and plants; Discuss the means of effecting this centralization, if it does not appear to the agents better to leave the ordering of seeds and plants to be done directly, as heretofore, by the agents.

"*Opinions, &c., of the Agents.*

"The agents of the conference at Clermont considered that the centralization would be very useful, and proposed, for this purpose, the mode of organization which seemed to them most convenient.

"The agents of the conference at Foix expressed the same opinion, and presented their proposals.

"At Carpentras the conference expressed the opinion, that, whenever the articles required can only be obtained by purchase, the centralization of the orders for these supplies will be more simple, more convenient, and more regular, but not always more economical: (1) because the seeds purchased are generally inferior in quality to those bought on the spot by the agents themselves; (2) because the expenses of carriage are great in the one case and nothing in the other. The centralization of orders does not then appear in all cases to offer the advantages to be desired, and ought to be restricted in the one case to species rare in France, such as cedars, Austrian pines, Corsican pines, and in the other, to supplies of seed which cannot be had in the locality.

"*Remarks, &c., of the Administration.*

"The ordering of seeds from merchants at a favourable time for the procuring of the supply, and the ordering of these in large quantities are favourable conditions for obtaining them on the most favourable terms possible. But notwithstanding this, the Administration does not intend to prevent in any way the agents from taking advantage of local supplies. To this end, at the periods for the despatch of the collective demands for aid, the agents will add to these demands the following information: (1) the quantity of seeds or plants of each kind necessary to meet the said demands, which can be delivered at their destination by the direct care of the conservator through the local resources; and let the destination of these seeds be stated; (2) Quantity of seeds or plants presumed to be necessary for the *reboisements obligatoires* during the season following the despatch of the information. In this let the quantity and species of plants and seeds to be sent by the Government be indicated, also the place of destination and time at which they are required; (3) Extract from the register of the nursery. Let each conservator state the number of plants required by him and their destinations; (4) Situation of the *sécheries*. And let each conservator report the quantity of seeds required by him, and their destinations.

" By help of this information the Administration will be able to give to the trade the necessary orders, and to provide for the direct transmission to their destination of the seeds and plants which cannot be obtained in the locality.

" VARIOUS QUESTIONS DISCUSSED BY THE AGENTS IN ADDITION TO THOSE SUBMITTED BY THE ADMINISTRATION.

" The questions discussed by the agents, in addition to the programme, do not appear to present in general other than purely local interest, and consequently it would be useless to reproduce the whole of them in the present summary. There are, however, some of those questions, which, on account of their wider interest, will be mentioned here.

" *Opinions, &c., of the Agents.*

" The agents of the conference at Carpentras have remarked, that the method of *reboisement* by strips and by clumps seems a desirable one to practise in certain regions, especially in the departments of the L'Izère and the Hautes-Alps. Clumps of larch would suit well for the high mountain pasture lands.

" *Remarks, &c., of the Administration.*

" This method of *reboisement* would only be efficacious in so far as it was practised concurrently with the works for improving pasture, and it is necessary to have it kept in mind, that the law for the *reboisement* of mountains limits its action to works of *reboisement* properly so called. Besides this, *reboisement* by clumps would have the effect of extending the defences to embrace very vast areas during the whole period of the first growth of the new plants, and during the successive periods of *reboisement*.

" The question of the mixture of different kinds of trees in the *reboisement* was under discussion at the conference at Carpentras, but the discussion elicited nothing new.

" *Remarks, &c.*

" There has not been obtained as yet a sufficiency of results to decide this question.

" At Foix, an agent said he had tried the effect of sulphur upon seedbeds of laburnum, of ailanthus, and of pines of Aleppo. It brought only to the laburnum a sensible augmentation in vigour of vegetation. The sulphur was only applied at the period of the August sap. The attention of the conference was called to the operation, which might be made the subject of interesting experiments.

" *Remarks, &c.*

" The operation of applying sulphur, which is pretty expensive, seems here to have had no other effect than to increase the power of vegetation in the plants of the seed-beds. It does not appear certain that important advantages result from its use. It will not be without use, nevertheless, to make experiments in this direction when a good opportunity may present itself.

" A proprietor of the department of L'Ariège had proposed to the Administration to grant to him land for the establishment of a central place for trials, experiments, and observations, in forest, pastoral, and hydrological matters, in relation to the *reboisement* of mountains.

"The agents deemed that such a field of experiments, however useful for the district in which it might be placed, would not be capable of furnishing certain information for other regions, and that, in this point of view, the proposed establishment does not present the promise of adequate general interest.

"*Remarks, &c.*

"The Administration shares the opinion of the agents at the conference at Foix.

"DESIRES EXPRESSED BY THE AGENTS ASSEMBLED IN CONFERENCE.

"The Agents at the conference at Foix expressed the wish that the Administration of Forests should be charged with the *mise en valeur*, or improvement, and the *reboisement* of the communal lands situated on the mountains.

"The execution of the law on the *mise en valeur* of the communal uncultivated lands lies with the prefects. There is nothing to prevent the execution of such works of this character, as have for their object *reboisement*, taking place under the care of the forest agents. It is thus that the law in question is applied in the department of the Vosges. The conservators will consider whether they ought not in this matter to advise with the departmental administrations.

"The agents at the conference at Carpentras have expressed the following wishes : (1) That in future the programme of the conference should be sent to each agent at least a month in advance ; (2) That the members appointed should bring with them samples, models of instruments, &c.; (3) That they should put down beforehand in writing, as far as possible, their answers to the questions on the programme, and their observations.

"*(Signed)* H. VICAIRE, Director-General of the Forest Administration.
"Paris, January 10, 1862."

The following is an Abstract of Report of operations in 1862 :—

In 1861 the expense of the *reboisement* of the mountains was settled as follows :—

Subventions for *reboisements facultatifs*, or sanctioned operations, granted to communes and private individuals for labour upon Crown lands, for nurseries, for *sécheries* (or places for collecting and drying seeds), for keeping up the works, and for various kinds of labour, - Francs, 548,855,30
Support of agents and overseers, - - - „ 46,718.94
Indemnities to agents, overseers, and assistants, „ 42,439,40

Francs, 638,013,64

A. *Reboisements Facultatifs.*
In 1862 the demands for subventions have been as follows :—
By communes and public establishments, 730 } 1428
By private people, - - - - 698 }
Of which only 40 have been rejected, as not falling within the scope of the law. These demands came from 39 departments in all. The areas reforested were, for what had been done by 742 communes and public establishments, 5,774н 58A ; by 394 private individuals, 1,714н 15A—total, 7,488н 73A 00. It was in the departments of Puy-de-Dôme and of Vaucluse that there was the largest extent of communal *reboisements.*

Subventions in money and in kind, amounting to 280,000 francs, representing about two-thirds of the total expense of the works.

The *reboisement* of private property has been most extensively carried out in the departments of the Gard and the Drôme, where the subventions amounted to 70,000 francs, representing about 40 per cent. of the total expense of the works.

The discovery, it is stated, is being made, that *reboisement* is an operation much more fruitful in immediate advantages than had been generally believed.

The report cites two examples of these advantages:

A commune possessed a district of 64 hectares covered with heath, which had not been sold, though in 1844 offered for sale at 7000 francs. At this time a sowing of *pin sylvestre* was undertaken, at the expense of the municipal chest; there was little additional expense; and now this district is valued at more than 70,000 francs. Another commune possesses a wood of 47 hectares in extent, planted with *pins sylvestres* about 15 or 16 years old. Lately a thinning was effected, which produced 16,000 francs.

These well-known facts have not a little contributed to the favour with which the works of *reboisement* are regarded in the departments in which they had been carried out.

Joint stock companies, or associations of capital, are now very advantageously employed for the *exploitation* of different branches of industry. The acquisition of mountain districts on generally very moderate terms, and their replenishment with wood by the help of large subventions, seems to form the basis of a speculation which is both productive and exempt from risk of loss.

The restocking of the mountain Crown lands had extended, in 1862, over 1866H. 03 ares, at an expense of 146,747 fr. 51 ct.

B. *Reboisements Obligatoires, or Enjoined Reboisements.*

In all cases in which public safety demands the creation of such hinderances as *reboisement* can offer to the irregular action of rivers or floods, or to the crumbling of the ground, and where the safety of the inhabitants, the condition of the roads, and the culture of the lower declivities, are most threatened by torrents and avalanches, the law has commanded the formation of woods; the extent of these is in proportion to the hydraulic effects they are designed to produce.

The Administration has carefully considered the condition of the districts where *reboisement* seemed to be most urgently required.

These careful inquiries in 1861 and 1862 extended over 21 departments, and gave rise to the origination of 269 enterprises, comprehending 136,756 hectares.

89 undertakings, extending over an area of 59,833H. 28A. have been officially inspected. The projects have encountered a good deal of opposition.

"It is necessary clearly to define the character of this opposition," says the Director General of Forests, "in order to exhibit the influence of the operation of *reboisement* upon the condition of the mountain population.

"In most cases the herds of cattle do not belong to the poorer inhabitants. The flocks of sheep belong to a certain number of local owners, who make all they can out of the communal lands, or to people from a distance, whose immense flocks, known by the name of *transhumant* flocks, every year cover the mountains leased from the communal bodies, at usually a very moderate rent.

"The operation of *reboisement*, far from introducing new restrictions into the already straitened circumstances of the poor inhabitants, would, on the ncotrary, be a source of numerous advantages. Without mentioning one of these advantages which can only be realised in the more or less distant future, there can be pointed out as an immediate and direct result, the unusual comfort diffused over these poor districts by the money devoted to the execution of the works in the form of wages, purchase of seeds and plants, and other outlays of various kinds.

"There is reason to think that the mountaineers, with their characteristic mental quickness, have already come to appreciate the operation of *reboisement*, and that the opposition which has arisen in several cases is only an expression of personal and isolated interests.

"It is, moreover, only through mistake that the pastoral population takes alarm at the undertakings of the Forest Administration, the greatest number have been in favour of this industry. Besides the immense tracts known by the name of pastoral mountains, which lie above the zone of forest vegetation, and of which the destination indicated by the nature of things cannot be modified, the actual sheep runs are in many cases not only preserved but improved from the double point of view of the pastoral interest and the preservation of the turf."

Nearly all the commissioners charged with the direction of these inquiries have earnestly approved of the projected *reboisement*, and in all cases the special commissioners, the Councils of the Arrondissement, the General Councils, and the prefects have adopted these projects. Besides, the General Councils have voted subsidies in favour of the *reboisements*. These subsidies, 20,000 francs in 1860 rose to 40,000 francs in 1861, and to 71,000 in 1862.

The gradual increase of these sums, more than their absolute importance, is an indication of the increasing movement of public opinion in favour of the operation.

Reboisements obligatoirs, or enjoined *reboisements*, had extended in 1862 into three departments, and into seven périmètres, or defined areas, of over 2061 hectares 87 ares, and has cost the State 69,576 francs 21 cents.

Thus the *reboisements* effected in 1862 amount in all to 11,416 hectares 63 ares.

C. *Sécheries and Nurseries.*

The *reboisement* in 1862 has been effected, by means of sowings, upon 8344н. 26а.—by means of planting, upon 3072н. 37а. For the sowings 95,403 kilogrammes of the seeds of different trees have been used. For the plantations there have been used 22,137,500 plants of different sorts of trees, besides those transplanted from the woods.

The *pin sylvestre*, or Scotclr fir, *épicéa*, or Norway fir, and the larch, have been most generally employed. Other trees have also been used—as the oak, the Austrian pine, the Mugho, the Aleppo pine, the cedar, the ailanthus, which are introduced gradually in proportion as new experiments are tried. The selection of trees has generally been directed by local indications.

Four *sécheries* were formed in 1861; two others were established in 1862. These supply from 15,000 to 20,000 kilogrammes of seed, corresponding to the *reboisement* of 2000 hectares. The outlay in 1862 amounted to 38,515,24 francs.

The Administration has, moreover, in 1862, set agoing the collecting of seed in the Crown forests, and has collected considerable quantities at a very moderate expense.

In 1861 there had been formed 473 nurseries, 330 hectares in extent, and capable of supplying about 60 millions of plants per annum.

In 1862 there had been formed 359 new nurseries, covering 272 hectares 96 ares, capable of producing annually about 40 millions of plants.

Many of these nurseries are of small extent, and are designed to supply plants for restocking the immediate vicinity. But in several cases central nurseries of considerable importance have been formed, situated in suitable districts, which had been bought or rented with a special view to the work in hand.

These nurseries have been the object of the greatest care, they are 14 in number, and are spread over ten departments. It is calculated that 5000 hectares is the extent annually reforested by means of planting, and that 40 millions is the necessary supply of plants. At the market price this quantity of plants would cost 240,000 francs.

In 1852 the expense for the formation of new nurseries, and the keeping up of the old ones, amounted to 153,772 francs.

D. *Administrative Measures.*

A new district under a forest conservator has been formed.

Annual conferences, attended by those taking a part in mountain *reboisement*, have been instituted.

The Director of the Government School of Forests has been appointed to visit the works of *reboisement* in the Pyrenees, the Alps, and the mountains of Central France. The object of this visitation is to encourage the efforts of those employed, to secure everywhere good methods of culture, and to report to Government upon the execution of the works and the result obtained.

According to the preamble to the law of 28th July 1860, the expense of *reboisement* was estimated at 180 francs per hectare. In 1861 and 1862 160,055 hectares 63 ares had been reforested. The expense during these two years had been 1,738,000 francs, or 180 francs per hectare, without taking into consideration the part of the above mentioned expense incurred in the preparation for enjoined *reboisements*, the formation of *sécheries*, the purchase of land, and other expenses not directly belonging to the work of replanting, properly so-called. The expense per hectare reforested will be still further reduced through use being made of the extensive resources supplied by the nurseries and *sécheries*, and of experience acquired by practice in the execution of the works.

ABSTRACT OF REPORT FOR 1863 :—

According to the detailed accounts of expenditure on the work in 1862, the expense has been as follows :—

Subventions,	*Francs*,	350,000,00
Purchase of Property,	,,	13,231,00
Execution and Superintendence of the Works,	,,	761,957,31
Total *Francs*,		1,125,188,31

A. *Reboisements Facultatifs, or Sanctioned Reboisements.*

Subventions have been granted to 450 communes, or public establishments, and to 983 private individuals.

These *reboisements* are extended over communal lands, 7,073h. 24a.; private property, 2,157h. 05a.; crown lands, 1,750h. 88a.; total 10,981h. 17a. Outlay, at the Government expense, 595,000 francs, besides the expense of keeping up the sanctioned *reboisements* of former years, 81,800 francs. When requested by proprietors, the Forest Administration have carried on the work of *reboisement* under the superintendence of its agents and guards, and they will continue to keep them up and develope them, so far as possible, till success appears assured.

The works have been spread over 40 departments.

The report specially mentions an experiment of *reboisement* in the Crown forest of the Luberon, situated on the formation called *neocomien*, belonging to the lower portion of the chalk formations, where the bare places are covered by enormous heaps of rocks, burnt by the sun, and entirely destitute of vegetable mould. Such is the district of which the forest agents have not been afraid to attempt the *reboisement*. Nothing has been neglected to ensure the success of this bold enterprise. After several attempts at plantation, which proved either fruitless or else too costly, the agents fixed on the method of replenishment by sowing, principally with the seed of the pine of Aleppo. The small quantity of vegetable mould still remaining between the rocks was gathered together into narrow ridges, and prevented from falling down by layers of stones. Upon soil prepared in this manner the sowings were most successfully effected. In the month of September, after the trial of an exceptionally dry and hot summer, the young plants appeared quite flourishing. To one who has seen the sowings of Luberon, (says the report) no *reboisement* will appear impossible.

Among private individuals the taste for forest improvement seems to have a tendency to increase wonderfully. The number of private proprietors who had received subventions, which was 394 in 1862, in 1863 amounted to 983.

B. *Reboisements Obligatoires, or Enjoined Reboisements.*

On the 1st January 1864, the districts comprehended in the Government undertakings were to be found in 23 departments.

Digests have been prepared of 264 enterprises, of which 77, embracing about 60,000 hectares, have been approved, with decrees declaring their public utility.

The works of *reboisement* then in course of execution in 26 circles extended over a surface of 1,853h. 57a.

The expense has been 154,850 francs, besides 13,100 francs for keeping up the works already effected. This outlay, however, is only a Government loan in terms of Articles 8 and 9 of the law of 28th July 1860.

In cases where direct *reboisement* did not present a sufficient probability of success, because of the absence of vegetable mould, it has been preceded by the restoration of the soil, by means of planting or sowing herbs or bushes. The principal plants growing spontaneously on the mountains are juniper, barberry, *l'argoussier*, *l'amélanchier*, which are chiefly found in the rockiest places, white fescue grass, whose luxuriant tufts appear on the steepest parts of the ravines, the sainfoin and lucerne, the long matted roots of which are well fitted to retain the soil upon the slopes.

"A considerable number of rustic barriers have been formed on the upper branches of the torrents. Instead of a large work of art constructed at the mouth of the ravine, which nearly always gives way, the system of defence consists in the formation of a number of dams across its ramifications throughout the upper portion of its course. The small furrows which form the highest branches of the torrents are dammed by simple faggots fastened with stakes. In the larger branches, and where the presence of rocks or any other circumstance appears to favour the construction, there are formed dams made of hurdles and boughs, or walls of stones placed one upon another strongly attached to the banks, or by rude carpenter work, the whole being completed by interweaving quantities of willow and osier wands into the banks, and into collections of earth which accumulate above the barricade.

"The success of these simple and economical works is remarkable. The water, arrested everywhere in its descent, falls with much less violence and rapidity; a great part of the materials which it bears along are stopped by the barricades, and no longer spread themselves over the villages and lands situated at the foot of the mountain; finally, the accumulation of these materials, joined to the quick growth of the willow wands, tends to efface the effects of the torrent between the successive barricades, and in some measure to lessen the torrent by stopping up the ramified furrows of which it was composed." Total extent of *reboisements* effected in 1863, was 12,834 h. 74 a.

C. *Preparatory Works.*

The question, which of the two methods of replenishment, sowing or planting, should be preferred, does not admit of absolute solution.

"In planting, two principal dangers are to be dreaded: The swelling up or upheaval of the soil in spring, produced by the alternations of frost and thaw, the effect of which is to lay bare the roots, and even to throw out the plants,—and the drought in summer. These dangers may often be averted or escaped, by placing, when circumstances permit, at the root of the plant one or two stones, intended at the same time to hinder the swelling of the soil and to keep the surface of the ground cool. When the trees are planted amongst herbage, after a turf is cut, for the purpose of planting a young tree, it is cut in two and placed at the root, either in the position which it occupied before the operation, or turning the grass side towards the ground.

"Autumn has been preferred for planting, spring for sowing; but experience has proved that sowings completed after the greatest heat of summer are the most successful; the young plant appears before the cold, then comes the snow to cover and protect it till the return of spring; it then resumes its scarcely interrupted growth, and when summer arrives it is robust enough to resist the heat."

D. *Nurseries.*

Since the beginning of the enterprise of the *reboisement* of the mountains, Government has felt the necessity of getting rid of the obligation to have recourse to purchase, because the outlay is great and the productions are not always to be depended upon. The harvest of the fruit of this foresight is now beginning to be reaped. Two kinds of nurseries have been formed. (1) Small nurseries, scattered over the district where the *reboise-*

ments are of small extent ; (2) More important nurseries, intended to supply subventions of plants to communes and to private individuals, as well as the *reboisements obligatoires* of specified périmètres.

The first-mentioned nurseries, containing less than 50 ares, are 355 in number, and cover in all 41 hectares 42 ares. The second, of greater extent, containing more than 50 ares, are 97 in number, and cover in all 144 hectares 72 ares. The expense of the nurseries, for establishment and keeping up, has been 163,000 francs.

The following is an account of the expense and returns of two nurseries of the second class, those of Arpajon (Cantal) and of Bourg (Ain):—The first, 7 hectares 43 ares in extent, has cost in all 51,252 fr. 60 ct. It has produced since its formation 4,365,310 resinous and broad-leaved plants, of the value of 42,712 fr. 60 ct., according to market value. The expense of keeping up will be from 10 to 12 thousand francs annually, and the return from 6 to 8 millions of plants, which, at an average price of ten francs per thousand, represent a value of from sixty to eighty thousand francs. The second nursery, 4 hectares in extent, has required an outlay of 29,107 fr. 53 ct., and it has supplied 2,050,000 plants, 20,000 francs in value. Its keeping up costs annually from 5 to 6 thousand francs, and it produces about two millions of plants, valued at 20,000 francs.

E. *Co-operation of the Departments.*

The General Councils have approved of the greater part of the projects submitted to them.

In 1863, 35 departments have granted sums amounting to 98,000 francs. These subsidies have been in 180, 20,000 francs.
,, ,, ,, 161, 40,000 ,,
,, ,, ,, 162, 71,000 ,,

Abstract of Report for 1864 :—

In 1863 the expense of the *reboisement* of the mountains amounted to 1,316,652 fr. 15 ct., apportioned as follows :—

Subventions to communes, to public establishments, and
private proprietors, - - - - - *Francs*, 494,000.00
Purchase of land, - - - - - ,, 23,879.13
Execution and superintendence of work, - - ,, 798,773.02

Total *Francs*, 1,316,652.15
In 1864 the expenses were, - - - ,, 1,401,822.48

A. *Reboisements Facultatifs, or Sanctioned Reboisements.*

Tracts of land belonging to communes or public establishments.—458 communes or public establishments have received, in 1864, grants of seeds and plants, or of money, amounting to 352,210 francs 15 centimes.

The tracts *reboisé* with the help of these grants were, 6,164 hectares 32 ares in extent. According to results determined by forest officers, the sowings and plantings succeeded in at least a proportion of from 60 to 80 per cent.

Land belonging to private individuals.—Government had in 1864 granted subventions to 739 private individuals for the *reboisement* of mountain territory, covering an extent of 1,601 hectares, dispersed among 28 departments.

Crown lands.—In the departments where the State possesses bare mountain territory, the Forest Administration has set the example of *reboisement* by sowing or planting every year areas more or less considerable. 1,834 hectares 70 ares of this kind of ground has been rewooded in 1864.

It is chiefly in the department of Ariége that the restockings of this kind have taken place during several years with remarkable success. Altogether the *reboisements facultatifs* of every class, in 1864, covered 4,743 hectares 90 ares.

B. *Reboisements Obligatoires, or Enjoined Reboisements.*

At the end of 1864 the number of projected enjoined *reboisements* were 322, covering a total extent of 168,300 ares. Of this number, 84, covering 61,814 hectares, have been at the same time subjects of a decree declaring them to be of public utility.

In the course of the same year, works have been executed in 65 périmètres. These works have consisted of *reboisements* properly so called, the keeping up of *reboisements* effected in preceding years, sowings or plantations of herbs or bushes, construction of dams, lopping trees, and enclosures. These have cost 249,000 francs.

The Government in 1864 has only had recourse to expropriations in the cases of two tracts of ground, extending to 25 hectares, for which the price paid amounted to 9,476 francs 47 centimes.

It is with the utmost reluctance that Government makes use of the privilege accorded by the 2nd paragraph of Art. 7 of the law of 28th July 1860. Since the law has come into operation, there have only occurred some three cases in which it was needful in the public interest to proceed to expropriation.

C. *Conferences.*

Conferences held in cantons to determine what lands should be replanted have continued to discharge this duty in a manner the most satisfactory.

D. *Résumé of Work executed in 1864.*

The total sum of *reboisements* effected in 1864 embraces an area of 12,193 hectares 32 ares.

E. *Kinds and quantities of Seeds and Plants used in 1864, and ways in which they have been apportioned.*

Of the 12,193 hectares 32 ares rewooded in 1864, 7632 hectares 44 ares have been sown with seed ; and 4559 hectares 88 ares have been planted.

The principal kinds of trees thus used have been, as in years preceeding, *le pin sylvestre*, or Scotch fir; *l'épicéa*, or Norway fir; *le pin maritime*, or maritime pine ; *le mélèze*, or larch ; *le pin noir d'Autriche*, or Austrian pine ; *le pin laricio*, or Corsican pine ; *le pin à crochets*, or Muglio pine ; *le sapin*, or silver fir ; *le chêne*, or oak ; *le châtaigner*, or chesnut ; *le hêtre*, or beech ; *le frêne*, or ash. It is with the greatest reserve that attempts have been made to introduce other kinds of tree, which are not indigenous, in such districts as have been rewooded.

Of the 161,260 kilogrammes of seed used in 1864, 137,028 kilogrammes have been supplied by *sécheries domamials*. But the Administration has not found any great advantage in preparing their own seeds.

In regard to plants, of 55,740,000, 49,334,000 have been reared in nurseries belonging to the State; and the others, 6,408,000 have been obtained from nurseries belonging to private parties.

The expenses of all kinds incurred in maintaining the State nurseries has amounted, in 1864, to 175,892 francs; and the value of the plants supplied from them in the course of the year, estimated at 6 francs per 1000, which is much below the average market price, amounts to about 300,000 francs. It is of some importance to add that, in regard to adaptation to their destination, the quality of the plants supplied from the State nurseries is in general much superior to that of the others; and, in the report it is added, there is reason to hope that the Administration will soon be in a position to dispense entirely with having recourse to purchase for the supply of plants.

The principal nurseries of the Administration collectively cover an area of 257 hectares 34 ares, and can supply 93 millions of plants per annum.

F. *Co-operation of Departments.*

The amount of subventions voted by the departments in 1864 is nearly the same as in 1863, being 81,104 francs, as against 78,000.

The following is a general *résumé* of what was effected in these first four years of the enterprise :—

In these years there were replanted with woods 41,083 hectares 26 ares. Of these there were *reboisements facultatifs,* or sanctioned

reboisements, on property belonging to private proprietors,	6056н 13а
to communes, -	21665н 84а
to the domaine, -	6853н 56а
	34575н 53а
Reboisements obligatoires, or enjoined *reboisements,* -	6507н 73а
	41083н 26а

The accomplishment of the work cost the State, in 1861, 638,013 fr. 64 ct.; in 1862, 1,125,188 fr. 21 ct.; in 1863, 1,316,652 fr. 15 ct.; in 1864, 1,401,822 fr. 48 ct.,—total 4,481,676 fr. 48 ct., being, on an average, 102 francs per hectare.

Thus far all seems to have gone on satisfactorily. Every thing had been done to carry public opinion, and the sympathies of those who were more immediately affected by the operations, with the enterprise. But it becomes necessary at this point to advert to the results of this commendable endeavour, and the supplementary legislation which this necessitated.

From the first the work had been prosecuted with vigour, and it had the support of many of the more intelligent inhabitants of the district; but after a time, as may be seen from these reports, a reaction began to manifest itself, and this became at length developed into strong opposition on the part of many.

"As may always be expected," says Cézanne, " difficulties which had not been taken into account began to make themselves apparent when the work was commenced. The word *reboisement* frightened the pastoral communities; there was promised to them herbage growing under the trees in about twenty years; but in awaiting this how were they to support the flocks, which supplied their only income? 'The operation,' cried they, 'is a flagrant injustice; they are ruining the mountains in order to enrich the plains.'

"The Administration saw that there was some foundation for this complaint, and they resolved to do what was right in the case; but the law spoke only of *reboisement*—their powers, and the funds placed at their disposal, related only to this; and something must be done to meet the case.

" It was thought at first that this might be effected by the law, *Sur la mise en valeur des biens communaux*, for the improvement of communal properties. The greater part of the lands to be replanted being communal lands, it was thought practicable to unite the two objects, and combine the two funds for a common action, and a mixed commission was nominated by the three ministerial departments interested; but it was found that the two laws which they sought to combine in joint action had two very different objects: the law on *reboisement* had for its object to secure the public safety, the other to promote the national wealth; the former acted on decrees with credits and subventions, the second by prefectoral resolutions granting simple advances; by the first the *Agents des Eaux et Forets* were charged with the reconstitution of communal property, to carry out the second the officials of *Les Ponts et Chaussées* labour to convert communal into national or personal property."

The Mixed Commission soon reported its powerlessness, and the Government had to follow up the law in regard to *reboisement* with one relative to *gazonnement*.

The following is a translation of the *Exposé des Motifs*, which accompanied the draft of this law, addressed to the councillors of State charged to support it before the *Corps Legislatif* :—

" GENTLEMEN,—When the law of the 28th July 1860, on the *reboisement* of the mountains, was submitted to the consideration of the *Corps Legislatif*, the honourable reporter, in the name of your commission, expressed himself in these words :—

"' It may be well, then, to recognise the fact that the *deboisement*, or destruction of woods on our mountains, is not the only cause, or even the principal cause, of the disasters produced by the ravages of the waters. Along with this, as still more hurtful, must be classed as a disturbing cause *degazonnement*, or the destruction of herbage.

"' In like manner, *reboisement* alone is not enough to remedy these evils. It would be impossible to replant with trees all the bare mountains, on account of the great expense. It would also be useless, as keeping up the turf is a sufficient preservative, the benefit of which has been proved by experience. It would also be difficult, looking at it from the stand-point of the wealth of the country, as it would substitute comparatively profitless forests for the magnificent pasturage, the destruction of which would ruin the population of the mountains.

"' But it is not the less true that, in conjunction with *gazonnement*, *reboisement* will have a most happy effect.

"' The present law will only produce all the good effects which may be expected when it shall be supplemented by *gazonnement*.

"'The experience and the investigations of engineers have shown that in certain cases it is indispensable to arrest a daily increasing evil, for only in this way can possibly be preserved certain districts unfitted for pasturage, and threatened with approaching destruction. *Reboisement* will create a great protection, preserving even the pasture lands, regulating the flow of the water, and preventing the formation of avalanches, and exercising certain specific effects during atmospheric perturbations.

"'The measure which is now proposed is truly a law for the public welfare, and has a right to all our sympathy; but it will not produce all the good that may be expected, until it shall be supplemented by measures for the protection of the herbage, and by measures repressing the increasing evils of depasturing.

"'The commissioners appointed by you pray earnestly for these measures, regarding which they have no power to take the initiative.'

"Goverment has not overlooked this view of matters in the preparation of the law of 1860. The *Exposé des Motifs*, or reasons assigned for this law, explained to the legislative body the various reasons which at that time led the Government to determine not to extend the action of the law to the restoration of herbage. One reads as follows, at page 17 :—

"'We do not conceal, that, even looking at it from the point of view of hydraulic results, which is the stand-point of the law, the restoration of the herbage is fitted to give important help to *reboisement*. At the same time, it does not appear possible to extend so far the operation of the proposed measures, and that for several reasons.

"' First, the financial resources which are at command are not adequate to meet the expense of the *reboisement* which it is desirable to encourage and execute, unless they be laid out with the greatest economy and wisdom; no part, therefore, should be diverted to works of a different nature, or inferior utility.

"' Second, *reboisement*, where executed intelligently, having solidified the soil, will also, in a certain degree, promote the natural restoration of herbage in certain places.

"' Third, there is room to hope that, having before their eyes the *reboisement* executed by or under the influence of Government, communes, to whom depasturing offers immediate and individual advantages, will be easily induced to undertake for themselves the restoration of their pastures, now that it has become more easy and sure of success.

"'Finally, the legislative body is engaged on a special law for bringing under culture communal lands, which will serve in cases altogether exceptional as a last resource.'

"The first reason which we have adduced still subsists; it is certain that it would be impossible to proceed with *reboisement* and with *gazonnement* simultaneously, and to a sufficient extent, with nothing but the resources created by the law of 1860; but the hindrances may speedily disappear should Government approve of the proposals which we shall shortly have the honour of presenting.

"As to the two last reasons, they rest upon conditions which it must be confessed have not yet been confirmed by experience. We shall later explain the causes which are opposed to the realisation of our hopes.

"It is right that we should furnish the *Corps Législatif* with a summary of the practical results of the law on *reboisement*. The success of this law may be confidently affirmed.

"*Reboisements facultatifs*, that is to say those set a-going by the simple encouragement of State subventions, have extended,—

   In 1861, to - - 3,237 hectares.
   In 1862, to - - 7,448 ,,
   In 1863, to - - 9,320 ,,

"*Reboisements* effected on the Crown lands have replanted,—

   In 1861, - - - 1,402 hectares.
   In 1862, - - - 1,866 ,,
   In 1863, - - - 1,750 ,,

"As regards *reboisement obligatoires*, that is to say, what is done in the périmètre, or boundary, the replanting of which has been pronounced necessary for the public welfare, the operations have been necessarily retarded by the fulfilment of legal formalities, but investigations and instructions have been carried on with activity.

"At the end of 1863, 264 undertakings, comprehending 140,000 hectares, were made the subjects of special consideration, and 77 had been the subject of special decrees declaring their public utility. The operations were being executed in 26 périmètres, and over an extent of 1,853 hectares.

"At least 40 Departments are profiting by the operation of the law.

"Several General Councils of Departments have desired to take part in the Government works. The sums voted have been,—

  20,000 francs in - - - - - 1860
  40,000 ,, ,, - - - - - 1861
  71,000 ,, ,, - - - - - 1862
  98,000 ,, ,, - - - - - 1863

"The import of such a constant and rapid progress has not escaped the notice of Government. The net cost of the operations has not been less satisfactory.

"The reasons assigned for the law of 1860, inspired by a very decided wish to avoid all illusions and chimerical promises, fixed the average expense of the work of *repeuplement* at 180 francs per hectare. What has been done in the average expense has not exceeded 108 francs, and the benefit of the law may thus extend to other and more extensive districts.

"Government has all along met with the greatest sympathy from the General Councils of Departments, Councils of Arrondissements, and Special Commissioners.

"But now, as we approach a very important and very delicate point—the moral disposition of the population towards the measures taken to carry into effect the law of 1860—we meet facts which have induced Government to think that it is opportune, and perhaps necessary, to complete this law in accordance with the wishes expressed by the Commission of the *Corps Legislatif*, at least in so far as relates to the communal pasture lands.

"The aim and consequences of the law relative to mountain *reboisement* has been but imperfectly understood and appreciated by Municipal Councils, and by the inhabitants of the communes most interested.

"Those who have the rights of pasturage, accustomed to the slender income derived from depasturing—and it must be admitted sometimes too poor to do without this, are disturbed by measures which temporarily restrain their individual privileges. Moreover, seeing that Government occupies itself exclusively with works of *reboisement*, they attribute to it the design of everywhere substituting forest for pasture, so as to progressively accomplish the suppression of depasturing.

"The consequences have been these: on one hand, a pretty large number of decrees proclaiming the public utility of the measures have been published, contrary to the advice of the Municipal Councils and to the wishes expressed by witnesses at the inquiry; on the other hand, the communes have refused to make the small sacrifices which would have been necessary for the restoration, in a future more or less distant, of pastures of which they believed they were destined to be deprived altogether.

"Independently of these obstacles and these misunderstandings, we ought to add, that many communes are to be met with which are really too poor to undertake the operations at their own expense, however inexpensive they may be, or to endure being deprived, even for a short time, of the incomes which a number of them derive in one form or another from the communal pastures.

"In this predicament Government has seen the necessity of intervention, and the pressing duty of enlightening the inhabitants, of reassuring them, and especially of meeting their real necessities and their just desires, by seeking to make compensation for the diminution of their privileges, looking to the valuable possessions of the pastures.

"Before having recourse to a new law, Government has tried what effects would be produced by the combined co-operation of the law *Sur la mise en valeur des biens communaux*, with the law on *reboisement*.

"A High Commission has been created 'for the purpose of finding out the best way to make the two laws co-operate towards a common end, and with the help of their mixed character to smooth away the difficulties which may arise between the two Ministerial Departments entrusted with the execution of these two laws—that of agriculture, commerce, and public works, and that of finance'.

"This Commission has acknowledged the impossibility, or at least the extreme difficulty, of reaching the same end with the two instruments at its disposal. And in order to attain this end it is necessary, in the first place, to find out the mountain districts, the consolidation of which is demanded by the public interest; to distinguish between ground which must be *reboisé* or replanted with woods, and ground which must be *regazonné* or planted with turf; to mark out the périmètres with reference to these; and in these périmètres to determine the number and form of massive woods which are to retain the floods and protect the pasturage. It is necessary that the enactments prepared by Government should be submitted, as a whole and in their harmony, to the various authorities, to the Councils, and to the Commissions, whose duty it is to give their advice in regard to the instructions issued. It is necessary that the subventions furnished by the State and the demands for local subsidies should be in proportion, on the one hand, to the public utility of the enterprise, and, on the other, to the advantages which would result to the local population; finally, it is necessary that the temporary privation of privileges should in certain cases be compensated by grants of money, at least to the poorest communes.

"It may be seen that this combination of circumstances, or conditions, can only be secured through a double operation—by means of two parallel codes of instructions, by means of two administrations, and by the application of two distinct laws. The difficulty is not wholly related to the truth, that to be useful an operation should have a single aim; but also to this, that the two laws, while presenting incontestable analogies, are, nevertheless, **distinguished by** notable differences.

"The principal stand-point of the law of *reboisement* is the public safety, the regulation of the water-flow, and the protection of the low grounds. Economy is only secondary.

"The stand-point of the law relative to bringing communal lands under culture is more especially economy, the improvement of the communal patrimony, the increase of the general food supply, and the increase of the municipal revenues.

"The formalities prescribed by these two laws are analogous; but they are not identical.

"According to the law of *reboisement* the initiative should be taken by the Central Government; according to the law *Sur la mise en valeur des biens communaux* the initiative is to be taken by the prefect.

"The law on *reboisements* sets agoing a very complicated machinery, more especially the special commissions; the law *Sur la mise en valeur des biens communaux* does not require the interference of these commissions.

"In another relation the law on *reboisement* offers two kinds of encouragement—fixed subventions and recoverable advances; the other law offers only recoverable advances.

"Finally, the law *Sur la mise en valeur des biens communaux*, conformably to the nature of its aim, tends to withdraw the p operty improved from the possession of the public; it formally authorises the State to enact that the improvements shall be consolidated; the law on *reboisements*, on the contrary, promises to throw open the ground for pasture whenever the trees are old enough, and the result expected from *regazonnement* is designed to be, as we have already said, to restore to the possessor a more valuable in the place of a more extended privilege.

"These differences will explain how it is the High Commission has been led to think that there is no hope of great and regular development, the necessity of which is now clearly shown, through the conjoint operations of this double initiative, of these double instructions, and these different tendencies, whatever may be done to organise the simultaneous application of the laws. The High Commission has unanimously acknowledged that the only practical efficacious means of obtaining the desired combination of *gazonnement* and *reboisement* in the *périmètres mixtes*, is to entrust the initiative and instructions to be given, and the execution of these, to one single administration under identical conditions, with the performance of the same formalities, with a single end in view, and that that which is arrived at by the law of *reboisement*, and of which a law simply supplementary to this would easily ensure the attainment.

"Government has adopted the proposal of the High Commission, and, the principle once admitted, the drawing up of the scheme presents few difficulties. Nothing is needed but to extend to the renewal of pasture in the *périmètres mixtes* the arrangements already adopted for the forest *repeuplement*, and to add to the funds created by the law of 28th July 1860 the necessary supplement.

"Gentlemen, we have very little to say upon the different articles of the scheme, and we shall advert to them here very cursorily.

"Art. 1, in reproducing the definition of art. 4 of the law of 28th July 1860, aims at defining the object of the new law, and at demonstrating that it is only a supplement to the older one. And it should be thoroughly understood that the action of the Finance Department can never take the place of the Agricultural, Commercial, or Public Works Department; that the Forest Administration will only exert its power to investigate or to execute operations for restoring the sward on hilly ground, looking exclusively to the public utility in the regulation of the water-courses, and the consolidation of the soil—that is to say, in conditions identical with those which led to its being invested with the power of encouraging or executing the *reboisements*.

"Art. 2, for the arrangement of preparatory formalities, refers simply to the law of 28th July 1860; it could not do otherwise, for in most cases the question is, how to accomplish the formation of mixed périmètres composed of woods and tracts of new turf; the directions ought to be combined for this double object, and consequently should be subjected to the same regulations.

"A temporary arrangement authorises the Forest Administration, in regard to ground included for the first time in the area of *reboisement obligatoire*, to substitute operations for renewal of the turf for operations for *reboisement* in such measure as they my deem fit.

"Art. 3 relates to art. 9 of the law of 28th July 1860. This last article enacts that, in cases where the State executes operations of *reboisement* upon communal lands, the communes may relieve themselves of the burden of repayment by giving up the proprietorship of half of the lands *reboised*. It has appeared that, when operations of *regazonnement* only are in question, the proportion of one-half is nearly always too great, taking into consideration the expense of the work. Hence, an arrangement by which the communes may always get exemption, by giving up land in proportion to the advances made for their benefit.

"Art. 4 fixes the different executive measures which should be specified by regulation of the public administration. This statute should determine the mode of certifying the advances made by the State, and the measures necessary for securing the repayment of these; it should also lay down rules for the allocation and settlement of grants of money which it may be necessary to allot, to communes which are too poor to submit to even a temporary deprivation of pasture, though in the view of its improvement,—grants of money, which, besides in certain cases, will constitute the greatest part, or even the whole, of the expense of restoration, and which should only be granted in cases of absolute necessity by a decree declaring the public utility of the measure.

"Finally, Art. 5 creates the financial resources required for the operation of the law. The Government is referred, for the specification of the nature of these, to the law of 28th July 1860, and in proposing to obtain these resources from extraordinary fellings it has only followed the line of action indicated by your Commission four years ago.

"Such, gentlemen, are the principal arrangements of the *Project de loi* which we have the honour to submit to you; they have a special reference to the mountain lands, because the Forest Administration, with whom the execution of them will lie, has only, in what relates to the regeneration of pasture lands which are not wooded, for its work, to carry out measures complementary to the law *Sur le Reboisement des Montagnes*; they have also a special reference to communal lands, because the improvement of meadows belonging to private proprietors have not appeared of a character to warrant either the application of coercive measures, or the employment of the funds of the State; these arrangements do not the less apply to numerous localities, and to areas of very great extent; they will produce no small effect by their physical action, and they will not be without interest, in more than one locality, in regard to their influence in pacifying the mind. We hope, gentlemen, that they will meet with your approval."

The draft, or *Projet de loi*, with such modifications as were proposed by the Commission to which it was submitted, was adopted by a unanimous vote of the *Corps Legislatif*, and was issued in the following terms:—

" Law of 8th June 1864, completing, in what relates to *gazonnement*, the law of 28th July 1860, *Sur le Reboisement des Montagnes*.

" Art. 1. Ground situated in the mountains, the consolidation of which is, by the terms of the law of 28th July 1860, recognised to be necessary on account of the state of the soil, and the dangers which may result to the lower ground, may be, according to the necessities of the public interest, either entirely returfed, or partly returfed and partly *reboiséd*, or entirely *reboiséd*.

" Art. 2. Applicable to the work of *gazonnement*, in so far as they contain nothing contrary to the present *loi*, are the Articles 1 to 8, and Articles 11, of the law of 28th July 1860, on *reboisement*.

" Everywhere, with regard to territory comprehended within the périmètre of obligatory *reboisements*, previous to the publication of the present law, the Forest Administration is authorised, after consultation with the Municipal Councils of the interested communes, to substitute *gazonnement* for *reboisement*, in such measure as they may judge necessary.

" Communes and public institutions, and private proprietors, may call for this substitution. In case of refusal by the Forest Administration, it shall be decreed by the prefect in council, after the fulfilment of the formalities enacted by 3 and 4 of the second paragraph of Art. 5 of the law of 28th July 1860.

" The decision of the prefect may be referred to the Minister of Finance, who shall make it law, after having taken the advice of the finance section of the Council of State.

"Art. 3. In every case, communes and public institutions may be released of repayment to the State by giving up at most the half of the returfed land, during the time necessary to repay to the State, both principal and interest, the advances made for useful works ; or they have the alternative of giving up entirely a part of the land, not to exceed one-fourth, all being specified by professional surveyors.

" Art. 4. There shall not be carried on the execution, at one time, of works of *gazonnement* and *enclosure*, on more than one-third of the surface to be *gazonnéd* in each commune, unless the Municipal Council shall authorise them being carried on over a more considerable extent.

" Art. 5. A proprietor expropriated by the execution of the present law has the right of recovering his estate after being *gazonnéd*, on condition of repaying the price of expropriation, and the expense of the operations, both principal and interest. He can exonerate himself from the repayment for the work executed by resigning one-fourth of his estate.

" Art. 6. An enactment of the public Administration shall determine (1) What measures are to be taken for selecting the portions pointed out in Art. 1 of the present law ; (2) Rules to be observed for the execution and preservation of the *gazonnement* ; (3) The mode of determining the grants made by the State, the measures necessary for securing the repayment of principal and interest, and the rules to be followed for the cession or resignation to the State of the possession and proprietorship of land ; (4) The mode of fixing and allocating grants of money, which, according to circumstances, may be allotted to communes in case of the temporary deprivation of them of pasture on the communal lands which are, for the time being, the subjects of *gazonnement* or *reboisement*.

"Art. 7. A sum of five millions is set apart to the payment of expenses authorised by the present law, amounting to 500,000 francs per annum.

"This shall be provided by means of extraordinary fellings in the Crown forests being made in aid of the ordinary resources of the treasury."

On the 10th November, in this year, was issued the Imperial Decree, embodying the regulation of the Administration for the execution of the laws of 28th July 1860, and the 8th June 1861, on the *reboisement* and the *gazonnement* of the mountains, in which, after the preamble stating what documents had been seen and considered, it is stated,—

"We have decreed and decree what follows :—

"TITRE I.—REBOISEMENTS ET GAZONNEMENTS FACULTATIFS.

"Art. 1. The proprietors of land situated on mountain tops or declivities, who may wish to benefit by the subventions to be granted by the State in terms of Arts. 1 and 2 of the law of 28th July 1860, and of paragraph 1 of Art. 2 of the law of 8th June 1864, should make their desires known to the *Conservateur des Forêts*.

"When a commune or public institution is in question, the request should be made to the prefect, who transmits it to the *conservateur*, along with his opinion and reasons attached.

"Art. 2. Ground belonging to communes or public institutions on which operations of *reboisement* or *gazonnement* are undertaken, with the aid of State subventions, are for the time submitted absolutely, the parts *reboiséd* to the forest *régime*, the parts *regazonnéd* to the pasture regulations prescribed by article 21 of the present decree.

"These operations, as well as the work necessary for preserving and keeping them up, are effected under the control and superintendence of the forest officials.

"Art. 3. If the ground belongs to several communes, and the success of the *reboisements* or *gazonnements* renders necessary combined operation, in accordance with Articles 70, 71, and 72 of the law of 18th July 1837, a Syndical Commission is appointed to attend to and carry on the operation.

"In any case in which the work has not been done, or has been badly executed, according to attestation of the forest officials, through the communes or the public institutions neglecting to conform to the decrees for the regulation of the right of pasturage, the prefect takes out a summons commanding the restitution of the subventions which have been allotted by the State.

Art. 4. The money premiums obtained by private individuals are paid after the work is performed, on presentation of a minute declarative of the works having been accepted, prepared by the local forest official in the form of the corresponding minute required on completion of operations for improving the crown lands, and on the advice of the inspector of the *conservateur*.

"A valuation is made of the subventions of seed or plants which are given to private proprietors before the beginning of the operations. This valuation is notified to the proprietor, and accepted by him. The amount can be recovered by the State in cases where the work is undone, where there may be an embezzlement of part of the seed or plants, or where the work is badly executed.

"Art. 5. All subventions exceeding 500 francs in value shall be decreed by our Minister of Finance; all subventions of the value of 500 francs and under shall be granted by the Director of Forests.

"TITRE II.—COMPULSORY REBOISEMENTS AND GAZONNEMENTS—SETTLEMENT OF THE PÉRIMETRES WITHIN WHICH REBOISEMENT AND GAZONNEMENT ARE NECESSARY.

"Art. 6. Whenever the Forest Administration deem it right to fix the *périmètre* of ground within which *reboisement* and *gazonnement* are required, the Director-General of Forests intimates to the prefect the names of the forest officials entrusted with the duty of drawing up the *procès verbal* of the survey of the grounds, the map of the district, and the plan of the projected operations.

"The prefect appoints the engineer of bridges and highways, or of mines, who is to lend assistance.

"Art. 7. The minute of *reconnoissance* is accompanied by an explanatory memorandum regarding the aim of the undertaking, and the benefits which are to be expected.

"The map of the district is prepared with the help of the registrar of lands. The number of the registral volume is given for each portion, also the extent, the proprietor's name; and in dealing with a commune or public institution, the total extent of land belonging to the commune or public institution.

"The périmètre is marked by a continuous border of a uniform bright colour. The grounds to be *regazonnéd* or *reboiséd* are distinguished by flat colours of different hues.

"The prospectus of the operations intimates what land is to be *reboiséd*, and what to be *regazonnéd*, it also fixes the period within which the whole should be completed, and contains,—(1) An approximate estimate of the expense, and a rough draft of the division of this expense among the different proprietors; (2) An indication of the subvention which should be offered to each proprietor; (3) An estimate of the actual value of each parcel, and its value in itself and in its superficies; (4) When necessary, a note of the indemnity which may be awarded to each commune in cases where there is a temporary deprivation of the pasturing on land included in the périmètre; (5) And all other necessary statistical information.

"Art. 8. The papers enumerated in the preceding article are forwarded by the Forest Administration to the prefect, who proceeds in each commune to make the enquiries prescribed by Art. 5 of the law of 28th July 1860, and paragraph 1 of Art. 2 of the law of 8th June 1864.

"The draft lies at the mayoral office for a month, at the expiry of which time a commissioner, appointed by the prefect, receives at the mayoral office, during three consecutive days, the depositions of the inhabitants as to the public utility of the projected operations. The month specified dates from the time when the project is advertised by proclamation and hand-bills.

"A certificate from the mayor attests the performance of this formality, as well as the publication of the prefect's decree requiring the members of the inquest to begin operations.

"After having closed and signed the register of declarations, the commissioner transmits it immediately to the prefect, along with his opinion and reasons annexed. He also sends the other papers which have served as a basis to the inquiries.

"Art. 9. The Municipal Council of each of the communes interested, called together by the prefect for the express purpose, shall examine the papers connected with the investigation, and at the end of a month shall give an opinion, by a resolution agreed to by them, along with the superadded assessors, in number equal to that of the acting municipal councillors. If it is necessary, this resolution shall declare it; if the Municipal Council authorises operations of *reboisement* to a greater extent than that fixed by Art. 10 of the law of 28th July 1860, also operations of *gazonnement* and of enclosure to a greater extent than that fixed by Art. 4 of the law of 8th June 1864, the minute of this resolution is added to the papers connected with the inquiry.

"Art. 10. The Commission ordained by par. 2 of Art. 5 of the law of 28th July 1860, and par. 1 of the law of 8th June 1864, is appointed by the prefect in all the departments traversed by the line of operations.

"The Commission assembles, in the place pointed out by the prefect, on the fourteenth day after he has given intimation. The papers giving directions are examined, also the declarations handed in to the clerk of the investigation; and after all necessary information has been collected from persons suitable to be consulted, the Commission gives its opinion, with reasons annexed, both concerning the utility of the enterprise and upon the different questions that have been submitted by Government.

"These different proceedings, from which the *procès verbal* or minute is prepared, should be completed within the course of another month.

"Art. 11. The prefect, after having taken advice from the Council of the Arrondissement, and from the General Council, shall forward all the documents, with his own opinion and reasons annexed, to our Minister of Finance, who, after having consulted our Minister of Agriculture, Commerce, and of Public Works, and also, when necessary, our Minister of the Interior, shall lay his report before us. We afterwards, in conjunction with our Council of State, shall decide upon the public utility of the operation.

"Art. 12. A duplicate of the decree declaring the public utility of the works is forwarded by the Director-General of Forests to the prefect, who is responsible for the performance of the formalities prescribed by Art. 6 of the law of 28th July 1860, and par. 1 of Art. 2 of the law of 8th June 1864. At the same time the Forest Administration intimates to the prefect, in regard to each registered lot, the operations to be effected, the conditions and time fixed for the completion of the offers of subventions by the State, or the advances they are disposed to give, and finally, if need be, the indemnities awarded for temporary deprivation of pasture.

"TITRE III.—THE EXECUTION AND KEEPING UP OF THE WORK.

"CHAP. I.—*Lands belonging to Private Proprietors, included in the Périmètres, fixed by decrees declaring their public utility.*

"Art. 13. At the end of one month, reckoned from the intimation made to him of the decree declarative of public utility, the proprietor of land included in the périmètre shall declare if he intends doing the work himself, or intends leaving it to the Forest Administration.

"Two copies are made of this declaration, and forwarded to the sous-prefecture of the locality, where they are registered.

"These copies are examined by the sous-prefect, who returns one to the proprietor, and sends the other immediately to the prefect.

"If the proprietor wishes to do the work himself, his declaration shall contain, moreover, some proof that he has the means of doing so.

"Art. 14. When no declaration has been made within the specified time, it is taken for granted that the proprietor refuses to undertake the work.

"Art. 15. The work completed by a private proprietor, with or without the aid of a subvention, shall be subjected to the surveillance of the Forest Administration.

"Art. 16. The Forest Administration shall proceed to the execution of operations to be effected on expropriated lands.

"The completion of the work is notified by the Forest Administration to the expropriated proprietor; this notification besides contains, (1) a detailed account, principal and interest, of the cost of works executed from the date of expropriation; (2) an estimate of the annual expense supposed to be necessary for their preservation and maintenance.

"Art. 17. When, in accordance with the Articles 7 of the law of 28th July 1860, and 5 of the law of 8th June 1864, the expropriated proprietor wishes to use his right of obtaining restitution, he makes a declaration to that effect at the sous-prefecture within five years of the notification having been made to him, in terms of the preceding Article.

"In this declaration he makes it to be understood whether he wishes to obtain restitution by repaying the money advanced by Government, or by giving up the half of his property if *reboisement* is in question, or the quarter if *gazonnement* has been effected.

"These declarations are registered and a deed is executed.

"Art. 18. If the proprietor decides on repaying the advances made by the State, he produces in support of his declarations necessary proof to establish that he is in a position to repay the expense of expropriation and the cost of the operations, both the execution and maintenance of them, both principal and interest.

"The declaration and attesting proofs are to be addressed, within a month, to our Minister of Finance, who decrees and fixes the formalities and the period within which the proprietor shall have his rights restored.

"Art. 19. If the proprietor offers to resign the half or the quarter of his property, according as the ground has been *reboiséd* or *regazonnéd*, a forest official and the proprietor, or his deputy, proceed to the division of the ground—that is to say, if it has been *reboiséd*, it is divided into two lots of equal value, and if it has been *gazonnéd*, into two lots, one being three quarters, the other one quarter, of the value of the whole.

"In case of dispute about the division of the lots, it is made by a third party, an *expert*, nominated by the President of the Tribunal.

"If one part of the work has been done by the proprietor, he is reimbursed by a proportionate deduction from the portion falling to the State.

"Where the land has been *reboiséd* the division is made by drawing lots when the parties disagree.

"CHAP. II.—*Lands belonging to Communes or Public Institutions included in the Périmètres fixed by decree declarative of public utility.*

"*Section 1st.*—Execution of works on ground belonging to communes or public establishments.

"Art. 20. Within a month, reckoned from the issuing of the decree declarative of the public utility, the communes and public institutions possessing land situated within the périmètres inform the prefects, by a resolution, with reasons annexed, if they intend to execute, at their own expense, the whole or part of the work on the conditions prescribed; or leave the State to do it at its own expense, subject to repayment; or finally, amicably to resign to the State the whole or part of the land included in the périmètre.

"When the commune or public institution fail to make known their intention within the above mentioned period, the State undertakes the work in accordance with Art. 8 of the law of 28th July 1860, and of paragraph 1st of Art. 2 of the law of 8th June 1864.

"Art. 21. Lands which have been *reboiséd*, or are to be *reboiséd*, belonging to communes or to public institutions included in the périmètres fixed by the decrees declarative of the public utility of the measure, are subject absolutely to the forest *régime.*

"Ground that has been or is to be *gazonnéd*, included in the same périmètres, falls under the application of the provisions and arrangements of the 8th section of *Titre* iii. of the forest code, and the 9th section of *Titre* ii. of the Act of 1st August 1807, which relate to the arrangements regarding pasture.

"Art. 22. When the commune or public institution has intimated its intention of carrying on the work, the Municipal Council, or the Administrative Committee, annually allots the funds considered needful, either for the execution of new works or for keeping up those already completed.

"Art. 23. The forest agents superintend the execution of the operations. In cases where the *conservateur* has proved that the work has been left undone, or badly done, a decision of our Minister of Finance shall decree that the State shall take charge of the operations, in terms of Art. 8 of the law of 28th July 1860, and paragraph 1 of Art. 2 of the law of 8th June 1864.

"When the ground belongs to several communes, and the success of the *reboisements*, or of the *gazonnements*, demands combined operations, if all the Municipal Council intimate their consent, a Syndical Commission is appointed to carry on the work in accordance with Art. 70, 71, and 72 of the law of 18th July 1837.

"*Section 2nd.*—Certification of the sums advanced by Government to the communes and public institutions, and the measures necessary to ensure repayment.

"Art. 24. When communes or public institutions intimate that they leave the operations to be performed by Government, the Forest Administration causes them to be done in accordance with the formalities used when the improvement of the Crown lands is in question.

"The statements of expense are prepared in accordance with the rules of liabilities of the Forest Administration.

"It is the same with the annual statements of the cost of maintenance.

"Art. 25. When several communes are interested in the operations, the division of expense is made in the way prescribed by Art. 72 of the law of 18th July 1837.

"Every year the parties interested receive a statement of the outlay on their behoof made by the State. After the completion of the works, the general account of the outlay is closed by the Minister of Finance, copies being delivered to the parties interested. The principal, forming the total of the amount, bears simple interest at five per cent. from the date of the completion of the works.

"Art. 26. The works effected by the State shall be kept up by the Forest Administration.

"The interest of the advances made by Government for this object, the account of which is closed annually by the Minister of Finance, is also five per cent. per annum. A copy of this account shall be delivered to all parties interested, along with a statement of the expense incurred.

"Art 27. Appeals for revision or rectification of the yearly accounts of expenses for the completion and keeping up of the operations shall, under pain of forfeiture, be laid before the Prefectorial Councils within six months from the notification of the said accounts. When this time has elapsed the accounts are confirmed.

"Art. 28. A statement of the produce, and one of the expenses incurred, shall be made and closed every year by the Minister of Finance, copies of which are sent to the parties interested.

"Within six months from this notification, parties interested may, as in the case of the expenses of the works, avail themselves of the privilege mentioned in the preceding Article.

"The value of the produce is deducted from the interest due to Government, and, in subordination to this, from the expense of the completion or keeping up of the works.

"Art. 29. When the advances made by Government are entirely repaid, either by the produce or by payments made by the parties interested, the latter are immediately put in possession of the ground managed for them by Government, under the restrictions resulting from their being subjected to the forest *régime* as regards the portion *reboiséd*, and with regard to the portion *regazonnéd*, subject to the regulation repeated in Art. 21 of the present law.

"If the communes and public institutions wish to repay the sum total of the Government loan, they must prove that they can do so, and execute the necessary commissions.

"*Section 3rd.*—Rules to be followed when communes and public institutions give up the enjoyment or proprietorship of grounds, as authorised by Art. 9 of the law of 28th July 1860, and Art. 3 of the law of 8thJune 1864.

"Art. 30. Should a commune or public institution wish to release itself from all Government claims by giving up either the proprietorship of the half of the ground *reboiséd*, or the use of not more than the half, or the proprietorship of a quarter at most of lands which have been *regazonnéd*, the Municipal Council, or the Administrative Commission, shall adopt a resolution relating thereto, with reasons annexed, which resolution shall be notified to the prefect.

"Art. 31. As regards land which has been *reboiséd*, when this is to be divided into two portions of equal value, this is done by an *expert* nominated by the prefect, and a forest agent nominated by the Forest Administration.

"The portions are assigned by lot, when the parties do not agree. This is done before the sous-prefect of the Arrondissement.

"If a part of the work has been executed by the commune or public institution, this is made up to it in the division by a proportionate deduction from the portion which falls to the Government.

"Art. 32. With regard to ground which has been *gazonnéd*, the division is made by an *expert* nominated by the prefect, and an agent appointed by the Forest Administration, according to the valuation of works of public utility effected by the State, and also the settlement of which portions of ground are to be given up to it altogether, or only for a time.

"When there is any dispute, this shall be done by an *expert* chosen by the President of the Tribunal.

"Art. 33. An account is kept by the Forest Administration of the produce of ground the use of which has been given up to the State. The enactments of section 2nd of chapter ii., *Titre* iii., of the present law are applicable to this account.

"*Section 4th.*—Method of determining and allocating indemnities, which may have been granted to communes, when there has been a temporary privation of the right of pasturage on communal land which has been subjected to *gazonnement* or *reboisement*.

"Art. 34. In cases where the right of pasturage on communal lands which have been subjected to *reboisement* or *gazonnement* has been withdrawn for a time, indemnities are granted in proportion to the resources, to the sacrifices made by the communes, to the wants of the needy inhabitants, and to the sums granted by the General Councils for *reboisements* and *gazonnements*.

"Regard is also had to any agreement made by any commune to suppress the keeping of goats, either wholly or in part.

"Art. 35. These indemnities are fixed by decrees declaring the public utility of the measure. They date from the day when the right of pasturage ceased, and they are paid into the communal treasury at the end of every year. These appear among the *extraordinary receipts*, under the name of accidental receipts, and the use to which they are to be put, regulated by the Municipal Council, in the form of sanctioned expenses *(dépences facultatifs)*.

"CHAP. III.—*General Enactments.*

"Art. 36. Before beginning operations within the limits of the périmètres fixed by Imperial decree, the limits of the périmètres, and if need be the boundaries of the said périmètres, must be determined at the expense of the State.

"Art. 37. Our decree of 27th April 1861, containing enactments of the Public Administration for the execution of the law of 28th July 1860, on mountain *reboisement*, is renewed.

"Art. 38. Our Ministers—the Secretaries of State in the department of Finance, in the department of the Interior, of Agriculture, of Commerce, of Public Works—are intrusted, each in his own sphere, with the execution of the present law.

"Given at Compiègne, 10th November 1864."

In 1865 there came into operation the supplemental law in regard to *gazonnement*, but circumstances which will afterwards be stated prevented the issue of the official report of operations at the usual time, and the reports for 1865 and 1866 were issued conjointly.

From these it appears that in these years nothing was done in *gazonnements* in connection with sanctioned works, or *reboisements facultatifs ;* but in connection with *reboisement* and *gazonnement obligatoires*, 37 new périmètres, embracing a total area of 25,916 hectares, had been considered, and 31 périmètres, in regard to which the prescribed formalities had been fulfilled, had been decreed of public utility. Works had been carried on during the same period over an area of 6491 hectares 3 arcs, embracing 120 périmètres, of which 41 were new ones. Of these 6491 hectares, almost all had been situated at great elevations, and 1613 hectares 88 arcs had been brought under *gazonnement*.

A tabulated statement of all the works executed in 1865 and 1866 is given. And the report goes on to say :—

" These works were executed at an expense of 924,122·64 francs, of which 55,978·34 francs were subventions, and 16,806·87 francs indemnities granted for temporary deprivation of pasturage, and the balance— 851,407·43 francs—might be considered money advanced to communes and public bodies which had given up to the Administration the execution of the works, subject to reimbursement, according to one or other of the modes of reimbursement specified by Arts. 8 and 9 of the law of 28th July 1860, and Art. 2 of the law of 8th June 1864.

"These indemnities for temporary deprivation of pasturage, provided for by Art. 6 of the law on *gazonnement*, were—in 1865, 4134·50 francs ; and 12,672·75 francs in 1866."

A tabulated statement, giving details of the expenses met by the State in 1865 and 1866 in these works of *reboisement* and *gazonnement* follows, and the report, summarising these, goes on to say :—

" The *reboisements obligatoires* in 1865 and 1866 extended over a total area of 5,919 hectares, of which 1,276·9 hectares have been *reboiséd* by proprietors by aid of subventions, and 4,624·91 hectares by the State in their stead ; *gazonnement* has been applied during the same time to 2,195·9 hectares; and 91,645 *barrages* have been constructed within the périmètres, and, in combination with these, numerous lateral hurdles ; the expense of the two operations amounts to the sum total of 924,192·64 francs, which has been created thus :—

"Subventions, in kind and in money, allotted for works
of *reboisement* and *gazonnement* decreed to be of
public utility, - - - - *Francs*, 55,978·34
" Advances made by the State for *reboisement* and
gazonnement, - - - - „ 485,219·56
" Advances by the State for works of maintenance, „ 169,641·27
" Advances by the State for *barrages* and various
works, - - - - - „ 196,546·60
" Indemnities allowed to communes for temporary deprivation of pasturage, - - - „ 16,806·87

Total, *Francs*, 924,192·64

" Besides these works decreed to be of public utility, there were other works sanctioned by the Administration.

"The sanctioned works, *travaux facultatifs*, of 1865 embraced a total extent of 7,734·15 hectares, and necessitated allocations on the part of the State to communes and private proprietors of subventions, in kind or in money, amounting to 374,772·32 francs, with an expenditure on the works of 169,776·26 francs for the *reboisement* of lands belonging to the State—in all, 544,548·58 francs.

"The subventions, in money and in kind, granted by the Administration for works of maintenance have amounted to a sum of 99,904·10 francs; the expense of maintenance of State *reboisements* has been 51,531·75 francs.

"Proprietors of different classes have further constructed, by means of the above-mentioned subventions, 5,804 new *barrages*, and have repaired 818.

"The whole expense to the State for sanctioned works of *reboisement*, in 1865, has amounted to 695,984·43 francs; and the expenses of every kind, incurred by communes and private proprietors, for the works for which subventions were granted, including funds granted by the General Councils of the districts, may be estimated at 650,000 francs.

"The extent of land belonging to communes, private proprietors, and to the State, over which *reboisement* was effected in 1866 was 5,697·80 hectares. The *reboisement* of lands belonging to the State cost 119,615·77 francs; the subventions, in money and in kind, granted by the State for *reboisements* by communes and private proprietors, amounted to 273,484·67 francs; the maintenance of State *reboisements* entailed an expenditure of 47,951·85 francs; the State contributed by subventions, in money and in kind, for the maintenance of *reboisement* by communes and by private proprietors, 112,573·75 francs; the total expense to the State for sanctioned *reboisement* in the course of the year amounted to 553,826·04 francs.

"There were constructed in the course of the year in the sanctioned *reboisements* effected with the subventions mentioned, 2459 new *barrages*, and besides these, 561 old ones were repaired. The quarter of the expense of the whole work born by private proprietors, by communes, and by departments, was estimated at 400,000 francs."

It has been stated, that circumstances prevented the issue of the report for 1865 at the usual time. In reference to this it is stated by the Director-General of the Forest Administration:—

"The report, relative to the operations carried on in 1865, could not be produced at the usual time in consequence of serious disturbances which the unexpected inundations in the autumn of 1866 occasioned in the greater part of the mountainous countries the regeneration of which has been entrusted to the Forest Administration. It might be expected that the restocking of forests, executed within a year before, could scarcely fail to have suffered much from such an outbreak of waters, and I thought it desirable, before making known the result of the works, to be fully and correctly informed on the extent of the evil done to them.

"Happily," he goes on to say, "the delay has, and has only, established the fact that any desolations which have occurred are trifling in importance compared with the calamities which have befallen the valleys and the plains.

"The deluges of rain which fell on the 23rd and 24th of September in the high lying regions of Auvergne, and of Vivarais, and on some spots in Savoie, transformed the most of the thread-like streamlets almost instantaneously into furious torrents, and raised in less than twenty-four hours, first the Lot, then the Aveyron, the Tarn, the river Arc, the Allier, and the Loire to a height which the greater part of these water-courses had never reached, even at the time of the floods of 1856.

"It may easily be conceived, that in such circumstances the works of *reboisement*, undertaken within a few years before on the brows and slopes of the mountains, could scarcely have any effect on the enormous masses of water, the impetuosity of which only the oldest woods could be of use in moderating. But if they have opposed no obstacle to the inundations, they have sustained perfectly the shock, and it may be affirmed that they have throughout exercised a happy influence.

"Thus in the Lozère, where the bridges carried away or damaged are reckoned by hundreds, where the valleys have been half-filled with sand and rocks, the *reboisements* and *gazonnements* executed on about 1700 hectares have perfectly maintained the soil on which they are situated, and protected the lower-lying grounds.

"In the périmètre of Chadenet, situated above the valley of the Crouzet, of 566 *barrages* which have been constructed, 2 only have been carried away, and the volume of earth and stones retained by the 564 *barrages* remaining standing is estimated at no less than 2000 cubic mètres, while, on the other hand, all the slopes rendered mobile by cultivation or by excessive depasturage have been cut up into ravines, and yielded up to the water-courses déjections which have increased considerably the disasters experienced.

"These results have been established by the prefect of Lozère in a discourse addressed to the Agricultural Society of his department.

"In the Cantal corresponding effects have been produced. On the slopes, stripped of woods, there are traces of torrential ravines to be met with at every step; at the base of these the meadows are covered with gravel and detached rocks; and the rocks and highways are cut up. But wherever the temporary prohibition of passage and pasturage has permitted vegetation to develope itself, and on spots on which *reboisement* has been carried out by the State, by communes, and by private proprietors—*reboisements* which cover a thousand hectares, there is no formation of ravines; and the lands and the lower-lying roads are untouched.

"The works executed in the Haute-Loire are much more important than those carried out in the Cantal. Besides the sanctioned *reboisement*, *reboisement facultatifs*, the périmètres, the *reboisement* of which was decreed of public utility, embraced in 1866 an area of nearly 5000 hectares, on 1650 hectares of which the *reboisement* has been effected. In the high mountains of Mézenc and of Mégal, where most of these périmètres are situated, none of the portions *reboisèd* or *regazonnéd* have suffered from the violence of the rains, whilst a contiguous mountain, that of Chaulet, which is being constantly traversed and broken up by the feet of sheep, has been ploughed up into deep ravines. Those good results established in the Mézenc and Mégal are due not to the action of the vegetation drawn over the denuded lands alone, but also to the restraining power of the *barrages*. Of 407 of these, constructed on the steep slopes of the Holme, nine only have given way before the impetuosity of the torrent of Ponteils.

"The department of the Ardèche has scarcely been affected by the storms of rain and the inundations, excepting in the north-west portions, and more particularly in the canton of Saint-Etienne-de-Lugdarès. This canton is situated on the plateau, at an elevation of 1200 mètres, surmounted by peaks of from 1400 to 1600 mètres. In this region is the périmètre of enjoined *reboisements, reboisement obligatoire,* of Borée, and the unoccupied domains of Mazan and those of Bonnefoi. Scarcely any traces of sand-hills are to be found in the portions which have been *reboisêd.* But it is not so with the adjacent lands, and more especially with the valley of Saint-Etienne-de-Lugdarès, the basin of the Allier. There, where terminate the slopes which have been *reboisêd,* the rock has been laid bare, deep excavations have been dug by the waters, and the valley has been covered with the material dug out and carried away.

" The department of the Gard, like that of the Ardèche, only suffered in its north-west portion, that is to say, in the Arrondissements of Alais and of the Vigan, in which are situated the principal périmètres of *reboisement obligatoire.* With the exception of two *barrages* carried away by the waters, the works which have been executed have stood wonderfully, and have at the same time protected all the lower-lying lands against erosion. But everywhere else, and notably at some distance from the périmètres of Montdardier, of Concoules, of Genolhac, and of Ponteils, new ravines and considerable accumulations of sand have been produced.

" In the department of the Puy-de-dôme, until 1866, enjoined *reboisement* had been carried out only in one périmètre, that of Clermont, about 400 hectares of which had been restored at the time of the inundations. But sanctioned *reboisements* had been there undertaken upon a great scale, and they extended over many thousands of hectares. Both have sustained perfectly the rude test of the deluges of the month of September 1866. In many places they have to some extent contributed to moderate the ravages of the waters. Thus, amongst the affluents of the Allier, the slopes of which have been happily protected by the recent replantings, may be cited the Couze-de-Chambon. This torrential water-course had always, when great rains fell, caused great havoc and desolation in the commune of Chambon d'Issoire. In 1866 the losses sustained, though still considerable, have been less marked than previously; moreover, the inhabitants convinced, as are likewise the agents of the Forest Administration, that the amelioration of the *régime* of the torrents ought to be attributed to the *reboisements* effected since 1862, on an area of about 200 hectares, have, without loss of time, hastened to offer contributions of day-labour towards the completion of these useful works.

It is befitting to make mention in the same way of the Puy-de-la-Chopine, or of l'Echorchade, the abrupt steeps of which were lately throughout a great extent denuded, and the *déjections,* spreading far, were augmented by every rain. Sowings and plantations of resinous trees, combined with *gazonnement* and a system of planting slips or cuttings, which might root, have completely changed the aspect of these grounds; and the storms of rain of the autumn of 1866, notwithstanding their extreme violence, have not effected any erosion of importance. In fine, in the basin of the Morge, the Arrondissements of Clermont and of Riom, the *reboisements* of Châtel-Guyons and of Royot, which are still of but limited extent, have given a striking illustration of the effects which works of this kind may produce. The old ravines are now stopped up, and the little rivers of the Grosliers and the Tiretaine, which in 1835, from the effects of rains like those which have led to the late inundations, would have ravaged and desolated the valleys, have not, it may be said, occasioned any havoc.

"In the departments of the Loire, the périmètres of which the *reboisement* has been decreed to be of public utility do not, as yet embrace more than an extent of 1700 hectares, of which about one-fourth part has been sown or planted. These works are evidently too restricted to be able to exert a really useful action on the neighbouring water-courses, the floods of which are so sudden and so disastrous. But it has been established that they have reduced, in some measure, the rapidity of the flow by the obstacles created, not only by the plants, but also by the dense herbage and bushes which have grown since the grounds were enclosed, or *mise en défends*.

"The numerous *barrages* erected in the department of Isère have acted well; they have prevented the crumbling down of the hills,—they have slackened the flow of the waters, and arrested on their way the enormous masses of earth and stone which previously would have precipitated themselves to the bottom of the basin. As for *reboisements*, properly so-called, they had not been undertaken to any extent previous to 1863. The works are thus of too recent a date to be able to modify the *régime* of the waters of the district; but the enclosures, *les mises en défends*, which have followed as a matter of course the declaration of the public utility of the *reboisements*, have had for effect, by covering again vast extents of ground with what may be called a spontaneous vegetation, to arrest the progress of *déjections* being carried away by the waters.

"The suppression, or rather the regulation, of the right of way, and the depasturing of these, has produced an almost immediate effect at Valbonnais, and on the eastern flank of the mountain of Connexe. The creation of ravines, previously so frequent, no longer occurs; and the old ravines have ceased to be a continuous menace to the population, or to the imperial road from Grenoble to Gap, which used to be cut up whenever a great flood occurred.

"The works of *reboisement* executed in the Maurienne, department of Savoie, extend only over 500 hectares; but old drains, transformed into dangerous ravines, have been stopped, and numerous *barrages* have been established on spots which were formerly more exposed to erosion. These works have stood well, and everywhere they have prevented the disintegration of the soil.

"The departments of the High Alps, and of the Drôme, did not suffer from the rains which caused such great disasters in Central France. Consequently, all the works undertaken in these regions by the Forest Administration remain uninjured.

"In the departments of the Lower Alps, and Vancluse, there have fallen only the usual rains, and there has been no general inundation.

"The trifling damages caused by the waters have been only local and accidental. But there may be collected, from the consequences of two days of rain which fell, a good many observations which tend to establish the efficacy of the works which have been executed.

"The ravines which furrow the *chantier* of the *reboisement* of Barrême formerly washed on to the imperial road immense quantities of material torn off from the mountain; now, the slopes are covered with numerous *barrages*, and there come to the road only small stones mixed with mud, which are easily stopped by the bordering ditch. In the commune of Saint-André, a mass of schistose granite, completely stripped of wood, and in full process of desintegration, has been almost consolidated through the effects of the *barrages* in combination with plantations. The principal ravine, by which formerly flowed torrents of black mud to the Verdon, which often spread themselves over the cultivated lands, is now cut up into sections by *barrages* of stone, and of facines; the willows, planted in the ground formed by the coming down of earthy material, fixed this mobile soil,—the bottom became level, and the slopes gentle,—and the ravine manifested a tendency to disappear altogether. The influence of *barrages* was equally shown in the *chantier* of Riou-Chanal, established to reduce the torrent of that name. It has been established that the Riou-Chanal, which formerly brought down blocks of from 10 to 15 cubic mètres, has been so subdued by the *barrages* that a foot-bridge, formed of a single plank at the embouchure of the ravine, at the height of a metre, 40 inches from the bottom of it, has not been carried away during many years; formerly, it would have disappeared after the first heavy shower of rain.

"The *gazonnements* being carried out now and for two or three years past in the lower Alps have given very beneficial results. Since the hills bordering the Labouret and the Seyne have been sown with sainfoin, there have no more been seen formed these numerous deep ravines which the waters dig out so easily in the disintegrated schists, of which the moutains in this region are composed. The simple prohibition of pasturing has frequently produced similar results; scarcely has the ground been shut up from the flocks than it covered itself again with a vegetation sufficient to extinguish the torrents. This fact has been established on the chantiers of Saint-André and of Castellane, and on many other spots.

"Not to multiply citations, which may be considered already too numerous, I shall now confine myself to indicating in a few lines the conclusions which naturally flow from the observations collected from all parts of France.

"These conclusions may be summed up thus:—The inundations of 1866 had for their point of departure the most elevated summits of the central plateau, they were too violent and too sudden to allow of the irruption of the waters into the low-lying valleys being retarded by the works of *reboisement* erected on only a few isolated spots.

"But if the works of recent creation, and the *barrages* which in completing the effect produced by these do not yet cover areas sufficiently extensive to cause them to modify perceptibly the *régime* of the great watercoures, they have exercised a very appreciable action on the spots subjected more immediately to their influence.

"They have not only slackened and divided the flow of the waters, but they have, beyond this, retained in their places enormous masses of earth and of rock which these waters would otherwise have swept away with them.

"This is one of the most indisputable and most useful of the effects of these works, for it must not be forgotten that the disasters, occasioned by the inundations, are not only those due to the elevation of the bed of the rivers, and to the flowing forth of their waters upon the plains; the desolations committed by them which are most difficult of cure proceed from deposits of pebbles and of sand, and these are the consequences of ravages committed by the waters in the higher-lying regions.

"When the rivers come down from wooded regions, which are thus protected from being cut up by ravines, their bed is regular and unencumbered with material in transit. If great rains do come, the river may overflow its banks, its waters may cover the plains, destroy some crops, and damage dwellings, but all of these damages are easily repaired, if they repair not themselves when the waters recede within their banks. Rivers like the Loire and the Allier, which come from granite mountains which have been for a long time stripped of woods, do not act so. At every flood they sweep away with themselves enormous masses of sand and of pebbles, which they spread over the cultivated fields, thus rendering them for ever unproductive. The bed of these rivers, constantly filling itself up with this debris torn from the mountains which they traverse, is of no depth; and their *thalweg*, being without any fixity, is displaced at every flood, passing into grounds which speedily disappear, carried away by the current. Now there is no better preservative of rivers against the filling up with sand than the fixation of the soil of the mountains by means of *reboisement*, or of *gazonnement*, or of *barrages*, and works tending to moderate the flow at the origin of the water-courses—that is to say, at the very source of the evil.

" The experiment has been made, and now we can foresee from the present that that day is coming when vegetation, drawn again over the slopes of the mountain, shall have consolidated the surface,—when the torrential water-courses shall have been diverted from these, and shall no more carry their *déjections* to the sea,—when all the old ravines shall have been stopped up, and the valleys and cultivated plains shall have almost nothing to dread from the violence of inundations."

In a subsequent part of the report, attention is called to the difference between the expense and the extent of the work of *reboisement* and *gazonnement* in different regions, which is pretty considerable; and the Director-General of the Forest Admistration goes on to say,—" These differences result generally from the nature of the works executed. Where the rocks which constitute the soil of the mountains present a sufficiently solid base, and where the water-courses do not charge themselves with great masses of disintegrated material, the operations ought to consist principally in the creation of vast extents of woods, or of dense herbage, destined for the retention of the vegetable earth, to fix it permanently, and as a consequence to control the *régime* of the waters. Then the artificial works are only accessory, and it sufficed, for the greater part of the time, to bring back again upon the slopes the vegetation which the abuse of pasturage had caused to disappear. But on other ground, as in the Alps in particular— where the grounds, devoid of consistency, are constantly being undermined by the waters, and in consequence crumble down on all hands—sowings and plantations would be insufficient to remedy the evil, if the consolidation of the soil were not previously secured by preparatory works, such as *barrages*, facinages, sustaining walls, and water-leadings from the torrents.

" These works, the complete efficacy of which has been demonstrated by six years' experience, occasion indeed a pretty considerable augmentation of the expense in the *reboisement* and *gazonnement* of those périmètres in which they are executed; but no outlay would appear to be more justifiable, if we take into account the vast extent of lands which are thus protected against the ravages of the waters, a good way beyond the boundary of the périmètres themselves.

" It would be superfluous work to go over the different proceedings adopted in the construction of *barrages*, and other artificial works, in regard to which the necessary details have already been given. I confine myself to point out the effects which have been produced by these works. A great number of small torrents have been extinguished; and villages, cultivated fields, and highways, which were severely threatened, have now been placed beyond risk of danger.

"Amongst places effectually protected may be named Sainte-Marie, a dependent village of the commune of Vars, in the High Alps, which a previous torrent had repeatedly invaded, and which was in imminent danger of immediate destruction; that of Chorges, in the same department, which is now traversed by a stream which has become inoffensive; the Bourg-d'Oisans, in the Isère, the existence of which was imperilled at every storm of rain by the muds brought down by the torrent of Saint-Antoine, the bringing down of which the *reboisement* may be said to have entirely suppressed; and a part of the town of Mende itself, against which the waters of a torrent, now extinguished throughout the whole of its course, were directing their flow.

"The great torrents require more time, and more especially more money to be spent upon them; nevertheless, the effect of the works is already making itself to be felt on many among them, the *régime* of which has been perceptibly improved; and there are some, even of them, which may be looked upon as extinguished. Amongst others may be mentioned that of Sainte-Marthe, in the High Alps, which was the terror of the valley of Embrun, and which is now so inoffensive that the inhabitants have in contemplation to bring again under culture all the lands previously abandoned on account of the ravages committed by the waters.

"In regard to the works of *reboisement* and *gazonnement*, strictly so-called, the results established, obtained on the occurrence of the inundations of 1866, admit no longer of any doubt being entertained in regard to the influence which they exercise on the *régime* of the waters.

"The test which these works have just sustained warrant the conclusion that the period of studies and experiments always needed at the commencement of such complicated operations, may be considered as having now come to an end. The forest agents are now satisfied in regard to the best method of procedure, and in regard to the most appropriate kinds of trees to employ.

"I shall not attempt to describe the methods which vary, it may be said, *ad infinitum*, with the regions, the lands, the exposure, the altitude, &c.; but it may be useful to make known the kinds of trees which have given satisfaction in the different mountain countries of France, and the degree of success which has followed sowings and has followed planting."

There is given a report relative to the different kinds of trees employed, and a tabulated statement of the degree of success which has followed all the operations of planting, of sowing, and of *gazonnement*, excepting on *reboisement facultatifs* of less extent than 50 hectares; from the commencement of these till 1866 inclusive, the success ranges from 1 per cent. to 100 per cent. The most frequently recurring figures are 60, 70, 75, 80, 85, and 98, and 100 per cent. or complete success is frequently reported.

A report of the seeds and plants employed in the works of *reboisement* and *gazonnement* follows, stating kinds, quantities, and prices, and the expense of maintaining *sécheries*, or places for drying seeds, where these had been erected.

This is followed by reports of grounds obtained by expropriation and otherwise, of subventions voted by the General Councils of thirty-six departments, of the Administration and surveillance, and of forest roads constructed or improved under the laws of 28th July 1860, and of 13th May 1863, from 1861 to 1866 inclusive.

In this there are given tabulated statements of 619,708 mètres of new forest roads executed, and of 817,547 mètres improved, the former at an expense of 2,209,753 francs, the latter 1,943,966; it shows a similar tabulated statement of 105,781 mètres of new roads, and 108,760 mètres of improvements, commenced in 1866 to be completed in 1867, upon which in 1866 there had been expended respectively 512,809 francs, and 186,932 francs, and a tabulated report of diverse works executed on other than forest roads, but also required for the getting out of the product of fellings in the forests, upon which had been expended 181,432 francs.

In August 1866 there was issued by the Director-General of the Administration of Forests, a circular, containing instructions and directions in regard to all matters pertaining to the work, arranged under the heads of (1) General disposal of business projects, specifications, estimates, works, and expenses; (2) works of restoration of forests after felling, fixation of dunes, *reboisement* and *gazonnement* of mountains and *sécheries*, for the preparation of seeds; (3) roads, bridges, and sustaining walls; (4) house-wells, and cisterns; (5) saw-mills; (6) ditches, enclosing-walls, and fences; (7) works executed for special payments; (8) works executed by forest warders; (9) works executed by brigadiers, and watchment of deposits; (10) works executed by concessionaries enjoying a temporary concession of advantages; (11) works executed by parties holding concessions of lands; (12) works executed by holders of concessions of lesser products; (13) works executed by insolvent delinquents; (14) works executed in repair of disintegration consequent on the felling or bringing out of timber, &c.; and forms of returns, accounts, and statements to be rendered to the Administration.

Though many of the instructions might prove suggestive to those who may encounter practical difficulties in carrying out similar operations, a translation of the whole does not appear to me to be necessary in a preliminary report on the subject.

The report of works executed in 1867 and 1868 is similar in character; but enough has been cited to supply data required for the formation of some idea of the nature, the extent, and the expense of the operation carried on.

This report, however, enters more fully into details of what has been done in different localities, and the beneficial effects which have followed. But these can be reported more satisfactorily in a separate chapter.

"The total amount of the expenditure, in relation to the laws of 28th July 1860, and of 5th June 1864, to the end of 1868," says M. Faro, Councillor of State and Director-General of the Administration of Forests, in this report, issued under the date of 30th May 1870, "is 10,187,240 francs 34 centimes. According to the provision of these two laws the expense might have risen to 10,500,000 francs in the time. There is then a balance remaining of 312,759 francs 66 centimes, which should be furnished by the produce of extraordinary fellings made, or to be made, within the limits specified by these special laws.

"To put before your excellency (the Minister of Finance) a complete *résumé* of the operations, from the commencement of the epoch when the law of 28th July 1860 came into operation, I have recapitulated in the following table all the areas *reboiséd* or *gazonnéd* to the end of 1868 :—

YEARS.	REBOISEMENTS FACULTATIFS, or Officially Sanctioned *Reboisements*.				REBOISEMENTS ET GAZONNEMENTS OBLIGATOIRES, or Enjoined Works.			Total Area of lands rewooded and covered with herbage.
	State Lands. Area.	Land belonging to communes and public bodies. Area.	Lands belonging to private proprietors. Area.	Total Area.	Reboisements. Area.	Gazonnements. Area.	Total Area.	
	h. a.	h. a.	h. a.	h. a.	h. a.	h. a.	h. a.	h. a.
1861	1,401 95	2,653 70	583 92	4,639 57	—	—	—	4,639 57
1862	1,866 03	5,774 58	1,714 15	9,354 76	—	—	—	11,416 63
1863	1,750 88	7,073 24	2,157 05	10,981 17	2,061 87	—	2,061 87	12,834 74
1864	1,834 70	6,164 32	1,601 01	9,600 03	1,853 57	—	1,853 57	12,192 92
1865	1,170 26	5,198 01	1,392 50	7,760 77	2,592 29	—	2,592 29	11,919 32
1866	986 09	2,909 42	1,739 75	5,635 26	3,107 96	1,050 59	4,158 55	9,590 85
1867	611 25	2,783 68	2,007 32	5,402 25	2,811 10	1,144 49	3,955 59	9,001 67
1868	216 50	2,663 20	2,129 63	5,009 33	3,263 94	335 48	3,599 42	8,108 80
					2,886 97	212 50	3,099 47	
Totals	9,837 66	35,220 75	13,325 33	58,383 14	18,577 70	2,743 06	21,320 76	79,703 90

" This table, in which are made reports of work done year by year, shows that the works have been extended over a total area of 79,703 hectares. The part referred to as ' Undertakings executed as Sanctioned Works ' is 58,383 hectares, that of ' Enjoined Works ' 21,320 hectares.

" Seeing that the work of 1869 is not yet closed, nor that of 1870 well begun, it may be anticipated that at the close of the current year the whole of the *reboisements* effected will embrace nearly 95,000 hectares, of which 25,000 at least have been restocked by *Reboisements et Gazonnements Obligatoires*. Such, approximately, will be the statement in the balance-sheet of the first decade, or 10 years work of the *reboisement* of the mountains, prescribed by the law of 28th July 1860. We must not, however, consider the actual influence of the works as limited to the areas which shall then come to be indicated.

" To compare the results obtained with the expenditure incurred, it is necessary to take into account not only the surveillance, and the regulation of the passage and pasturage of the portions of the périmètres not replanted, but also, and more especially, the numerous *barrages* which have caused their influence and protective action to be felt at very great distances, as well as in the valleys."

PART IV.

PAST, PRESENT, AND PROSPECTIVE ASPECTS OF THE WORK.

HAVING passed in review the evils which were devastating valuable land, by torrents washing away mountain sides and depositing the detritus in the valleys and on the plain; the remedial measures which have been at different times proposed; the legislative measures by which *reboisement* and *gazonnement* have been enforced and regulated; the practical measures which have been adopted; and the change which has thus been effected, both on the face of nature and in local popular opinion; we are prepared, with advantage, to cast a glance over the whole field, that we may see what has been, what is, and what is to be.

In the Introduction, I have stated some facts in regard to the ravages which were committed by torrents some forty years ago. But, to a full realization of the state of the case, it is necessary that something should be known of the previous history of Alpine torrents, and scarcely less necessary, with a view to preventing misapprehension, that something should be known of the vast extent to which the mountains ravaged by torrents, and now subject to *reboisement*, were and are covered with primitive forests; and while this may be necessary to prevent misapprehension, it may at the same time bring into view what evils have resulted from what are at most but partial clearings.

CHAP. I.—PAST HISTORY OF ALPINE TORRENTS.

Washington Irving represents Knickerbocker, in his *History of New York*, as deeming it proper to give the history of New York from the very founding of the city, and, to enable him to do this satisfactorily, to cite and demolish or sustain the various schemes of cosmogony which learned men had proposed to enable them to account for the creation and existence of the world. I do not propose to myself to go so far back as this; but I know that the expositions of the physical geography of France, and of what are generally reckoned geological phenomena, by Cézanne, and by Costa, and by others, are not out of place in their treatises in this department of hydrology, and by referring to these I may enable many who never saw, and are never likely to see, the works of *reboisement* and *gazonnement* of which I write, to form a more correct idea of what is being done, and has been done and accomplished by the works referred to.

The views of Labéche, cited by Surell, have been given in full, in so far as they related to pre-adamic torrents and their effects; and the geological doctrines of MM. Cézanne and Costa de Bastelica, in regard to pre-adamic torrent action, have also been given.

According to the views advanced by Cézanne, extensive districts of France owe their existing surface, composition, structure, and contour, to moraine-like deposits by far-extending glaciers, existing and flowing during what is generally spoken of by geologists as the glacial or as the drift period, a period long anterior to the 6000 years which constitute what we may designate the historical era of the world's existence; and to deposits on a stupendous scale of what were the *lits de déjection* of torrents—compared with which the torrents of the present day are as tiny streamlets—the *régime* of which followed close upon the glacial era, succeeding it apparently immediately, and giving occasion for the designation the *torrential era*, intermediate, there at least, between the eras of glacial and of alluvial deposits.

M. Costa advances similar views in regard to the character of the geological formation upon which M. Cézanne founds his theory, but he considers that the moraine-like deposits, which M. Cézanne attributes to glacial action, may have been, and probably were, like the others, the *lits de déjection* of torrents, and he alleges that, compared with the *régime* of torrents, the *régime* of glaciers is temporary, local, and accidental; while this is universal, extending to all lands, if not also to all worlds, and extending over all time.

We are thus by both carried back to a time in which, if the earth was not without form and void, the mountains then were naked and bare.

Observation shows that, now at least, soil capable of nourishing plants when exposed naked and bare is soon covered with vegetation. A little decaying cheese, or fruit, or damp bread, so exposed is soon covered with mould. In experiments designed to test the hypothesis of spontaneous generation, ingenuity seems to be baffled in the endeavour to devise a crucial experiment which shall either establish or disprove the hypothesis. Under conditions the most unlikely, the simpler organisms make their appearance; to prevent this has hitherto proved impracticable, if it be not impossible; and what is seen thus in the laboratory on a simple scale is seen on a large scale taking place everywhere.

"Whenever a tract of country, once inhabited and cultivated by man," says Marsh, "is abandoned by him and by domestic animals, and surrendered to the undisturbed influences of spontaneous nature, its soil sooner or later clothes itself with herbaceous and arborescent plants, and, at no long interval, with a dense forest growth. Indeed, upon surfaces of certain stability and not absolutely precipitous inclination, the special conditions required for the spontaneous propagation of trees may all be negatively expressed and reduced to these three: exemption from defect or excess of moisture, from perpetual frost, and from the depredations of man and browsing quadrupeds. Where these requisites are secured, the hardest rock is as certain to be overgrown with wood as the most fertile plain, though, for obvious reasons, the process is slower in the former than in the latter case. Lichens and mosses first prepare the way for a more highly organised vegetation. They retain the moisture of rains and dews, and bring it to act, in combination with the gases evolved by their organic processes, in decomposing the surface of the rocks they cover; they arrest and confine the dust which the wind scatters over them, and their final decay adds new material to the soil already half-formed beneath and upon them. A very thin stratum of mould is sufficient for the germination of seeds of the hardy evergreens and birches, the roots of which are often found in immediate contact with the rock, supplying their trees with nourishment from a soil deepened and enriched by the decomposition of their own foliage, or sending out long rootlets into the surrounding earth in search of juices to feed them.

"The eruptive matter of volcanoes, forbidding as is its aspect, does not refuse nutriment to the woods. The refractory lava of Etna, it is true, remains long barren, and that of the great eruption of 1669 is still almost wholly devoid of vegetation. But the cactus is making inroads even here, while the volcanic sand and molten rock thrown out by Vesuvius soon become productive. Before the great eruption of 1631 even the interior of the crater was covered with vegetation. George Sandys, who visited Vesuvius in 1611, after it had reposed for several centuries, found the throat of the volcano at the bottom of the crater 'almost choked with broken rocks and *trees* that are falne therein.' 'Next to this,' he continues 'the matter thrown up is ruddy, light, and soft: more removed, blacke and ponderous: the uttermost brow, that declineth like the seates in a theater, flourishing with trees and excellent pasturage. The midst of the hill is shaded with chestnut trees, and others bearing sundry fruits.'"

He adds in a foot note,—"Even the volcanic dust of Etna remains very long unproductive. Near Nicolosi is a great extent of coarse black sand, thrown out in 1669, which, for almost two centuries, lay entirely bare, and can be made to grow plants only by artificial mixtures and much labour.

"The increase in the price of wines, in consequence of the diminution of the product from the grape disease, however, has brought even these ashes under cultivation. 'I found,' says Waltershausen, referring to the years 1861-62, 'plains of volcanic sand and half-subdued lava streams, which twenty years ago lay utterly waste, now covered with fine vineyards. The ash-field of ten square miles above Nicolosi, created by the eruption of 1669, which was entirely barren in 1835, is now planted with vines almost to the summits of Monte Rosso, at a height of three thousand feet.'"

To the spread of vegetation and the growth of trees is attributed the extinction of the primitive torrents,—to the destruction of forests, which had protected the land for ages, is attributed the reappearance of them in our day,—to the spread of forests over denuded ground is attributed the extinction of some which seem to have originating in later times,—and to aid in this work is the object of the *reboisement* and *gazonnement* which are being carried out. The whole process is thus sketched by Marschand, in his work entitled *Les Torrents des Alpes et le Paturage*:—

"After their elevation, the Alps presented everywhere abrupt crests, separated by deep rents. Physical and chemical agencies disintegrated the rocks everywhere naked, and formed of their accumulated debris the first slopes of crumbled materials. The waters flowing on these extremely steep lower slopes, gnawed them away little by little, and levelled up the bottom of the valleys. At this epoch all the water-courses must have had a character essentially torrential; they carried away immense quantities of materials, which have formed the beds of alluvial deposits, the thickness of which is at times so considerable.

"But soon a powerful vegetation came to cover the upper slopes, and to arrest, or rather to retard, the great work of levelling. When one pictures to himself what must have been then the configuration of these mountains formed of rocks without consistency, he is led to suppose that the power of vegetation must have been then much greater than in our days, for it is doubtful if, in existing circumstances, forests could be produced by spontaneous growth on the very steep lower slopes of unstable soil, where we find them to-day.

"From the time that the mountains were covered with woods, the torrents took a more regular course; to the primitive disorder succeeded a life more calm, more regulated. The destruction continued, but with less rapidity; the very steep lower slopes were reduced by local landslips, erosions led to the formation of torrents, which, after having had their period of ascendancy, by degrees became extinct, and clothed themselves with woods. We are promoting every day this slow and measured destruction of the mountains; we meet in all chains of mountains with these erosions, with bare mountains, and in a word, with torrents, but their number is very limited, and their development inconsiderable. When the upper slopes are covered with forests, as in Styria or in the Afenin Engadine, the wooded curtain in which torrents occur arrests their overflow, and generally prevents them from becoming formidable.

"Unhappily man, improvident and avaricious, has frequently destroyed the forests, that he may thereby get possession of the soil; he has hindered forests from forming themselves; he has substituted for them pasture grounds often but ill maintained. With the ruin of the soil begins that of the people. The more unhappy a people are the more selfish do they become, and the more they destroy; so that, from the time that the evil begins it cannot but go on increasing.

"In restoring to the mountains their ancient forests, we have for our end and design to arrest the disorders which have appeared on the deforested lands,—in a word, to maintain on all the lower slopes their fixity, and all this is in augmentation of the public wealth."

The views expressed by M. Marschand are in accordance with views expressed by other students of the subject, and I know of no writer of the present day on the subject who has advanced views conflicting with them.

Cézanne, writing in regard to facts underlying such views, and which are the facts upon which they are based, says,—"These facts which had some novelty in 1840 are to-day, in 1870, above and beyond all dispute. They have been verified throughout a great extent of the Alps.

"M. Gras, *Ingénieur en chêf des mines*, has confirmed them satisfactorily in an interesting memoir, published in 1848. According to this geologue, the formation of the torrents now extinct must have followed close upon the epoch of the glaciers and of the erratic boulders; the Alps must have found themselves at that time completely denuded by the cold and the protracted continuance of the ice.

"At length," says he, "the productive powers of nature restored vegetation to the bosom of the Alps, and came to cover them with thick forests. This *boisement* greatly modified the *régime* of the water-courses, which lost their torrential character, and the deposit of material on the beds of *déjection* was extinguished.

"When man, in process of time, began to inhabit the Alps, he destroyed a part of the forests and extended cultivation over the flanks of the mountains. The clearings have re-awakened to some extent the destructive action of the torrents, and given a new life to their deposits; these have re-appeared in a great many places, without becoming, however, so numerous and so extended as aforetime."

And on *reboisement* and *gazonnement*—the means by which the greater and more destructive torrents of pre-adamic times were extinguished—being employed artificially, at great expense, but on a corresponding magnitude, to bridle and subdue, and if possible utilize their successors of the present day, and to cause them to minister to the promotion of industrial operations which they have disturbed and destroyed, they have proved efficient.

Chap. II.—Existing Forests.

While the torrents which have committed such ravages and devastations in France are attributed, and justly so, to the clearing away of forests, it must not be supposed that the forests have been utterly and everywhere absolutely destroyed. According to a valuable paper addressed to the Academy of Sciences in 1865, by M. Becquerel, it appears from official statements that France has an area of 52,768,610 hectares, or 131,921,525 acres, of which 8,804,554 hectares, or 22,011,376 acres—or about a sixth part of the whole surface of France—are covered with forests, and of this the new plantations constitute but a fractional part.

M. Marschand, in a work I have cited, gives the following picture of the extent to which natural forests are still extant on the Alps :—" The Alpine mountains may be considered as divided into three great zones: at the summit, around the rocks and glaciers, are the pastures; lower down are the forests; the bottoms of the valleys, where the villages are usually situated, is cultivated. This division is necessary, and wherever it has been disturbed the greatest misfortunes have followed these infractions of the laws of nature.

"The zone of pastures, or *alpages*, consists usually either of valleys or of high acclivities; its existence is due to this, that at such heights, where forests and cultivation do not flourish, herbage grows spontaneously—thanks to the fertility of the soil, enriched and improved by the great quantity of snow which covers it during the winter. This zone exists everywhere, and everywhere can be modified. In many countries the abuse of the pasturage by overstocking has impoverished, and sometimes ruined, the higher pastures; but this is a local, and generally a temporary, evil, which wise regulations and skilful labour can remedy.

"Below the *alpages* are situated the forests; this is their natural position, here they grow to the greatest perfection—in a word, nature has placed them here to protect the valley by arresting the torrents which flow over the *alpages*, and the avalanches which slide down the higher slopes. Their salutary influence extends to the climate, and to the production and regulation of a slow and measured flow of water in the spring. In a word, without these forests the Alps would only present an immense ruin, threatening all the districts traversed by the floods which flow down its sides.

"I should add, that there is often too much dogmatism displayed in fixing the altitudes of the different zones, especially the forest zone. I have often heard the erroneous argument,—' I have seen at such a place a grove of larches, or of some other tree, situated at a great height—2800 mètres for example. I therefore conclude that we may plant the kind of trees I have seen at the same altitude.' In opposition to this opinion I would remark, that the upper limit of the forests rises with the bottom of the valleys; that at the entrance to some of these—that of Urbage, for example—it does not exceed 1400 or 1500 mètres (the forest of St Vincent), whilst higher up, round the lake of Parouard, it reaches about 2500 mètres.

"The limit is not always the same, but it varies in the same valley following its direction—in a word, the upper limit of forest vegetation is regulated by the local climate, and will change with it. In general, it may be admitted that the summits of the Alps are bare, exposed as they are to all the winds; that the upper limit of forest vegetation rises wherever the forest is sheltered from the north, and where it can receive the influence of the south wind.

"I now come to the cultivated zone. In the French Alps this has suffered the most; to the primitive gentle slopes have succeeded, at many points, more or less terrible erosions, which increase every year; it is in this zone that the most awful havoc is caused by torrents descending from above. It is there that villages and fields are sometimes carried away, and sometimes buried under the mud, to which by analogy the name of lava is given. In a word, it is the cultivated zone which suffers the havoc caused by the blindness and apathy of the dwellers in the higher districts. It is here that the torrents may be obstructed by artificial obstacles, doubtless insufficient to arrest their ravages permanently, but which may permit us to await without great danger the restoration of the upper districts."

From this it will be seen that forests still exist, and that to such an extent as to be the characterestic feature of a broad and widely-extended zone in the Alps, where there are forests of an extent of which few untravelled students of arboriculture can form any conception. In the Vosges the extent and the conservative influence of the forests is such, that we have seen M. Surell boldly declaring that a writer on torrents, familiar only with torrents as seen there, evidently did not know what torrents were.

Chap. III.—Laws Regulating the Reboisement Effected and Measures Adopted.

The work of *reboisement* which has been and is being carried on in mountainous regions of France must not be confounded with the work of *sylviculture* in the Landes, and in the district of the Gironde. The object aimed at, and the system of operations adopted in each of these enterprises, is different from those of the other. The latter, advocated by Brémontier, was begun in 1787; it was interrupted in 1789, resumed in 1791; abandoned in 1793, and begun again in 1801, from which time it has been prosecuted without interruption, and with most satisfactory results. The object aimed at was to arrest and utilize the dunes, or drift sands of Gascogny and adjacent lands. Though still commanding attention, it may be spoken of as the work of the first half of the present century; the work of *reboisement* has been the work of the latter half of the century. Originating from the publication of the *Étude sur les Torrents des Hautes Alpes*, by M. Surell, it has for its object to arrest the destructive effects of these.

As stated by M. Magne, in his report to the Emperor, of which a translation was previously given (ante pp. 147–152) between 1843 and the date of his report, February 1860, "sixty-three general councils have urged the necessity of measures being taken for the reforesting of the mountains. A report and a *projet de loi* were prepared by the Director-General of Forests in 1845. This *projet de loi*, remitted for examination to a Commission composed of forest administrators and distinguished *savants*, was amended in many parts and submitted to the Chamber of Deputies in the session 1847." But, he adds, nothing came of this law.

This law, M. Cézanne alleges, became abortive, through its being too radical in its enactments: it subjected to the forest *régime* all lands on which were to be effected the reproduction of forests or of pasture lands. With the law of 1860 it was different. The law of 1860, it is stated by him, limited the action of the Government to the *reboisement*, strictly so-called, and to this in périmètres, or specified areas, embracing only portions of the country at large and of the localities, and it provided for the proceedings being carried on principally by subventions to the proprietors themselves, in the form of money grants, or of the grant and delivery of seeds and of plants. There was required the sanction or approval of the Council of the Arrondissement and of the General Council of the district, and, in fine, that of a mixed Commission, composed of the prefect, of members of the General Council, and of the Council of the Arrondissement,—of an *Ingénieur des Ponts et Chaussées*,—of an official of the forest service,—and of two landed proprietors: and only after these had been obtained could be obtained a decree to determine the extent of the périmètre, or specified area given up to the Forest Administration for *reboisement*. Art. 4 provided, indeed, that recourse might be had to expropriation, but only in cases in which this is required in the interests of the community in consequence of the condition of the soil, and the dangers resulting from this to the lower-lying grounds. And Art. 5 increased the number formalities of which had to be observed before anything could be done by the State in carrying out the works; it required, besides a public enquiry, the judgment of the Municipal Council.

The law prescribed the course to be followed in regard to proprietors and to communes, property belonging to whom might be included in périmètres, and who either could not or would not execute the works themselves; the law determined, also, in what cases it was competent to the State to make pecuniary advances in aid of the works, and to what extent the State should in such cases participate in the benefits resulting from the operation.

The Minister of State estimated at that time the total area of lands susceptible of *reboisement* at 1,133,000 hectares; and the Commissioners calculated, from data supplied by work done, that it would cost upon an average 180 francs per hectare to do the work; and they estimated that, making allowance for what portion of the expense might be met by proprietors and communes, 80,000 hectares might be replanted by means of the credit of ten millions, decreed by the Government for the prosecution of the work.

This credit was spread over a period of ten years, and was designed to be, to the extent of five millions, covered by the sale of forest lands; while the remaining five millions were to be met by extra fellings, and by the ordinary resources of the Government.

At this rate it was reckoned that it would take 140 years to complete the *reboisement* of the mountains; and it was considered that this was not an unreasonable time to be required to undo the work of twenty centuries; but it was arranged that the first effects should be directed towards the points which were most threatened; and this has been done.

The works by which it has been accomplished have been mainly those recommended by Surell :—

First, the formation of *zones de défense*, or *zones de défends*. Zones along the main channel of a torrent, and ramifications of this in the basin drained by it—the former enclosed, the latter simply protected by prohibition of trespass by pasturing sheep or cattle thereon.

Second, *boisement*—the planting of some of these zones more or less extensively with trees and shrubs.

Third, *gazonnement*—or the creation of what, in contradistinction to dead, crumbling, denuded slopes, were called *berges vives*, by promoting the growth of a dense covering of herbage.

Fourth, the construction of *barrages*, or wears, generally of facines, &c., but in some cases of stone, to arrest the current and so to prevent erosion, and to arrest detritus in its progress towards the valley.

At first, and for a time, it was intended that the zones should be enclosed; but, with the modification of the law of 1860, introduced by that of 1864, which substituted to a great extent *gazonnement* for *boisement*, it was considered enough, except in some special cases, to prohibit the admittance of flocks within the area of operation. And it was found, that the zone left to itself, or sown broadcast with appropriate seeds, by degrees became covered with a natural turf of herbage and bush.

The *berges vives* generally take the form of rounded, elongated banks, which become in like manner clothed with verdure. The object aimed at is to give to a transverse section of the valley a stable waving outline, by the bringing down of unstable elevations, and filling up with the debris intervening depressions, converting the acute, projecting, and retiring angles which they generally form into connected curves. This operation was sometimes effected on an extensive scale by blasting, but in general the pickaxe could do all that was required.

Simultaneously with this operation has been carried on the erection of *barrages* where necessary.

The *barrages*, or barriers or wears, were designed to arrest and retain gravel which might fall or be washed down slopes, and so to prevent its reaching the *cone de déjection* to add to accumulations there. The structure of them varied with the material at command, and the requirements of the situation. There are *barrages* of large blocks of stone in solid masonry, held together by iron clamps; there are dry stone dykes; there are some *barrages* formed of stakes, and of wickerwork or hurdles; and in acute angled beds of currents, there are laid in the bottom beds of fagots and stone, over which are spread the debris of demolished surmounting hills; and sometimes a simple gabion is placed in the *thalweg*, with its mouth directed up stream, and left like a bow-net in that position to be filled with earth by the flood occasioned by the first storm of rain which may come; and sometimes for this there is substituted a tree or a bush half buried, with its branches and roots in the bed of a stream. All that is sought to be done is, with the readiest materials to adapt the barrier to the requirements of the locality, giving it strength proportionate to the strain; and in some cases a turf or tuft of grass may suffice.

The importance of giving fixity to the hills, in combination with the arrest of debris by barriers adapted to the locality, can, perhaps, only be realised fully in view of the landslips, of a greater or less extent, constantly occurring. In view of these, the arrest of the debris by the *barrage* was primarily designed; and it was sought to effect this by so regulating the flow of the water as to diminish its power of erosion, and of undermining the confining banks of the torrent. With this effected, it was considered that the extinction of the torrent would follow in the course of time, but that by *reboisement* the work might be so expedited as to accomplish within a few years what otherwise it might have required decades to effect,—and within a decade what might have required a century. And with this in view, the production of timber, and the securing of large and early pecuniary returns from the sale of forest produce, were considered of but secondary importance compared with the prevention of the crumbling away of the mountains. In the prevention of this, and of direful consequences which might follow—the destruction of houses and lands by the ravages of the torrents, the devastation of the valleys, with their houses and fruitful fields, by covering them with the debris, and the inundations of the plains below, destroying life and property to an extent to affect materially the national resources—was compensation looked for. And in accordance with this view of the matter have the plantations been managed throughout the period embraced by this report. They have been and are being managed in accordance with the most advanced forest science, and with a businesslike view to the reduction of expenditure and the increase of returns; but all this has been and is being done in subordination to the great objects aimed at—the regulation of the flood, the extinction of the torrent, the conservation of the mountains, and the preservation of the plains; and the proceeds obtained are looked upon as a set off against the expenditure incurred rather than as a reimbursing revenue.

In the prevention of the evils referred to there was expected, and in the prevention of these evils it may be said there has been obtained, a return satisfactorily compensating the outlay of thought and labour and money demanded by the enterprise.

CHAP. IV.—DEVASTATIONS OCCASIONED BY TORRENTS WHICH IT WAS SOUGHT TO ARREST AND PREVENT, AND MEASURES EMPLOYED.

It was the devastations in the valleys and plains, which were occasioned by torrents, which first caused attention to be given to the subject of *reboisement*; and for a time it was the allegation of the mountaineers that the operations were begun and carried on solely in the interests of the dwellers in the plains.

Details already given show, as was subsequently seen by the mountain population, that they had a beneficiary interest in these operations, as real as that of those for whom they had thought that their interests were being sacrificed. A few more of these may be given.

M. Ladoucette, in his *Histoire, &c., of the High Alps*, writes,—"Saint-Eusèbe, a village of 603 inhabitants, built on an argillaceous layer at an elevation of more than 200 mètres, or 650 feet, above the Drac, and at a distance of a kilomètre, about two-thirds of a mile, from the river, presents the phenomenon, at once curious and alarming, of the sinking down of this layer. It is seen distinctly from the royal road along the Drac; while for some years it was entirely concealed by the elevation of the ground before it. The crevasses, which exist along the whole line, give the unwelcome intimation and assurance that the successive overturns of its terrible neighbour will end in withdrawing its foundation, and bringing about its total ruin. The prospect is one which it is frightful to contemplate."

M. Cézanne, in his continuation of the work of Surell, writes,—" There may sometimes be seen, at a certain distance from the hills bordering a torrent, a series of parallel fissures, between which the land appears to be dislocated by unequal landslips. It is not seldom that these landslips extend over a vast surface, and in such a way that the cultivated fields and even detached houses of a higher lying plateau have evidently sunk to a lower level.

" When, by such fissures, a mountain is thus cut up into prisms unstably balanced, it is enough that some day, on the occurrence of a storm of rain, the gushing waters find their way into a cleft and lubricate the surfaces of these, to occasion a landslip. If the mass of material which crumbles down be considerable, it temporarily bars up the ravine, and soaking itself by degrees with the water it becomes soft ; it then gives way all at once, and precipitates itself in a rush of viscous lava, the impulsive force of which is more formidable than that of pure water, for this lava, not having the fluidity of water, cannot, like it, flow round a resisting object without dragging it along with it. Such appears to be the origin of these *débâcles* of mud, observed by Saussure, and described in his *Voyage dans les Alpes*, and which he in like manner attributed to the rupture of a *barrage*. And it is only by the supposition of such temporary *barrages* giving passage to a *débâcle*, that certain phenomena which accompany torrent floods can be explained and accounted for.

" A proprietor on the bank of the torrent of Sainte-Marthe states that many a time, while he heard with anxiety, in the silence of the night, the grand roar of the torrent which was eating away his domain, he has remarked distinctly a time of arrest of this—a sudden quieting—which he attributed at once to a sudden cessation of the flood ; but after some minutes the uproar recommenced with greater force than ever ; and anew the tumult of blocks of stone striking against each other gave response to the bellowing roar of the waters."

M. Scipion Gras relates a case which may be considered characteristic, —" On the 4th June 1827 the village of Goncelin, not far from Grenoble, was suddenly threatened by a torrent flood, the inhabitants in alarm ran up on to the embankment; but the waters subsided, the flood seemed to have passed, and, reassured, they retired from the embankment, when, all at once, they saw issue from the gorge a mountain of water which precipitated itself upon them with fury. Forty-two houses were engulfed or overthrown, twenty-eight people were surprised and drowned ; and half of the village, buried under a layer of mud, of stones, and of rocks, had to be rebuilt on this mass of ruins. How can such phenomena be explained if not by the formation and giving way of a temporary natural *barrage.*

" The Secheron, a torrent of the Tarentaise, flows between two schistose hills, otherwise firm, but which, since 1853, at which time they were stripped of wood, have been subject to movements which have been rather disquieting. In April 1869 this torrent, stopped by a quantity of earth which had crumbled down, threatened to overwhelm two villages ; the tocsin sounded throughout the valley, assistance was organised, the engineers and the soldiers hastened to the spot, and it required all their exertions to divert the issue of the waters, and prevent the Isère itself from being stopped in its course."

M. Cézanne cites, in further illustration, the case of the Lake Saint-Laurant, already detailed (ante p. 81), and goes on to say,—" These cases are not extraneous to the matter in hand, but they show how it becomes of importance to fix and consolidate the crumbling hills bordering a torrent; and they make intelligible how, in certain special cases, by the suppression of *débâcles*, the extinction of a torrent, of which time would appear to be a necessary element, may be in fact the immediate and decided result of some artificial operation." And he refers in illustration to the case of the torrent of Vachèrio, which I shall afterwards cite.*

* I have not myself seen much of landslips in France; but I have visited the scene of such in the Notch of the White Mountains in New England, and seen their disastrous effects. These mountains are the loftiest in the United States east of the Rocky Mountains; several of the summits, Washington, Jefferson, Adams, and Madison, tower 5350 feet, 5261 feet, 5383 feet, and 5039 feet, above the level of the Connecticut river; and Munroe and Quincy rise to the approaching height of 4932 and 4470 feet. I travelled from Burlington, it was in 1834, long before railways had been introduced into the locality, and I can still reproduce the feeling of solitude which stole over me in the midst of the mountains and forests, with no one near me but the driver of my light conveyance, and a feeling of shrinking, when hungry and weary, from what seemed to be a realization of the poets wish for some vast wilderness and endless continuity of *solitude*. But having to trust to the accuracy of my recollection, I prefer giving details in the words of another to giving them in my own:

"The first view of the White Mountains, as distinguished from the multitude of peaks and summits which meet the eye in every direction, is obtained a short distance from Littleton; but Mount Washington is not seen till arriving near to Crawford's. The first view of these mountains is magnificent, and as they are approached they become more and more so, until the bare bleak summit of Mount Washington, rising far above the immense piles which surround it, strikes the traveller with awe and astonishment. But the emotions which one receives from the grand and majestic scenery which surrounds him here are utterly beyond the power of description. There is no single object upon which the eye rests, and which the mind may grasp, but the vast and multiplied features of the landscape actually bewilder while they delight.

"These mountains are the loftiest in the United States east of the Rocky Mountains; and their heights above the Connecticut river have been estimated as follows:—Washington, 5350 feet; Jefferson, 5261; Adams, 5383; Madison, 5039; Monroe, 4932; Quincy, 4470. From the summit of Mount Washington, the Atlantic ocean is seen at Portland, 65 miles S.E.; the Katahdin Mountains to the N.E., near the sources of the Penobscot river; the Green Mountains of Vermont on the west; Mount Monadock, 120 miles to the S.W.; and numerous lakes, rivers, &c., within a less circumference. The *Notch* or *Gap* is on the west side of the mountains, and is a deep and narrow defile, in one place only 22 feet wide. A road passes through, which is crossed by the river Saco; into which several tributary streams enter from the mountain heights, forming beautiful cascades. Lafayette Mountain is situated in the northeast part of the township of Franconia, nearly equidistant from Mount Washington at the northeast, and Moose-Hillock at the southwest, being about 20 miles from each; and it is obviously more elevated than any other summit in sight, except the White Mountains.

"At the *Franconia* Notch, near the road leading from Franconia to Plymouth, and about three miles south of Mount Lafayette, a foot-path has been cleared out from the road to the top of the mountain. The point where the path commences is six miles from the Franconia iron works, and the length of it from the road to the summit is three miles; and throughout this distance it is almost uniformly steep. The ascent for the distance of about two miles is through a thick forest of hemlock, spruce, &c. Higher up, the mountain is encompassed with a zone, about half a mile in width, covered with stinted trees, chiefly hemlock and spruce. Above the upper edge of this zone, which is about half a mile from the top, trees and shrubs disappear. The summit is composed chiefly of bare rocks, partly in large masses, and partly broken into small pieces.

"The view from the top is exceedingly picturesque and magnificent. Although it is not so extensive as that from the summit of Mount Washington, yet owing to the more advantageous situation of Lafayette, being more central as it respects this mountainous region, it is not inferior to it either in beauty or grandeur. The view to the northeast, east, south, and southwest, is one grand panorama of mountain scenery, presenting more than fifty summits, which when viewed from this elevation do not appear to differ greatly in height. Some of these mountains are covered with verdure to the top, while the summits of others are composed of naked rocks; and down the sides of many of them may be seen *slides* or *avalanches* of earth, rocks, and trees, more or less extensive, which serve to diversify the scene. The only appearance of cultivation in this whole compass is confined to a few farms seen in a direction west of south, on the road to Plymouth, extending along the Pemigewasset branch of the Merrimack. To the west is seen the territory watered by the Connecticut and the Ammonoosuck.

"At a place in the road through the Franconian Notch where the path up the mountain

On the subject of landslips, there is valuable information supplied by Mr Marsh, in his treatise on *The Earth as Modified by Human Action.* He says,—" Earth, or rather mountain slides, compared to which the catastrophe that buried the Willey family in New Hampshire was but a pinch of dust, have often occurred in the Swiss, Italian, and French Alps. The commences, is exhibited to the view of the traveller, on the mountain opposite to Lafayette, the *Profile* or the *Old Man of the Mountain*, a singular *lusus naturæ*, and a remarkable curiosity. It is situated on the brow of the peak or precipice, which rises almost perpendicularly from the surface of a small lake, directly in front, to the height (as estimated) of from 600 to 1000 feet. The front of this precipice is formed of solid rock, but as viewed from the point where the profile is seen, the whole of it appears to be covered with trees and vegetation, except about space enough for a side view of the Old Man's bust. All the principal features of the human face, as seen in a profile, are formed with suprising exactness. The little lake at the bottom of the precipice is about half-a-mile in length, and is one of the sources of the Pemigewassat river. Half-a-mile to the north of this there is another lake, surrounded with romantic scenery, nearly a mile in length, and more than half-a-mile in breadth. This is one of the sources of the southern branch of the Ammonoosuck, which flows into the Connecticut. These lakes are both situated in the Notch, very near the road, and near to the point where the steep ascent of Mount Lafayette commences. The northern lake is 900 feet above the site of the Franconian iron works, and the highest point in the road through the Notch is 1028 feet above the same level. Other curiosities in this vicinity are the *Basin* and the *Pulpit*.

" A portion of the Gap, including the Notch in the *White Mountains*, which is the most sublime and interesting, is about 5 or 6 miles in length. It is composed of a double barrier of mountains, rising very abruptly from both sides of the wild roaring river Saco, which frequently washes the feet of both barriers. Sometimes there is not room for a single carriage to pass between the stream and the mountains, and the road is cut into the mountain itself. This double barrier rises on each side to the height of nearly half-a-mile in perpendicular altitude, and is capped here and there by proud castelated turrets, standing high above the continued ridges. These are not straight, but are formed into numerous zig-zag turns, which frequently cut off the view and seem to imprison the traveller in the vast gloomy gulf. The sides of the mountains are deeply furrowed and scarred by the tremendous effects of the memorable deluge and avalanches of 1826. No tradition existed of any slide in former times, and such as are now observed to have formerly happened, had been completely veiled by forest growth and shrubs. At length, on the 28th of June, two months before the *fatal* avalanche, there was one not far from the Willey house, which so far alarmed the family, that they erected an encampment a little distance from their dwelling, intending it as a place of refuge. On the fatal night, it was impenetrably dark and frightfully tempestuous; the lonely family had retired to rest, in their humble dwelling, six miles from the nearest human creature. The avalanches descended in every part of the gulf, for a distance of two miles; and a very heavy one began on the mountain top, immediately above the house, and descended in a direct line towards it; the sweeping torrent, a river from the clouds, and a river full of trees, earth, stones, and rocks, rushed to the house and marvellously divided within six feet of it, and just behind it, and passed on either side, sweeping away the stable and horses, and completely encircling the dwelling, but leaving it untouched. At this time, probably towards midnight, (as the state of the beds and apparel, &c., showed that they had retired to rest,) the family issued from the house and were swept away by the torrent.

" Search, for two or three days, was made in vain for the bodies, when they were at length found. They were evidently floated along by the torrent and covered by the drift wood. A pole, with a board nailed across it, like a guide post, now indicates the spot where the bodies were found. Had the family remained in the house they would have been entirely safe. Even the little green in front and east of the house was undisturbed, and a flock of sheep (a part of the possession of the family) remained on this small spot of ground, and were found there the next morning in safety—although the torrent dividing just above the house, and forming a curve on both sides, had swept completely around them, again united below, and covered the meadows and orchard with ruins, which remains there to this day. Nine persons were destroyed by this catastrophe, and the story of their virtues and their fate is often told to the traveller by the scattered population of these mountain valleys, in a style of simple pathos and minuteness of detail, which has all the interest of truth and incident of romance in its recital.

" The number of visitors to the White Mountains has been considerably increased, on account of the interest excited by these *avalanches*. The most sublime views of them, (several of which are nearly equal to the memorable one which swept away the unfortunate Willey family), may be seen all along for several miles, in passing through the Notch. They are also observed from various points in the country around, extending down the sides of many of the elevated mountains, and the astonishing effects of this extraordinary inundation are also witnessed in the great enlargement of the channels of the streams which rise in these clusters of mountains. This is the fact especially with regard to the channel of the principal branch of the Ammonoosuck, which rises near the summit of Mount Washington," landslip which overwhelmed and covered to the depth of seventy feet the town of Plurs, in the valley of the Maira, on the night of the 4th of September 1618, sparing not a soul of a population of 2,430 inhabitants, is one of the most memorable of these catastrophes, and the fall of the Rossberg, or Rufiberg, which destroyed the little town of Goldau in Switzerland, and 450 of its people, on the 2nd of September 1860, is almost equally celebrated. In 1771, according to Wessely, the mountain-peak Piz, near Alleghe in the province of Belluno, slipped into the bed of the Cordevole, a tributary of the Piave, destroying in its fall three hundred and sixty lives. The rubbish filled the valley for a distance of nearly two miles, and, by damming up the waters of the Cordevole, formed a lake about three miles long, and a hundred and fifty feet deep, which still subsists, though reduced to half its original length by the wearing down of its outlet.

"The important provincial town of Veleia, near Piacenza, where many interesting antiquities have been discovered within a few years, was buried by a vast landslip, probably about the time of Probus, but no historical record of the event has survived to us.

"On the 14th of February 1855, the hill of Belmonte, a little below the parish of San Stefano in Tuscany, slid into the valley of the Tiber, which consequently flooded the village to the depth of fifty feet, and was finally drained off by a tunnel. The mass of debris is stated to have been about 3,500 feet long, 1,000 wide, and not less than 600 high.

"Occurrences of this sort have been so numerous in the Alps and Apennines, that almost every Italian mountain commune has its tradition, its record, or its still visible traces of a great landslip within its own limits. The old chroniclers contain frequent notices of such calamities, and Giovanni Villani even records the destruction of fifty houses, and the loss of many lives, by a slide of what seems to have been a spur of the hill of San Giorgio in the city of Florence, in the year 1284.

"Such displacements of earth and rocky strata rise to the magnitude of geological convulsions, but they are of so rare occurrence in countries still covered by the primitive forests, so common where the mountains have been stripped of their native covering, and, in many cases, so easily explicable by the drenching of incohesive earth from rain, or the free admission of water between the strata of rocks—both of which a coating of vegetation would have prevented—that we are justified in ascribing them for the most part to the same cause as that to which the destructive effects of mountain torrents are chiefly due—the felling of the woods.

"In nearly every case of this sort, the circumstances of which are known —except the rare instances attributable to earthquakes—the immediate cause of the slip has been the imbibition of water in large quantities by bare earth, or its introduction between or beneath solid strata. If water insinuates itself between the strata, it creates a sliding surface, or it may, by its expansion in freezing, separate beds of rock, which had been nearly continuous before, widely enough to allow the gravitation of the superincumbent mass to overcome the resistance afforded by inequalities of face and by friction; if it find its way beneath hard earth or rock reposing on clay or other bedding of similar properties, it converts the supporting layer into a semi-fluid mud, which opposes no obstacle to the sliding of the strata above.

"The upper part of the mountain which buried Goldau was composed of a hard but brittle conglomerate, called *nagelflue*, resting on an unctuous clay, and inclining rapidly towards the village. Much earth remained upon the rock, in irregular masses, but the woods had been felled, and the water had free access to the surface, and to the crevices which sun and frost had already produced in the rock, and, of course, to the slimy stratum beneath. The whole summer of 1806 had been very wet, and an almost incessant deluge of rain had fallen the day preceding the catastrophe, as well as on that of its occurrence. All conditions, then, were favourable to the sliding of the rock, and, in obedience to the laws of gravitation, it precipitated itself into the valley as soon as its adhesion to the earth beneath it was destroyed by the conversion of the latter into a viscous paste. The mass that fell measured between two and a half and three miles in length by one thousand feet in width, and its average thickness is thought to have been about a hundred feet. The highest portion of the mountain was more than three thousand feet above the village, and the momentum acquired by the rocks and earth in their descent carried huge blocks of stone far up the opposite slope of the Rigi.

"The Piz, which fell into the Cordevole, rested on a steeply inclined stratum of limestone, with a thin layer of calcareous marl intervening, which, by long exposure to frost and the infiltration of water, had lost its original consistence, and become a loose and slippery mass instead of a cohesive and tenacious bed."

He then goes on to say,—" In Switzerland and other snowy and mountainous countries, forests render a most important service by preventing the formation and fall of destructive avalanches, and in many parts of the Alps exposed to this catastrophe the woods are protected, though too often ineffectually, by law. No forest, indeed, could arrest a large avalanche once in full motion, but the mechanical resistance afforded by the trees prevents their formation, both by obstructing the wind, which gives to the dry snow of the *Staub-Lawine*, or dust avalanche, its first impulse, and by checking the disposition of moist snow to gather itself into what is called the *Rutsch-Lawine*, or sliding avalanche. Marschand states that the very first winter after the felling of the trees on the higher part of the declivity between Saanen and Gsteig, where the snow had never been known to slide, an avalanche formed itself in the clearing, thundered down the mountain, and overthrew and carried with it a hitherto unviolated forest to the amount of nearly a million cubic feet of timber. Elisée Reclus informs us, in his remarkable work *La Terre*, vol. i. p. 212, that a mountain, which rises to the south of the Pyrenæan village Araguanet in the upper valley of the Neste, having been partially stripped of its woods, a formidable avalanche rushed down from a plateau above in 1846, and swept off more than 15,000 pine-trees. The path once opened down the flanks of the mountain, the evil is almost beyond remedy. The snow sometimes carries off the earth from the face of the rock, or, if the soil is left, fresh slides every winter destroy the young plantations, and the restoration of the wood becomes impossible. The track widens with every new avalanche. Dwellings and their occupants are buried in the snow, or swept away by the rushing mass, or by the furious blasts it occasions through the displacement of the air; roads and bridges are destroyed; rivers blocked up, which swell till they overflow the valley above, and then, bursting their snowy barrier, flood the fields below with all the horrors of a winter inundation."

And he adds in a foot-note,—" The importance of the wood in preventing avalanches is well illustrated by the fact that, where the forest is wanting, the inhabitants of localities exposed to snow-slides often supply the place of the trees by driving stakes through the snow into the ground, and thus checking its propensity to slip. The woods themselves are sometimes thus protected against avalanches originating on slopes above them, and as a further security, small trees are cut down along the upper line of the forest, and laid against the trunks of the larger trees, transversely to the path of the slide, to serve as a fence or dam to the motion of an incipient avalanche, which may by this means be arrested before it acquires a destructive velocity and force.

"In the volume cited in the text, Reclus informs us that 'the village and the great thermal establishment of Barèges in the Pyrenees were threatened yearly by avalanches which precipitated themselves from a height of 1,200 mètres and at an angle of 35 degrees; so that the inhabitants had been obliged to leave large spaces between the different quarters of the town for the free passage of the descending masses. Attempts have been recently made to prevent these avalanches by means similar to those employed by the Swiss mountaineers. They cut terraces three or four yards in width across the mountain slopes, and support these terraces by a row of iron piles. Wattled fences, with here and there a wall of stone, shelter the young shoots of trees, which grow up by degrees under the protection of these defences. Until natural trees are ready to arrest the snows, these artificial supports take their place and do their duty very well. The only avalanche which swept down the slope in the year 1860, when these works were completed, did not amount to 350 cubic yards, while the masses which fell before this work was undertaken contained from 75,000 to 80,000 cubic yards.'"—*La Terre*, vol. i. p. 233.

In many cases such as are cited the evil may be traced to the infiltration of water upon argillaceous beds, such as are referred to, which thus become lubricated, and so admit of the sliding over them of thick beds of superincumbent earth, bearing with them, it may be, houses, and trees, and cultivated fields; in other cases, the infiltrated water comes upon beds of materials the disintegration of which leads to similar results.

M. Marschand—after describing a deposit on which is situated the village of Meyronnes, and its lands in the upper part of the valley of Barcelonette, in the Lower Alps, which deposit was then in movement in one mass throughout the whole extent, from Saint-Ours to the Ubayette, a distance of about 3½ kilomètres, or two miles and a half, threatening direful consequences, which he details—states, that any one may see at a glance from a road on the Sylve, a mountain situated on the other side of the valley, that this movement is manifestly attributable primarily to the waters of a stream, the sources of which are, at Fous-Vive and at Saint-Ours, being absorbed largely by the ground which it traverses, which is thereby softened,—and secondarily to the percolation of water produced by the melting of the accumulated snow on the southern slope of the mountain of Saint-Ours,—and, in fine, to the meadows covering the ground being extensively irrigated, and an additional percolation of water resulting from this irrigation. The cohesion of the mass was being thus destroyed, and the base of the mass was being at the same time undermined by the waters of the Ubayette; and it was manifest that the catastrophe threatened must happen sooner or later.

A similar case is reported by him, as having been seen by him in Tessin, imperilling the village of Campo. In this case the process was more advanced than in that at Meyronnes; the river Rovana, an affluent of the Maggia, having attacked the mountain on which the village was standing.

In neither of these cases was the ground in movement wooded. But he mentions also the crumbling of a portion of a forest of Norway firs, 120 mètres long by 90 broad, at Güruigel Bruck, on the mountain of Güruigel, on the east slope of which is the source of the torrent Gürbe; and he states that in the forest of the Gürbe are immense heaps of rubbish, the remains of former landslips, while at a higher level are the denuded mountain sides, whence the material has slid. But it is mentioned by him that above the land bared by the landslips specified, and distant only a few mètres from the summit of the mountain, are to be seen numerous springs. And to water thus supplied may be attributed the landslips which have occurred.

Such is one of the aspects of the enterprise; it is to prevent landslips as well as to preserve the lower-lying valleys from *déjections*.

In all such cases as these last cited there is required drainage and desiccation as well as *reboisement* and *gazonnement*.

"When the landslips are occasioned by infiltrations of water into ground which retains it in great quantity—as, for example, at Meyronnes, or at the Gürbe—it is of primary importance," says M. Marschand, "to cut off these waters. How this is to be done must be determined by a careful study of the ground, which must be brought, if possible, into a healthy state—(1) By turning off and leading away to a distance from the lands in movement all the streamlets flowing thither; (2) Causing all waters which traverse these lands to follow the line of most rapid declivity; (3) Searching out the sources which feed the mud, and draining them by carrying off the water by ditches and by tile-drains; and (4) Suppressing irrigation on or above the lands in movement."

The work of *reboisement*—applying that term, as is often done, to the whole of the operation of reforesting the denuded mountains—is thus found to embrace in practice a variety of operations over and above the mere sowing of seeds and planting of trees. The object aimed at is the extinction of the torrent, and nothing tending to the accomplishment of this is neglected, and there is a pleasurable excitement experienced in observing how this is done.

"The extinction of a torrent," says Cézanne, "is a struggle with a formidable foe, in which are called into exercise the same qualifications which command success in war,—bravery, energetic perseverance, and that sage tact which discerns the weak point in an enemy and carries the attack direct to the heart of his defences." And he cites the torrent of Vachères as an interesting illustration of the stratagetic skill with which the works employed have been distributed in the basin of the torrent.

"The torrent of Vachères, (says he,) one of the very worst in the Alps, is on the left bank of the Durance, over against Embrun, and in the plain its vast cone of sterile gravel presents a sad contrast to the rich cultivated grounds by which it is bordered, and which by it are menaced. This torrent is in reality a small mountain river, its *bassin de réception* covering an area of 7000 hectares, or well nigh 3000 acres, embraces excavations in three mountains; and some of the affluents are 20 and 30 kilomètres, or 14 and 20 miles, in length. In all the upper parts of these, which may be considered the sources of the torrent, the ground is comparatively firm, and the water limpid; but when it approaches the gullet by which it debouches into the plain, it traverses an extensive bed of detritus, apparently a formation of the glacial period, an old moraine it may be, a confused mixture of mud, and sand, and erratic blocks, torn from the far off summits of the High Alps. In this quarter the torrent is enclosed in precipitous banks, to the depth of 100 mètres, or nearly 350 feet, and these banks are being eaten away unceasingly at their base, and are in a state of the most complete instability.

"Here the torrent at once changes its character; clear thus far, here it loads itself with muddy *déjections*—the everchanging divarications begin—and the least storm of rain causes the hills to crumble down, and gives rise to the most violent effects of a *débâcle*, or breaking up of a dam. Near to this spot the principal torrent receives two considerable affluents: on the right, the torrent of the Grande Combe comes down from the mountain Saint-Sauvier; while, on the left, the torrent de l'Homme tears and eats away, in a great fan-shaped basin, the mountain of Baratier. These two torrents wear down a black schistose earth of the worst kind, and, between the two, a single flood suffices partially to dam up the principal torrent, which is then driven against one or other of the confining unstable banks. Thus all the producing causes of disaster meet within a space of about a kilomètre, or two-thirds of a mile square; for this reason those desolate spots have been chosen by M. Costa as the field of battle, while he is satisfied with simply prohibiting the access to flocks in the upper part of the basin.

"The bold plan, which is in course of execution, consists in breaking the living force of the principal torrents by a massive wall, behind which the water-course will accumulate its *déjection* in such a way that the base of the existing banks forming the gullet will be covered deep by these; the crest of these banks will then be broken down, and a gentle and regular slope will replace the torn and ruinous surfaces which they now present. While these works are in course of execution, a longitudinal dike, built higher up the *barrages*, in a situation happily chosen, prepares for the torrent an artificial bed, into which it will be cast when they shall have banked up the ancient bed.

"One of the two affluents mentioned above, de l'Homme, already extinguished by planted banks and small *barrages*, has become innocuous. And its counterpart, coming in an opposite direction, the torrent of the Grande Combe, which now, after having flowed for some distance parallel to the principal torrent, falls into this below the wear, will, by means of a cutting, the locality for which is indicated by a natural depression in the ground, be brought into the torrent above that *barrage*. And by the change thus made the torrent of the Grande Combe will be led away to a distance from the black schist, with which it now charges itself to repletion; the bed which is now hollowing out will be filled, and the hills between which it now flows will be laid out in banks and subjected to the usual treatment.

"M. Costa hopes, that by the new channel which he has in view for the torrent of the Grand Combe, the extinction of it will be brought about as by stage effect, for in the course of a few hours the muddy waters of to-day will have given place to a sheet of limpid water, flowing into the torrent above the *barrage*."

While all the credit given to M. Costa by M. Cézanne is justly due to him for devising and executing such works, it is also due to M. Surell to mention that the sufficiency of such measures in some cases was not unforeseen by him. In view of the whole subject of torrents, he remarks,— "When a torrent is examined with attention, it may be seen that all its parts are not equally hurtful. The mischief is often committed by but one branch of it, and the others contribute but little thereto. It would, then, be useless to apply the same treatment to all without discrimination; the attack must be made on the devastating branch, and that once extinguished, the ravages will be found to have ceased."

From what has been stated, it will be seen that the work assumes a variety of forms; but *reboisement* seems still, as from the first and all along, to be considered the most important, if not the most necessary, of the various forms which the work of extinguishing torrents assumes, or *reboisement* and *gazonnement* in combination; and I would now report how the work is being executed in the High Alps, where the importance of this enterprise in all its magnitude has been realised by all classes of the population.

There, over the whole surface of the *berges vives*, sloping but often very steep banks to be covered with vegetation, are traced horizontal level banks, about 6 or 7 feet broad, with a slight inclination towards the mountain, designed to give to the water facilities for collecting and remaining there. Towards the edge of these banks, where previously the earth has been loosened to a considerable extent by the pickaxe, they plant broad-leaved trees of three or four years growth in such proximity to each other that the extreme branches touch, and in such a way that the collet of the root is buried some eight inches under the surface of the ground as a security against drought. The stem is pruned to the level of the ground, that too rapid vegetation may not exhaust the plants; and the pruning is repeated until the vigorous appearance of the young trees testifies that their roots have at length reached a moist subsoil capable of supplying them with nourishment.

These embankments are made at distances from each other varying with the degree of slope; and the intermediate strip is sown broadcast with forage plants, or plants chosen from amongst those which grow spontaneously on the mountains. Sometimes, midway between the rows of trees, are planted other hedges of trees which receive less attention, but which grow pretty well in favourable spots; and between these trees, and midway between these rows, there may be planted lines of lucerne, while on the space between grow herbs of various kinds. This is the case where the main lines of trees are from 20 to 100 feet apart.

When the slope is very precipitous, the embankments are sustained by stones or hurdles; and when on steep declivities the soil is so disintegrated as to be unable to withstand the violent impinging of the rain-drops in a storm, the ground sown is sometimes protected by a covering of straw, or with cuttings of herbs, &c., which the growth of the herbage underneath soon renders unnecessary.

M. Marschand gives the following instructions, and refers to the *Traité Elémentaire de Sylviculture* of M. Franckausen, translated into French by M.

Amyot, as containing valuable details in regard to the measures adopted—both in sowing and planting—in the mountains of the Oberland. Of the preparation of the soil he says,—When the surface of the soil is bare, and of too great a declivity to give any certainty of stability, the first thing to be done is to fix it, which may be done by means of hurdles.

The soil being fixed, it is next requisite to prepare it for the reception of seed or plants; and in reference to this he quotes a proverb, current in the south of France, to the effect that good weeding, hoeing, or digging, may count for a watering. And he goes on to remark, that plantations should only be made in ground well broken up and well wrought, any danger of such soil being carried away being met by the hurdles employed.

The digging and breaking up, he recommends, should penetrate to a depth of from 16 to 20 inches, and should be accompanied by the removal of stones, and the filling up of the hollows they created with the good superficial soil surrounding them.

When the soil between the hurdles has been thus broken up, the location of the plants must be determined by the nature of the soil. On calcareous rubble, the plants must be set immediately below the hurdles, for such ground being constantly falling they will thus be protected from injury by the falling stones. But on ground more stable—as on marls, for example—they may be planted in the middle of the bands between the hurdles, or even immediately above these, the earth which may accumulate from the continuous falling being too little to destroy or injure the plants.

When the surface is covered with vegetation, and stable, hurdles are unnecessary; but it may be well to break up the ground in plots on places so narrow and steep that there may be some danger of the falling down of the earth occasioning erosion. These plots may be from 16 to 40 inches square, or in the some cases the ground may be broken up in horizontal strips 3 feet or more in breadth, and 12 or 15 feet long, at such a distance from each other that the branches of the trees to be planted may touch when they have attained to the state of perches, a distance varying with the kind of trees planted from a fathom to 20 feet.

When the work takes this form of strips, it is necessary to make the surface as horizontal as possible; otherwise, the earth may be swept to the lower edge of it by the first storm of rain which may occur. In many cases, the lower sides of such strips may be sustained by low walls composed of the stones taken out in breaking up the ground. In stony ground, such walls are built on the upper border of the strips, in such a way as to arrest rolling stones, and so keep these from falling against the plants. This system has been employed with the happiest results by M. Demontzey in the southern Alps, securing at the same time other advantages besides that referred to.

When it can be done, it is well to leave the broken-up ground for some time exposed to atmospheric influences—allowing a winter, or at least some months, to intervene between the preparation of the soil and the planting of the trees.

With regard to the method of *reboisement* by sowing seeds of trees, he writes,—" This method of *reboisement* it is not in general advisable to adopt, as it rarely gives satisfactory results; while a considerable gain of time is secured, with greater probability of success, by planting young trees. But there is a method of sowing frequently adopted where the ground is prepared in strips, which—thanks to the good preparation of the soil, and the relatively pretty large extent of the ground broken up—succeeds well.

"There are made of these a kind of *pépinières volantes*, or temporary nurseries, in the centre and on all points of the lands to be replanted, in which may be found, at befitting times, and at little expense, plants with which to supply void spaces, or even to carry on the *reboisement* over the entire surface, when a small number of strips may have been sown for this purpose."

Of plantations, he says,—" The success of these depends on the plants employed, and on the time at which the operation of planting is performed. There has been much discussion on the question, whether this should be done in spring or in autumn. Spring is preferable in the Alps, as the frost of winter extrudes the plants from the ground, and destroys many of them. And there it is necessary to plant as much as possible after rain, while the ground is moist, thus giving the best security for their success; and where this is made a point of some importance, the *pépinières volantes* are of great service—they allow of young plants being had on the ground at any time; while the difficulties of procuring plants in sufficient quantity, at a given time, at a great many different places, often prevents their arrival at the time required, and is otherwise prejudicial to the work.

"It is scarcely necessary," says he, " to add, that in planting great care must always be taken to place the best soil finely comminuted around the roots, which should be placed and disposed with consideration and attention,—to heap up the soil and press it down with the foot, &c.,—to take, in short, all the care recommended for plantations in general, and which it is unnecessary to repeat."

In regard to the choice of plants, he says,—" In the Grissons, where the mountains are calcareous, and where the climate in summer is extremely hot, they can use only plants which have been transplanted in a nursery. The plantations always succeed, and the inspector of the forests of the canton, M. Coaz, attributes the success which has been obtained solely to this use of retransplanted plants.

"In the Oberland, in like manner, they employ only retransplanted plants, and rarely do these perish. There are thus obtained indisputable facts, over against which can only be set the fact of the success of some species of resinous trees particularly robust, such as the Austrian pine and the larch, transplanted or sown; but even these, however, when first transplanted in the nursery, and then replanted, are unquestionably superior to those which have not been so treated.

" The objections which may be raised to the employment of retransplanted plants, are the pretty high price of them, and the difficulty of procuring them in great quantities. But in reply," says M. Marschand, " to the first of these objections, I have no doubt that the final result will be generally obtained at less expense with retransplanted plants than with others, taking into account the interminable labour required with these in supplying the places of dead plants—works often more onerous than the original planting. And I can adduce in support of this opinion," says he, " a great many examples of this having been the case in the French Alps.

"The second objection may be easily met—it is only requisite to extend the nursery proportionately with the area to be replanted; if there have been made *pépinières volantes*, in strips within the périmètres, there may be transplanted thence the young plants, and those which are not required may be left there, where they will not fail to grow." And he goes on to say,—" It may be superfluous to add, that the superiorty of transplanted plants over others consists in this, that within one or two years after being transplanted they have acquired body, have become more densely branched, and have formed more tufted and branched roots, and so can better adapt themselves to transplantation to the place destined for their growth.

"Plants obtained from the strips where they have been reared are, on the contrary, always rather slender and poor, because they have grown up in a very crowded condition."

The kinds of trees best suited for the work of *reboisement* is the next subject to which attention is given by M. Marschand. "The choice of the kind of tree to be planted," says he, "ought always to be made with great care; and if it have been practicable to make trial of different kinds in the locality, never should extensive works be attempted with any but the kinds the success of which has been made certain." And he goes on to say,—"I have seen the most beautiful *reboisements* obtained by means of—(1) the Scotch fir, (2) the Austrian pine, (3) the Siberian pine, (4) the larch, (5) the Norway fir.

"The *pin sylvestre (pinus sylvestris)*, or Scotch fir, transplanted and replanted, succeeds always; it is employed in the *Contre-fort* of the Alps, which constitutes the principality of Lichtenstein. This tree, which does not grow well but on deep earth, covers in the Alps immense areas, but it becomes remarkable there for its poor and stunted appearance; and this variety, which offers no redeeming advantage, should be rejected.

"The *pin noir d'Autriche*, or Austrian pine, is very robust, and may be considered the pine of calcareous lands; it has almost everywhere given very good results; its qualities and its products make it valuable, and it is not without reason that day by day the adoption of it is spreading on all hands.

"The *pin à crochets*, the Mugho or dwarf pine, is common in the Alps; the greatest mass of this in growth which I know extends from the *Engadine* to the Munster-Thal; it is the forest of Offen, more than 50 kilomètres, or about 35 miles, long. The tree—now creeping on the ground, now shooting up—presents everywhere a poor appearance, and yields wood fit only for fuel.

"This tree rarely attains to great dimensions; it is well, therefore, to be chary in the employment of it—it should be consigned to dolomite chalks, and pebbly ground, unsuitable for all other kinds of forest vegetation. Yet I have often seen this tree in demand in the Alps, though often enough it possessed no claim to be classified with forest trees.

"The *pin Cembro*, or Siberian pine, is a tree growing at great altitudes; it is not much employed, nor has it generally succeeded well on the Alps. In the Grisons and the Oberland, it is considered one of the most robust of trees, the success of which when planted is most certain.

"The seed of it should be gathered in autumn; during the winter it is kept in sand or in saw-dust, in a place slightly moist and of mild temperature, such as a cellar or stable; or the cones may be left spread on hurdles in such places. In spring the seeds are slightly watered daily for a fortnight, at the end of which time they are taken out and sown. Unhappily the mice are very fond of this seed, and scarcely have they been committed to the earth when they are devoured; in Engadine they surround the seed-beds with frames of planks, sunk about 16 inches, and covered with wire-cloth—and thus the mice are kept out. M. Coaz, by successive waterings, causes the seeds to germinate in the boxes in which they are kept during winter, he then sows them on the ground; and the mice do not attack these.

"The young trees are transplanted when one or two years old, according to their strength, to be planted out two years later, when three or four years of age.

"The *reboisements* executed in the environs of Stalla, below the pass of Juliers, at an altitude of about 1800 mètres, or 6000 feet, have succeeded perfectly, not a plant has died; but it may be doubted whether the Siberian pine will have a rapid growth at such altitudes.

"The finest masses of this tree which are known to me are situated in the environs of Saint Moritz, Upper Engadine, at an altitude of 1800 mètres; they are very compact and complete, and of various ages, and are beautiful forests.

"The *melcze*, or larch, is the most robust and valuable of the trees of the Alps, and is the one which it should be sought to multiply and diffuse as much as possible. It succeeds pretty often when sown, but *always* when planted; and its growth is rapid enough to produce quickly good results, in fixing the soil and regulating the water-flow.

"The *épicéa*, or Norway fir, is not held in high estimation in the Alps. In Switzerland and in Austria it is much employed, even in southern climates; in general it is planted out after transplantation."

With regard to deciduous or broad-leaved trees, M. Marschand says,— "I am myself no advocate for the employment of these in *reboisements* on the Alps. The resinous trees have been located by the Creator on the great mountains, because they possess, in view of the general *règime* of the waters, properties which the broad-leaved trees do not.

"But I may add," says he, "that in the level lands of lower-lying spots, extending to 1200 mètres, or 4000 feet, in the southern Alps, the acacia succeeds well; the ash and the sycamore equally well; and, in fine, as a bushy growth giving a first shelter, I have seen employed with success the plum tree of Briançon, and the variety of willows which cover calcareous slopes; on the calcareons coasts of the Adriatic, they employ as a first shelter the juniper.

"In conclusion, I repeat," says he, "that since with care a direct *reboisement* may be obtained by means of resinous trees, recourse should never be had to provisional protection excepting after the most manifest failure with these."

It may be desired to compare with these matured opinions, deliberately expressed by M. Marschand, the opinions which have been expressed by others. To facilitate this being done I may repeat here that, at the first of the annual conferences of agents employed in the works, instituted by Ministerial appointment, and held in 1861, it was stated that the kinds of trees which up to that time had been employed, had been chiefly the *épicéa*, or Norway fir, the Scotch fir, the Austrian pine, the Aleppo pine, the Corsican pine, the larch, the ailanthus, the acacia, the Mount Atlas cedar, the white oak, the *ilex*, or evergreen oak, the cork tree, the chestnut, the willow, the white poplar, and the birch; and of shrubs—the filbert the shumack, the hazel, &c. But this referred to a much wider range of country than the High Alps alone, to which M. Marschand's remarks refer.

The opinions expressed by the agents employed, in regard to the adaptation of these several kinds of trees and shrubs for which they had been selected, and in regard to localities for which one and another of them were appropriate, and the annotations of the Administration on the opinions expressed, have been given (ante pp. 177-207).

In the report of operations in 1865-66, it is stated that the kinds of trees most extensively diffused in the region of the Alps—including the Isère, the High Alps, the Lower Alps, and the Drôme—in the order of most importance, were the Austrian pine, the Scotch fir, the Norway fir, the larch, the oak, the Corsican pine, the alder, the ash, the silver fir, and the Mugho, or dwarf pine. And in this region much use has been made of suckers and twigs of willows and poplars and of herbaceous plants.

In the Pyrenees—including the Eastern Pyrenees, the High Pyrenees, the Lower Pyrenees, Aude, and Ariege—the kinds of trees most common were the Scotch fir, the Mugho, the Austrian pine, the larch, the chestnut, the maritime pine, the oak, and the acacia. And in the region of the Cévennes, and the central plateau—including Ardèche, Grand Lozère, Hérault, Puy-de-Dôme, Cantal, and High Loire—there were employed principally the Austrian pine, the Scotch fir, the Norway fir, the oak, the maritime pine, the Aleppo pine, and the ailanthus.

It is then stated generally that, in the selection of trees, the maple, the acacia, and the filbert, were preferred for unstable ground, on account of their rapid growth and their roots sending forth numerous suckers. The oak and the walnut were reserved for strong, dry, solid grounds; while in the moist depths of the ravines the alder, the poplar, the ash, the osier, the white willow of the Alps, &c., were made use of.

These are, in some soils, preferred to the coniferae, in view of the object aimed at; and some other trees have been employed experimentally. But a preference or prejudice has been expressed in favour of indigenous trees.

Amongst the bushes cultivated may be mentioned the black thorn, the bramble, the myrtle, the juniper, the *hippophoi*, and above all the barberry. This last, by virtue of its strong root, was formerly spread over the country from the valley to the mountain summits, but the root being in demand as a dye, this led to reckless destruction of it, and it had almost entirely disappeared.

Amongst the herbs employed are the sainfoin, the lucerne, and the restharrow, a plant indigenous to the *combes*, which may be seen suspended over the edge of the precipice, the crumbling crust of which it holds and retains as with the grip of despair.

In regard to most of the trees, of which mention has been made as used in the work of *reboisement*, much information may be found in almost any English work on Arboriculture, Forests, or Forest Trees. But when I was engaged in the study of this subject I failed to obtain the kind of information I required, to enable me to learn for myself, and to give to others, counsel, in regard to measures to be adopted in carrying out works of such magnitude as would be requisite in some of our colonies, if it were attempted to prevent by sylviculture the devastating effects and consequences from torrents, and from inundations from which occasionally they suffer. This information is to be had at command in France. In a work entitled "*Cours élémentaire de culture des Bois créé a l'Ecole Forestiere de Nancy, par M. Lorentz, Directeur-Fondateur de cette école, ancien Administrateur des Forêts, Officier de la Legion d'honneur, Membre Correspondent de la Societe Imperiale d'Agriculture, &c., complété et publie par A. Parade, Conservateur des Forêts, Directeur de l'Ecole Imperiale Forestiere. Cinquième Edition, publicé par A. Lorentz et H. Nanquette, avec une preface par L.*

Tassy, 1867," there is abundance of such information, which I purpose embodying in a separate volume on Forest Science and its practical application in the forest economy of France.

Chap. V.—Devastations and Restorations.

There are still extant forests of great extent on the mountains of France, but there have been extensive clearings. And while we picture to ourselves mountains begirt with forests, it is expedient with the object we have in view to introduce into the picture such scenes as have been described in the Introduction, as presented by Devoluy, by the vicinity of Embrun, and by the valley descending from the col Isoard, and others which have been given of the ravages and devastations wrought by the torrents which owe their birth to the clearings which have been made in these ancient forests. Elsewhere it is the same ; and the study of this will show what evils have resulted from what may be considered but partial clearings.

M. Cézanne follows up a lengthened lucid and instructive exposition with the statement,—" These long explications which have been given can give but a very inadequate and incomplete idea of the treatment applied to torrents ; on the other hand, it suffices to visit any one of the périmètres, and cast a glance over the whole, to receive a convincing demonstration of what is being done, and to be imbued with absolute confidence in the efficacy of the cure. If then the Administration of Forests desires to form at any time a special service for the artificial extinction of torrents, the best measure to take would be to send their agents on a mission into the High Alps, as *L'Ecole des Ponts et Chaussées* send their students to visit the works in course of execution."

I am aware of the importance of this suggestion, and I would make the same to those whom I desire to move to the adoption of like measures. Meanwhile, without detriment to this suggestion, I can produce statements innumerable, and of unquestionable authority, descriptive of what was, within the last twenty years, the condition of various localities, and what is the condition into which they had been brought by *reboisement* and *gazonnement*, and *barrages*, at the time when operations were interrupted by war.

To students of Forest Science, information embodied in the official documents, of which translations have been given, may suffice to enable them to form a definite idea of what has been done, and enable them, perhaps, by a vivid fancy, to reproduce the past, to picture the present, and to imagine what the future is likely to be ; but others may prefer being supplied with less formal and more detailed information—and such is at command.

Sect. I.—*The High Alps.*

The state of desolation to which this region had been brought has been again and again brought under notice ; but other details are not awanting.

Of that desolation some idea may be formed from the following account of the vicinity of Embrun, given by Surell,—" In going from Gass towards Embrun, following the highway numbered 94, more than a fourth of the journey is made on the beds of torrents. They are seen scattered over the whole country, inundating all the valleys, and furrowing all the slopes, and hence comes that air of desolation so peculiar to the country, which at once strikes strangers on their crossing these mountains for the first time.

"The multiplicity of torrents in this department is a fearful scourge, it is like a leprosy which has seized upon the soil of the mountains. The torrents eat into the sides of these, and, dejecting on the plains heaps of debris, by a long-continued succession of accumulation they have created enormous beds of *déjections* which are ever increasing and extending. They threaten to overwhelm everything. They doom to perpetual sterility the soil which they bury beneath their deposits. Every year they are swallowing up some additional estates. They intercept communication between different parts of the district, and hinder the establishment of a good system of roads. And these ravages are to be deplored all the more because they take place in a country which is very poor, and is devoid of manufactures, and one in which arable ground, which is the only resource of the inhabitants, is rare. These, it often happens, succeed in creating a small field, but only after prodigies of labour and perseverance, and then comes the torrent unexpectedly and deprives them, it may be in one hour, of the fruit of ten years of labour and toil.

"The dread which these torrents inspire appears in the names which have been given to them. Thus is it with the torrent *l'Épervoir*, the hawk, and with the torrents *Malaise*, ill at ease, *Malfosse*, evil pit, and *Malcombe*, *Malpas*, *Malattret*,—all names speaking of evil. Some bear the name of *Rabioux*, the enraged; several others that of *Bramafaim*, howling hunger. There are some which seem ready to swallow up entire villages and even market towns; and there a dark cloud hovering over the sources of the torrent is sufficient to spread alarm over a whole community."

From this statement some idea may be formed of what some thirty years ago was the state of things there. The passage is cited in the official report of works executed in 1867 and 1868, and with it the following statement by M. Surell in regard to what influenced him in doing what he did in the matter is given:—

"There was yet another consideration which determined me to undertake this study, and I must say that it is this which all along has given direction to me in my work. This wretched department going fast to ruin, and the Administration, whose duty it is to look to the conservation of its territory, not having yet tried to put forth the least effort to avert the coming evil, it appears to be high time to call the attention of the Administration to the state of this country. It seems to be ignorant of the extent of the evil, and it is my belief that in throwing light upon this plague, and showing what might be done to cure it, I am discharging a sacred duty."

"As may be imagined," writes M. Faré in the report cited, "a state of things such as this has commanded the most serious attention of the Forest Administration from the time they were intrusted with the execution of the law of 28th July 1860."

A summary is then given of the extent to which *reboisement* and *gazonnement* had been effected, and the Director-General goes on to say,—"I have cited above some of the statements made by M. Surell, which bring into prominence the imminence of the danger with which the French Alps were being threatened. In further reference to the sad picture thus presented, and to make apparent the results already produced by the works of restoration executed by the Forest Administration, I shall confine myself to reprinting, from a report presented in 1869 to the *Conseil Général des Hautes Alpes*, by M. Gentil, *Ingenieur en chéf des ponts et chaussées*, the following passage :— ' Torrents are one of the most disastrous plagues of the High Alps ; the *cones de déjection* invade the valleys, bury under their heaps the cultivated ground, end in annihilating every kind of cultivation, and hunt the inhabitants away from the country ; and at the same time the erosions occasioned by them destroy the sides of the mountains ; and thus is destroyed at one and the same time all the value of the mountain and the value of the plain.

" ' The embankments attempted on the *cones de déjection* at the issue of the gorges, by which come down the materials carried off by the waters from the higher-lying lands, have always failed, or at best the effects produced by them have been but precarious. The dikes in a few years have disappeared under the rubbish from the mountain.

" ' But the Forest Administration has succeeded, by the consolidation of the soil, in the creation of a robust vegetation on the flanks of the *bassins de réception*. The results are assured : the case of the works at La Batie, at Sainte-Marthe, at Resail, has demonstrated most manifestly and most indisputably, that it is quite possible not only to arrest *déjections*, but also to re-establish vegetation on mountains the most ravaged by these torrents.

" ' It is not required of me here to show by what means the Forest Administration has succeeded in extinguishing the torrents. I confine myself to specifying the results of these operations. These results, in regard to the valleys, to the lands there, and to the roads by which they are traversed, are remarkable in the extreme, and it is now required of me to point out these to the Departmental Administration.

" ' From the time that the soil in the *bassin de réception* is consolidated, and by plantings and sowings and works of the Forest Administration the soil is fixed, material is no longer torn away and thrown into the current which transports it to the lower-lying parts. The waters assume in some measure a regular *régime*, they come clear and free from mud upon the *cones de déjections*, they dig out there a stable bed for themselves by carrying away the less ponderous material ; at this stage embankment becomes possible in the valley, and it is practicable at little expense to keep in one unchanging direction the flow of waters which no longer carry away the stones. Properties along the banks are then securely protected ; they are no more exposed to a sudden disaster such as those of which we have so many examples ; they recover with this security their money value ; and the population reassured may count upon their harvests.

" ' On the other hand, the fixing of the bed of the current permits the erection of bridges and aqueducts on the roads and highways ; communication is protected against the frequent interruptions to which it was exposed when the torrent was in full activity ; and, in fine—nor is this the least important result of the regeneration of the basin of reception—the principal rivers no longer receive the masses of *déjection* which encumber their beds and create confusion in times of flood. In illustration of these results, which have been thus referred to in a general and summary way, may be cited the following facts :—

" ' The torrent of Sainte-Marthe, near Embrun, was threatening to extend its *déjections*, so as to cover the imperial road, No. 94. A proposal to construct a dike on the left bank had been formally discussed ; the expense of this was estimated at about 45.000 francs, and it was considered that it should be met in part by the State, and in part by the proprietors on the river bank. But since the execution of the works of *reboisement*, in the basin of Sainte-Marthe, by the Forest Administration, this water-course has lost its torrential character, and has settled its bed in the *cone de déjection*, the embankment has become useless, and the project which had been under discussion has been entirely abandoned.

"'The torrent of Palps, in the commune of Risoul, was threatening both the departmental road, No. 4, and the imperial road, No. 94. In 1865 the draft proposal of work to be executed to cause the torrent to be conducted directly into the Guil, and to settle the bed at the end of the departmental road, had been discussed and presented for approval. The execution would have entailed an expense of 30,000 francs. But the Forest Administration has consolidated and covered with herbage the *bassin de réception* of this torrent, and they have been able to leave the waters to flow on in the course they have taken, and to construct a simple aqueduct under the imperial road, No. 94.

"'The torrent of Riou-Bourdoux was noted as one of the formidable torrents of the High Alps; the quantity of material which the waters put in movement at every flood, had, in some measure, led to the abandonment of the construction of a bridge for the passage of the imperial road, No. 94; the Forest Administration has enclosed, *mis en defens*, the basin of reception, and executed some works of consolidation and of *gazonnement*. The *régime* of the torrent has been in consequence so far changed, that, at little expense, the bed on the cone can be definitively settled, and a bridge erected for the imperial road.

"'I might bring forward other examples of what has been effected; those which have been given may suffice to make appreciable how complete and efficient are the results obtained.'"

This testimony is endorsed by the Director-General, who says,—" I have nothing to add to this report of the eminent engineer-in-chief of the department of the High Alps, save to say, that I fully share his firm conviction that it is practicable to arrest the déjections of the torrents, and to re-establish vegetation on the most ravine-furrowed mountains."

He subsequently, in another connection, cites the following statements from a report which had been presented to the Council General of the High Alps in 1868, by a Commission appointed by themselves:—

"Our Forest Commission felicitates itself in having to present to you a most satisfactory picture of the works of *reboisement* and *gazonnement* undertaken in the High Alps.

"The two distinguished engineers who were appointed along with us, more competent in many respects than the members of our Commission, have expressed with warmth the satisfaction afforded them by their visit to these immense and interesting works; and as for the Commissioners themselves, the same who in 1866 had seen the works in an embryotic state, they know not whether they should praise most the admirable harmony which is characteristic of the works as a whole, or the wonderful results already obtained.

"*Gazonnement* is substituted for *reboisement* wherever the *boisement* is not indispensably requisite to consolidate the ground, and there have been planted scarcely any but broad-leaved trees in active growth, which are becoming speedily defensible in such a way as to permit of the early restoration to the depasturing of sheep of the grounds thus reconquered.

"By these results the most active resistance has been deadened, and in many localities, where the mere apprehension of works to be undertaken had created the most violent complaints and the keenest opposition, the agents of the Administration are to-day loaded with praise, and those who in the beginning showed the greatest hostility, come forward of their own accord to solicit of the Commission the extension or the completement of the works in course of execution.

"We can certify, and do, that in many communes the evidence of the results obtained has allowed syndicates of dikes to give up their works of defence as being rendered superfluous by the consolidation and *gazonnement* of the grounds in the higher basin of the torrents which they proposed to embank; and we certify what we have ourselves seen, that in the périmètre which we have visited, works undertaken only two years ago have sufficed literally to extinguish completely many ravines, dangerous affluents of the torrent of the Sasset, affluents which were producing the greatest masses of déjections of earth and blocks of stone which were encumbering the bed of the torrent; we certify, in fine, that in spite of the many terrible storms of rain which this year have desolated our country, and in particular the quarter of Sasset, not a *barrage*, not a dike, not one of these verdant strips which give to the périmètre, as a whole, the aspect of a smiling parterre, has been cut in upon, and that the growth of the bushes and of the herbaceous plants has attained an unlooked for development.

"The results, of which we have submitted to you a report, lead us to invite the Council to solicit the immediate publication of decrees of public utility for the périmètres approved at its first sederunt.

"The Commission, penetrated with the greatness of the interest which the whole department has in the continuation and in the development of the works of *regazonnement* and *reboisement*, propose to you to vote for this source, as in former years, a subvention of 500 francs."

Of this region, it was considered originally that, by reason of the extent of the evil, little could be reckoned on proprietors and communes taking the initiative in the work, and that subventions, however extensively they might be distributed, would be altogether inadequate to bring about a restoration of the mountains, from which vegetation had almost entirely disappeared; but it has been accomplished, and details are given which may enable any one to fill up the outline of the picture which such general statements may suggest of what has been effected.

Appended to the report cited are numerous monographs on the work done in different périmètres of *reboisement* in different parts of France; and, amongst these, are monographs on what has been done in several périmètres in the High Alps.

The first of these is on the works executed in the périmètre of Sainte-Marthe, in the valley of the Durance. This valley of the Durance has at all times been deemed one of the most convenient highways between France and Italy. By the valley of the Durance, says Cézanne, passed Hannibal, Caesar, Charlemagne, Charles VIII., Louis XII., François I., Louis XIII. Victor Amédée, Duke of Savoy, invaded and devastated it in 1692. Lesdiguières, Catinat, Berwick, Vittars, carried on campaigns in the High Alps; there is not a gorge in the department, nor a pass, which has not become famous as a battle field; and there the Vaudois and the Protestants formerly found refuges in which they were safe.

The Col du Mont Genèvre, upon which the valley of the Durance abuts, is less elevated than those of Mount Cenis, of the Great and the Little St Bernard, of the Simplar, and of the St Gothard; sheltered as it is against the north wind, it presents exceptional advantages in the inclement season; and the pass is not, like the others, a desert and inhospitable *col*, but is a cultivated and inhabited plateau, in which there are two villages—the principal villages of two communes.

The Dauphines of a former day, more especially Humbert II., and Louis II., and subsequently Louis XI., took a special interest in this region; they dug channels of irrigation, raised dikes, and founded useful establishments. The memory of them is still maintained among the mountain population.

By order of Louis XIV., Vauban, after having completed the fortifications of Briançon and Embrun, of which the Romans had laid the foundation, constructed in all its parts the stronghold of Mont Dauphin.

Previous to the revolution, Dauphiny was considered an independent State connected with France; not incorporated with the kingdom, but governed by the king under the title of Dauphin of the Viennois. It embraced the territory now forming the departments of the Isère, the Drôme, and the Higher Alps. Until then the Higher Alps, being an integral portion of Dauphiny, shared the revenues of a wealthy province; and they were dependent on a central administration not far distant, where their requirements were known, and where an interest was taken in efforts the inhabitants had to put forth. Subsequent to the division of France into departments it was otherwise. But Napoleon I., having his attention directed to the subject by M. Ladoucette, who was prefect of the High Alps, constructed, by the Durance and the Mont Genèvre, the great military road from Spain to Italy, and caused the route from Paris to Nice to pass by Gass. And the department was enriched by Napoleon with many vitalising institutions which were subsequently suppressed.

At the time of the reverses experienced by the nation in 1815, and in despite of orders received from the authorities in France, the inhabitants of Briançon, of Mont Dauphin, and of Fort Queyras, refused to open their gates to the Austrio-Sardinian army; and by this patriotic conduct there was preserved for France a great quantity of munition and war stores which the army had taken to Italy.

But from that time onwards the High Alps have exhibited a continuous decadence. Through the policy of centralization the business of the department, in common with that of the whole country, removed to Paris, failed to receive from the Administration the attention given to districts less inaccessible and more out-spoken; and depopulation followed, apparently in consequence of the continuous impoverishment of the land.

M. Léonce de Lavergne, in a treatise on *Économie Rurale*, writes,—" The two departments of the High and the Low Alps present a sad contrast to the other parts of this region. *They have retrograded instead of advancing.* It is the least wealthy portion of the district; it has only 22 inhabitants to the hectare, while even Corsica has 27. For more than a hundred years past all who have known the French Alps speak of the coming destruction of the whole of the vegetable soil by the periodical ravages of the torrents."

The population of the High Alps, it is stated by Cézanne, increased by 14,000 from 1806 to 1846; but it diminished by 11,000 from 1846 to 1866; and he gives official returns in proof of this fact, with official returns showing the continuous increase of the population of France throughout the whole of the time specified. "We have lost," says he, " in twenty years what we had gained in forty; and the public wealth has experienced a similar falling off. If we look to the production of timber, this from 1831 to 1847 increased over the whole of France 44 per cent., but in the High Alps only 30 per cent. From 1847 to 1868 it increased over the whole of France 103 per cent., but in the High Alps only 12 per cent., and he cites official returns in proof, so that, whether tested by population or by wealth, the High Alps only followed at a distance in the general advancement of France till 1847, and from that time onward it has, in the words of Lavergne, *retrograded instead of advancing.*"

M. Cézanne goes on to say,—" The smiling fertile valleys of this region are narrow; the greater part of the villages extend up the steeps to the region of a trying climate, where even the sweat of man cannot make the ground productive. There the peasants live on the verge of the habitable parts of the land. The least physical accident—frost, drought, or rain—is disastrous to them; and the roar of the torrent in their immediate vicinity is always threatening them. They toil and they suffer without complaint; they neither blame the Government nor God; but when the misery becomes unbearable they depart, they yield to the greater attractions which invite them to the opulent cities of Marseilles and Lyons and Paris. It is to the betterment of their circumstances indeed; but it is a cruel wrench; and the nation suffers.

"There are some who have said,—'Let the mountains crumble into ruins if they cannot sustain themselves against the effects of the weather!' But others have said,—' These are a part of the ramparts of France! and what will it advantage us in the day of danger to have fortresses on the frontier, if behind these strong places there be only a desert, supplying to the army of France neither woods, nor fields, nor railroads, nor population?' There were others who had their attention called to the subject, and more especially to the importance in many respects of the *reboisement* of the mountains; and the reaction commenced; but the state of the land was not more encouraging than was the condition of the people."

I have given a translation of M. Surell's account of Dévoluy; not less saddening is the account given by him of the valley descending from the Col Izoard, which he cites as a typical specimen of the *bassin de reception* of a torrent. He says,—" The aspect of this monstrous channel—a gorge which serves as the common point of accumulation and discharge of several lateral torrents—is frightful. Within a distance of less than two English miles, more than sixty torrents hurl into the depths of the gorge the detritus torn from its two flanks. The smallest of these secondary torrents, if transferred to a fertile valley, would be enough to ruin it."

And of this torrent of the Col Izoard, he says in an appendix to his treatise,—" This gorge, dreadful in appearance as it is, is nevertheless the most convenient route there is leading to the valley of Queyras à Briançon. The bed of the torrent serves as a district road; and from this one may judge of what the district roads of the department are. The traveller who should be caught by a storm of rain in the midst of this defile would there infallibly lose his life. Where could he find refuge from the dangers pressing upon him on all sides? The soil sinks under his footsteps; if he remains in the bed he is engulfed by the torrent; if he try to climb the mountains he is crushed by the blocks of rock, and by the clods of earth, which are there tumbling down from all parts; consequently the inhabitants, when they venture on this route, have to take care and see that there be no threatening of bad weather.

"The torrent of Labcoux, which leads into the Dévoluy, presents nearly the same features with the same consequences. The district road is the bed of the torrent, and the mountains which enclose it are in many places so steep, or they are composed of so crumbling a material, that it would be difficult to find refuge on them in the event of a storm of rain suddenly swelling the torrent, and effacing under a body of furious waters all traces of the road."

Such was the state of things, and such were the evils, which were fearlessly attacked by the Forest Administration and the hydraulic engineers of France with the simple appliances of *reboisement* and *gazonnement*, or the planting of trees and the fostering of the growth of herbage.

It was like David going forth with his sling and his five smooth stones in his shepherd's wallet to kill Goliath of Gath; but, preposterous as seemed his scheme, and ridiculously inadequate his provision for what he proposed to do, *he did it;* and so likewise was it here.

Of what was done the following account is given :—" The torrent of Sainte-Marthe is situated on the right bank of the Durance, and has for its origin the brows of the Mount Saint-Guillaume, the height of which is 2,500 mètres, well nigh 8,350 feet. The point at which it reaches the river after a course of 8 kilomètres, above 5 miles, is about 2 kilomètres, about a mile and a third, below Embrun, at an altitude of 700 mètres, or well nigh 2,350 feet. The difference of level between the origin and the embouchure of this torrent is about 1,800 mètres, or 6,000 feet. The slopes of the *thalweg* are consequently very great.

"When the works were commenced on the *bassin de réception*, the surface of it was absolutely bare and everywhere cut up by ravines. But as this upper part is formed of sandstone and of pretty hard compact limestone, the disintegration was only superficial.

"The *canal d'écoulement* is a narrow gorge, and has an extremely steep descent, all along which exist *berges vives* in a tumbling down condition. The upper half is formed of earth, stones, and blocks of rock which have been borne thither; the lower half traverses black marl almost in a state of clay or mud.

"Everything to produce the well-known effects of torrents is found in this torrent. The *bassin de réception*, entirely denuded of vegetation, forms a funnel in which the waters, at the time of storms of rain, rush to a common centre almost instantaneously. The mass of waters precipitates itself on the steep declivities of the *thalweg*, from the first tearing away from the flanks of the upper hills large quantities of stones and of rocks of all sizes; lower down the flood mixes up with itself the black mud furnished by the washing away of the lower-lying hills; and then, like an avalanche, which in some respects it resembles, it precipitates itself with a violence which nothing can resist, and debouches at the bottom of the valley at the extremity of the gorge which forms the summit of the *cone de déjection*. Fine properties in the environs of Embrun, of a value of at least 300,000 francs, an imperial road, with a bridge and dike belonging to the State, of a value of more than 200,000 francs, and a district road of great importance, were all being threatened with destruction. Dikes had been constructed along the side of the torrent to protect the plain; but the bed of the torrent rose higher and higher still. It had been necessary to meet this by raising higher and higher the embankment; and it had now come to pass that the torrent was several metres above the level of the property along the banks. Although it was imprudent to elevate still higher the torrent, a new scheme of embankment, which it was estimated would cost 45,000 francs, had been formally discussed; and it was about to be carried out.

"It was in these circumstances that the torrents of Sainte-Marthe was attacked in 1865; from 1863 the whole of the basin, which measures 530 hectares, upwards of 200 acres, had been enclosed, *mise en défens*, with the consent of the Municipal Council of Embrun, though it had been opposed by the inhabitants of the hamlets on the sides of the torrents. These had in reality the greatest interest of all in the execution of the works, as their dwellings and their crops, dragged along with the general movement of the soil, were tending towards engulfment in the torrent.

"The works began with an improvement of the basin. Two years of enclosure had prepared the ground. All the ravines were cut up into portions by more than 200 *barrages*; channels to lead off and disperse the water were cut; and seeds of forage plants were sown over places which required them.

"Attention was then given to the consolidation of the hills bordering the *canal d'écoulement*. With this view there were constructed, first, strong *barrages* in the upper-lying parts of these. The years 1865, 1866, and 1867, were employed in securing the command of the head of the torrent, and diminishing the violence of the flood. It would have been imprudent and almost impossible to construct *barrages* in the middle of the black slime of the lower-lying portion of the *canal d'écoulement*, inasmuch as the force of the flood would not have been sufficiently reduced at that time.

"In 1868 it was considered that the last part of the work might be taken up with some chance of success. If matters had been less pressing this might have been deferred for one or two years more; but it was deemed of importance that the results should be made apparent.

"In constructing the lower series of *barrages*, the work was begun anew from below, instead of being continued from above. First, there were planted at the lower extremity strong *barrages* capable of withstanding the strongest floods. Others were then constructed successfully further up the torrent, and pretty near to each other, that each might give support to the one above it. And in proportion as land was gained by each *barrage*, the hills were cut into shape by the pick-axe to give them the angle of stability.

"In the same time that the principal *thalweg* was thus being consolidated, *boisement* and *gazonnement* were carried on on the lateral slopes. The ravines were choked with small *barrages* of stones, with hurdles, and with *fascines;* and the ground was drained at spots where infiltrations of water were producing subterranean disintegration.

"These works, carried on in combination with each other, have proved completely successful. The torrent is now [1870] extinguished. For two years the greatest storms of rain have deluged the basin, but have had no other effect than to occasion a moderate increase of the flow. This has carried off no material, nor has the stream overflowed its banks All danger to the plain has disappeared. THE SYNDICATE HAS DISSOLVED ITSELF.

The new scheme of embankment has been abandoned. The proprietors have again brought under culture all the lands previously invaded, and a few years ago they planted vines and orchards within the very embankments of the torrent formerly reared. These facts are patent to all; and they have been officially certified by the *Service des Ponts et Chaussées*.

"The expenditure, including that of 1868, has been 91,134 francs 24 centimes. The number of *barrages* constructed is 759. The total length of the *barrages vivants* and the hurdles is 32,270 mètres. The length of roads, 9,400 mètres. The length of channel to carry off and disperse the water, 1117 mètres. The extent of ground regenerated and restored is 400 hectares. The extent of what may be considered as regained and maintained is 300 hectares."

Such is the report of the Administration in regard to the extinction of the torrent of Sainte-Marthe. In accordance with this is the statement by M. Gentil, previously cited. And in accordance with both are statements which have been subsequently published.

In the same valley is the torrent of Vachères, in regard to which corresponding details are given, the torrent of Chague, the torrent of Riou-Bourdoux, the torrent of Reallon, the torrent of Valoria, the torrent of Trente-Pas, the torrent of Lhubac, in regard to all of which are given similar details not less romantic, but true. *Ex uno disce omnes.*

And there are other, and many other, valleys in these same High Alps in which corresponding work has been done, varying only as it varied with varying circumstances in different localities here.

Statements scarcely, if at all, less interesting and encouraging than this I have quoted in regard to the valley of the Durance are given in regard to operations in the basin of the Drac, including those in the Communes of Saint-Jean-Saint-Nicoals, of Orcieres, and of Champoleon, which two latter give their names to two branches of the river, the Drac d'Orcieres, and the Drac de Champoleon.

From a *Note sur les Dessèchements, &c.*, by M. Montluissant, in the *Annales des Ponts et Chaussées* for 1833, it appears that the Drac has when in flood poured into the Isère, at its confluence with this river a little below Grenoble, 5,200 cubic yards of water per second, such is its importance.

M. Cézanne intimates that nothing could be more profitable to a student of such works than would be a visit of inspection to the works of *reboisement* in the High Alps, if there could be secured the advantage of making such inspection under the guidance of a man who has organised them amidst difficulties, which description can only imperfectly bring before the mind, and such a one, says he, is M. Costa de Bastelica—a man heart and soul devoted to the work. Whilst some functionaries sent to the Alps, says he, from the day of their arrival live in thoughts of the day when they shall be able to leave the miserable country, M. Costa has attached himself to those mountains by a protracted residence. Daily witnessing the ravages of the torrents, and imbued with the ideas of M. Surell, he has had faith in success, and when the law of 1860, and the confidence in him justified by his administration, gave him the means and opportunity of action, he gave himself to the work with an enthusiastic interest, to which his country, witness of his efforts, renders with unanimity its homage.

The time is not yet long gone past when the scared mountaineers were threatening his house and his person; he sees at length the success crowning his efforts, and every spring decking a new group of *combes* until then naked and desolate, while the torrent subdued clears for itself a channel in the old *déjections*; and in manifestation of his victory, M. Costa, as if in defiance to his enemy, has here and there thrown *passerelles*, or foot-bridges, at a height scarcely exceeding a mètre above the water, and takes pleasure in showing that the torrent, Sainte-Marthe for example, which but lately rose above its embankments and carried off high bridges, has respected for years this feeble barrier.

But if I should tell all that is told of what has been done in one of the périmètres in these, I should feel strongly disposed to tell with similar prolixity of all that has been done in all: I should feel like a boy running down a steep declivity unable to stop till he has reached the bottom. A stand must be made somewhere—I make it here. In language suggested by the Bible, I leave off before I begin, that I may report what has been done elsewhere, and the changes which have been effected there by the works executed, and their results.

Sect. II.—*Department of the Isère.*

Nowhere, perhaps, are torrents to be seen acting with such fury as they have displayed in the department of the High Alps; and it was in the arrondissement of Embrun, more especially, that they were to be found in greatest numbers, and in their most terrible forms. In proportion as we recede from this district, which may be regarded as the centre of their action, they are seen less and less violent, and are more and more rare, until, at a great distance, their characteristic peculiarities finally disappear; but as is thus indicated, they do not all at once cease, nor are they confined to the region of the High Alps alone.

Throughout a great part of the former independent State of Dauphiny they have committed ravages.

The following account of Dauphiny and Provence is given by Mr Marsh, the facts stated in which were supplied by the work by Charles de Ribbe entitled *La Provence au point de vue des Bois, des Torrents et des Inondations*:
—" The provinces of Dauphiny and Provence comprise a territory of fourteen or fifteen thousand square miles, bounded northwest by the Isère, northeast and east by the Alps, south by the Mediterranean, west by the Rhone, and extending from 42° to about 45° of north latitude. The surface is generally hilly and even mountainous, and several of the peaks in Dauphiny rise above the limit of perpetual snow. Except upon the mountain ridges, the climate, as compared with that of the United States in the same latitude, is extremely mild. Little snow falls, except upon the higher mountains, the frosts are light, and the summers long, as might indeed be inferred from the vegetation, for in the cultivated districts the vine and the fig everywhere flourish; the olive thrives as far north as $43\frac{1}{2}°$, and upon the coast grow the orange, the lemon, and the date-palm. The forest trees, too, are of southern type, umbrella pines, various species of evergreen oaks, and many other trees and shrubs of persistent broad-leaved foliage, characterising the landscape.

"The rapid slope of the mountains naturally exposed these provinces to damage by torrents, and the Romans diminished their injurious effects by erecting, in the beds of ravines, barriers of rocks loosely piled up, which permitted a slow escape of the water, but compelled it to deposit above the dikes the earth and gravel with which it was charged. At a later period the Crusaders brought home from Palestine, with much other knowledge gathered from the wiser Moslems, the art of securing the hillsides and making them productive by terracing and irrigation. The forests which covered the mountains secured an abundant flow of springs, and the process of clearing the soil went on so slowly that, for centuries, neither the want of timber and fuel, nor the other evils about to be depicted, were seriously felt. Indeed, throughout the Middle Ages, these provinces were well wooded, and famous for the fertility and abundance, not only of the low-grounds, but of the hills.

"Such was the state of things at the close of the fifteenth century. The statistics of the seventeenth show that while there had been an increase of prosperity and population in Lower Provence, as well as in the correspondingly situated parts of the other two provinces I have mentioned, there was an alarming decrease both in the wealth and in the population of Upper Provence and Dauphiny, although, by the clearing of the forests, a great extent of plough-land and pasturage had been added to the soil before reduced to cultivation. It was found, in fact, that the augmented violence of the torrents had swept away, or buried in sand and gravel, more land than had been reclaimed by clearing; and the taxes computed by fires or habitations underwent several successive reductions in consequence of the gradual abandonment of the wasted soil by its starving occupants. The growth of the large towns on and near the Rhone and the coast, their advance in commerce and industry, and the consequently enlarged demand for agricultural products, ought naturally to have increased the rural population and the value of their lands; but the physical decay of the uplands was such that considerable tracts were deserted altogether, and in Upper Provence, the fires which in 1471 counted 897, were reduced to 747 in 1699, to 728 in 1733, and to 635 in 1776."

As an example and illustration of what has been done in the department of the Isère, to the north of the High Alps and of the department of Drôme, I take at hap-hazard the monagraph given by the Forest Administration on the works executed in the périmètre of the Bourg-d'Oisans:—

"The territory of the Bourg-d'Oisans has been formed by the union of the two communes, formerly distinct, of the Bourg-d'Oisans and of Gauchoirs; it lies with a general exposure to the northeast, and comprises two distinct valleys, which meet at the confluence of the Vénéon and the Romanche; the one (Bourg-d'Oisans) is throughout a cultivated plain of an average extent of 3 kilomètres, or nearly 2 miles, traversed by the Romanche throughout its entire length, and includes all the escarpments which overlook it on the southeast to an altitude of 1800 mètres, or 6000 feet; the other (the Gauchoirs) lies on the left bank of the Vénéon, and extends to the upper ridge of the mountains, the dominating peaks of which attain a height of 2900 mètres, or upwards of 10,000 feet. The slopes of the mountainous region of the territory present abrupt declivities of from 45 to 60 degrees; with the exception of the dominal forest of Riou-Péroux, and of some masses of resinous trees of no great extent, they are entirely denuded and furrowed by numerous ravines, amongst which may be mentioned the torrent of Saint-Antoine.

"This torrent has hollowed out for itself a vast notch in the very flank of the mountain, and by this notch, which is almost vertical, it was enlarging itself unceasingly by the tumbling down of the upper parts. It was transporting on to the cultivated fields, and even on to the highway of the imperial road No. 91, at each storm of rain, rocks and stones mixed with mud; and it was threatening immediately the Bourg-d'Oisans, which it was necessary to secure by the construction of a strong dike designed to alter the direction of the current of dejections. This constant *écoulement*, or coming down of rocks and earth, was still further augmented by the proprietors of adjacent lands, who, finding this mode of getting out timber cost less, projected trunks and whole trees from the forests by the friable slopes and precipices of these declivities.

" The vegetable soil, kept on the surface by some bushes and tufts of turf, which the teeth and the treading of the flocks were destroying, no longer offering resistance, was converted into mud by the waters, and washed away on all hands; the argillaceous marls and slates which constituted the subsoil of the escarpment over which the torrent rolled its water, divided into thin layers, cleft and fragile, fell into a state of desintegration with the greatest rapidity, and filled up the basin of the Saint-Antoine with their débris continually renewed. On the other hand, the depasturing of flocks on the mountain carried to excess, in completing the ruin of the pastures, was extending the ravages of other torrents, and was hindering in the plain the fixation of the gravels of the Vénéon above the dikes constructed to confine it, and was destroying the vegetation which, by spontaneous or natural growth, would have opposed serious obstacles to the violence of the current.

" In order to fix the dry accumulations of alluvial deposits brought down by the Vénéon, to arrest the disintegration of the declivities, and to protect the houses which compose the Bourg-d'Oisans, it was becoming of importance to subject to a rigorous regulation the exploitation of the communal property, and to fix in a secure manner the hills of the most dangerous torrents.

" It was with this design that was proposed, by reports of the 31st March 1864, and 12th March 1865, the *reboisement* of the périmètre of Bourg-d'Oisans.

" Notwithstanding the energetic opposition raised by the inhabitants of these lands, which are essentially pastoral, who saw in the subjection of their mountains to a protective *régime* a great disturbance to their sole industry, the works were declared to be works of public utility by a decree of the 4th April 1866.

" The périmètre comprises 893 hectares 50 ares to be rewooded. Of this 73 hectares 42 ares belong to individual proprietors; the remainder to the commune; and 994 hectares 17 ares to be *regazonnéd*, belonging to the section of the Gauchoirs, in regard to which was pending a lawsuit between that section and the commune of Villard Eymoud.

" The individual proprietors, with the exception of one whose land was already wooded, refused either to undertake the work at their own expense, or to leave it to the execution of the Administration. Almost all of these properties are wooded. They are situated on the basin of Saint-Antoine, and have been included in the périmètre in order that sentence of expropriation might be recorded in regard to them : for it is essential to the success of the works to be executed in this basin, and for the securing of the results obtained, to prevent not only the grubbing up of woods, but *les coupes à blanc etoc*, or the felling of them with a clean sweep, and even partial exploitations, which, by denuding the ground, would give birth immediately to new ravines and, as a consequence, to new dangers.

"The Municipal Council of the Bourg d'Oisans, by a decision of the 10th May 1866, decided to give up to the Administration the execution of the works of *reboisement* and *gazonnement;* the State, moreover, remains charged with the *gazonnement* of the lands in dispute with the commune of Villard-Eymond, this last having refused to make known its intentions.

"Face to face with the increasing devastations, and with the urgency of a prompt execution of the works designed to avert them, the Forest Agents undertook from 1866 the different operations of acknowledged necessity. They constructed on the torrent of the Saint-Antoine four *barrages* of dry stone masonry, measuring together 460 cubic mètres. They dug *un canal de dérivation*, or channel for leading off the stream, 240 mètres long. And that they might be able to fix, by rewooding, the abandoned river-bed of the Vénéon, they covered with *fascines* of willow wands a surface of 5 hectares, composed of mobile earth, on the precipitous slopes, and planted 45,000 alders, or birches, and 240,000 resinous trees, on an area of 16 hectares.

"Prosecuted with vigour in the subsequent years, the works executed up to the 1st January 1869 may be summed up thus :—

"The plantation, on 43 hectares 11 ares, of resinous trees, alders, willows, and birches;

"The sowing, on 43 hectares 11 ares, of 590 kilogrammes of seeds of the same kinds of trees;

"The construction of 20 *barrages* of stone, comprising 1920 cubic mètres;

"The opening of a *canal de dérivation* of 240 mètres;

"The employment of more than 300,000 willows in hurdles and *fascines* for the fixing and consolidation of shifting land.

"The whole expense of the work, inclusive of material, transport, and labour, amounted, for the works of all kinds, to the sum of 21,206 francs 30 centimes,

"If it were required to specify the failures which have occurred in the works of *reboisement*, properly so called, sowings and plantations, notwithstanding the care which the Agents have bestowed upon them, it might be affirmed that the works in *barrages* and willow *fascine* works have succeeded in a very remarkable way, and have offered an effectual barrier to the torrent of Saint-Antoine. And although the *barrages* and *fascine* works, acknowledged to be necessary, have not been entirely completed, the torrent appears subdued, the considerable mass of material which it was carrying away has been arrested, the tumbling down of earth does not produce more; and, on the imperial road, No. 91, from Grenoble to Briançons, communication is no longer intercepted.

"But here, as in other périmètres, the plantations of Austrian pine have only given mediocre results: the plants of these brought from remote nurseries got heated in the transport, and arrived more or less damaged; the Norway firs and the larches have succeeded to the extent of 50 per cent.; the willows, the alders, and the birches, planted in the moist alluvia of the Vévéan have succeeded to the extent of 90 per cent. The sowings have generally failed; they have been unable to withstand on these denuded lands the vigour of the winter and the sun's heat during summer. The want of success should be in part attributed to certain mistakes made in the distribution of the different kinds of trees, to exposures, and to elevations which are unsuitable to them.

"The works of *gazonnement* have been carried on cotemporaneously with the works of *reboisement*. They consisted in 1867 of a sowing of 600 kilogrammes of fescue grass, on 70 hectares 58 ares, and of the construction of five stone *barrages* of 58 cubic mètres. In 1868 they consisted in enclosing the same lands.

"But the sowing executed in 1867 is far from having given satisfactory results. The Administration could supply only fescue seed, and experience has shown that, on the grounds of the Alps, *gazonnement* can only be obtained by the simultaneous employment of many kinds of seeds of forage plants.

"The five *barrages* have stood well, and have accomplished the purpose for which they were employed, arresting the sweeping away of material into the bed of Vénéon.

"The total expense for material, transport, and labour, for the works of all kinds connected with *gazonnement* has been 752 francs 95 centimes.

"The surface of the périmètre of the Bourg d'Oisans is too extensive for the subjection of it to forest *régime* not to have excited certain discontents among a population of which the principal industry is the rearing of cattle. To meet the reiterated protests and complaints of the communes interested, and to appease this opposition, the conservator, after having visited the places, has proposed to maintain in the *périmètre de gazonnement* unenclosed properties between the Gauchoirs and Villard-Eymond, and about 450 hectares situated in the canton of Cornillon.

"To resume, notwithstanding the checks experienced in the works of *reboisement* on 86 hectares and 23 ares, recorded from 1866 to 1869, there are only spots of limited extent on which the regeneration sought to be effected has not been accomplished.

"As a consequence of the enclosures, the *gazonnement* straightway ensuing, and some seeds of resinous trees, carried by the wind, have naturally reclothed the soil; and the whole surface may be considered in good keeping. When the local nursery, established within the périmètre, shall have provided the requisite plants, it will become an easy matter, and by no means costly, to complete the existing stock of trees."

Thus are failures and success alike reported. Similar monographs are given on the works executed and difficulties overcome, and results obtained in the périmètres of Roissard and of Cornillon in the same department, in which operations were being carried on at the same time on fourteen other périmètres, the area of the whole comprising upwards of 7,600 hectares.

SECT. III.—*Department of the Drôme.*

The former independent State of Dauphiny included, besides what are now the departments of the High Alps and the department of the Isére, the department of Drôme. In this department there were, within the first decade of these operations, seventeen périmètres in which works of *reboisement* or *gazonnement obligatoires* were being carried on. Not the least important of these, but not the most important of them, was the périmetre of Luc.

"The périmètre of Luc, in the Arrondissement of Die, lies at an altitude of 800 mètres, on a mountain side, of which the general inclination is to the north-west.

"The area is 815 hectares 72 ares, of which 593 hectares 83 ares belong to the commune of Luc, and 222 hectares 40 ares to different individual proprietors. The soil is compact argillaceous marl, the surface is deeply ravined by the torrent of Luc and its numerous branches.

"The works of *reboisement* declared to be of public utility, by two decrees dated respectfully 11th February 1863 and 12th August 1865, were begun in 1863. During all the period embraced by this monograph, they had to be confined within the communal part of the périmètre, awaiting while the proprietors of private lands opposed every work of restoration. Of them, 110 hectares may be considered as perfectly improved, *remis en valeur*; at least they will now require nothing more than works of maintainance of little importance. The ground there has been consolidated by means of numerous *barrages* of stone and of *fascines*. The soil there has been cultivated in horizontal strips and garnished with young plants of the oak, the Austrian pine, the Scottish fir, the Norway fir, the ash, and the accacia. In the beginning recourse was had to sowings, but this method of restocking the ground not having given good results, it was not long ere it was abandoned. Between the cultivated strips there have been scattered seeds of forage plants; herbs bruised by the teeth of cattle in browzing have been topped; and a few slips and suckers of *bois blanc*, more particularly of the willow, have been planted in the new grounds which were being formed above the *barrages*.

"The expense, borne entirely by the Treasury, had amounted altogether in 1869 to 49,097 francs 80 cents., exclusive of a sum of 1081 francs allotted to the inhabitants of Luc as compensation for temporary deprivation of pasturage.

"It was estimated that the works remaining to be executed would entail an additional expenditure of about 120,000 francs.

"But by this time the good effects of the works were beginning to make themselves felt, the torrent of Luc no more rolled down water charged with materials; its *régime* had become more regular, its bed had become hollowed out, it no longer threatened, as previously, to carry away the upper parts of the village, and thus the population had come to desire that the works might be speedily happily completed, not only because they had ascertained what good results had followed, but because, besides this, the wages paid had contributed to diffuse comfort amongst them."

SECT. IV.—*The Lower Alps.*

While, as has been stated, Embrun might be considered the centre of torrential phenomena, which gradually diminished in violence as distance from the spot increased, we have found these ravaging and devastating lands in the departments of Isère and Drôme. Throughout a great extent of the Lower Alps they have told with crushing effect. Mention has been made in preceding pages oftener than once of the valley of L'Ubaye, in which is situated the town of Barcelonette; there they have ravaged and devastated the land in a way and to an extent which make it terrible to contemplate.

In this department there were, when the report of operations in 1869 was prepared, ten périmètres in which works of *reboisement*, or *gazonnement obligatoires*, had been decreed, but in three only had they then been begun.

Of what was done in the périmètre of Labouret, it was reported that this périmètre, in the Arrondissement of Digne, was composed of 113 hectares

28 arcs, belonging to the two communes of Beaujeu and of Vernet. It extends over two very precipitous slopes, the one with a north-west, the other with a south-east exposure. The soil, belonging to the jurassique formation, is composed of black marls and calcareous rock, which easily disintegrate under the action of meteoric agents. The imperial road, No. 100, traverses the périmètre on an embankment of about 3 kilomètres, along the right bank of the torrent of Labouret, and leaning on the left hand against a very abrupt declivity of rocks in a state of decomposition. Formerly numerous transhumant flocks passed hither and thither in the bed of the torrent in going to the mountain pastures, where they spent the summer, or in returning thence to winter in the plains of the Crau. These devoured any little herbage which had come to fix itself on the hills between the two roads. When the rains came, if these were abundant and sudden in the region of the Lower Alps, the waters flowing freely on a col, completely denuded, and accumulated rapidly in the ravines, sweeping away with them earth and stones, and forming quickly a blackish mud which precipitated itself with violence towards the foot of the mountain, and in doing so destroying the imperial road and covering with a layer of gravel the cultivated grounds of Beaujeu and of Javie.

The *reboisement* of the périmètre of the Labouret was declared of public utility, after the legal formalities, by a decree of 18th June 1862. The works were begun in 1863. They embraced the construction of *barrages* as well as the sowing and planting of forest trees or of forage herbs.

"The *barrages* constructed up to 1869 were in number 2,139:—viz., 210 of stone, 503 of wood, and 1,426 in planted *twigs* or *fascines* of willow. The wooden *barrages* in general offered but little resistance; the greater part have been destroyed and swept away by the first floods of the torrent. Those of stone and *fascines* have done better; they have rapidly consolidated the ground and facilitated the subsequent execution of works of sowing and planting. One year after their erection, when the earth washed down arrested by them had reached about the level of the *barrage*, no time was lost in fixing this mobile earth by planting it with shoots and cuttings of quick-growing wood, especially willows. The steeper hills were also fixed by the growth of fescue grass.

"The kinds of trees which have been employed are the cedar, the Austrian pine, the Scotch fir, the Mugho or dwarf pine, the Corsican pine, the ash, the willow, the acacia, the ailanthus, the *hippophaë*, the walnut, the cytisus, the *alisier*, the maple, &c.

"There have been sown 4,616 kilogrammes of seeds of resinous trees, and 332 kilogrammes, of broad-leaved trees. There have been planted out besides this 200,729 plants of resinous trees, and 439,261 plants of broad-leaved trees, including 10,000 shoots or cuttings, and further, there have been sown 8,495 kilogrammes of the seed of sainfoin, or of *fénasse*.

"In the first years of these operations the seed was sown broadcast without any previous preparation of the ground, but this yielded no results. So also the first plantations made in the bed of the torrent before the consolidation of the hills were rapidly swept away by the waters. From 1867 the works have been better directed. *Barrages*, and more especially those constructed of *fascines*, have been multiplied in the birth-place of the ravines to stop the washing away of mud, and the sowings, as well as the plantations, have received all the care required.

"The total expense up to this time [1869] amounts to the sum of 42,221 francs. Thanks to these works, the périmètre has completely changed its aspect; 100 hectares are rewooded, and now require only the labour needed for their maintenance. The remaining 13 hectares represent calcarious precipices, on which one can scarcely dare to hope to fix some herbacious plants by persevering efforts. Formerly, after the least storm of rain, the torrent rolled away masses of mud, of stones, and of fragments of rocks, which covered anew the imperial road as well as the lower-lying cultivated fields. Now the waters, the current of which has been retarded, deposit all such materials above the *barrages*, and the imperial road is no longer flooded. If the Treasury have imposed on itself great sacrifices, it will not be long in reaping the benefit of these; for as soon as the vegetation shall be completely fixed on the Labouret, the expense of the maintenance of the road, which has already been reduced to a marked extent in consequence of the works, will not exceed those of a mountain road of average conditions."

A corresponding report was given of what had been done in the périmetre of Seyne, in the same Arrondissement, a périmètre of 1,250 hectares, situated at an altitude of from 1,400 to 2,400 mètres, and formerly covered with forests, but which in 1861 presented the desolate aspect of a vast desert ravaged by torrents. These, five in number, are tributaries of the torrent De la Blanche, which precipitates itself into the Durance. But adherence to the principle adopted, of citing details of only one périmetre in each locality, in illustration of what has been and is being done, forbids this monograph being also translated.

M. Cézanne intimates, that in the Lower Alps there had been victories gained which may be classed with those of M. Costa, already noticed. "Previous to the law of 28th July 1860," says he, "slight *barrages* of hurdles, combined with the reclamation of waste ground, had fixed the lands of the Mollard, near Sistéron, the detritus of which was covering the site of the town. The torrent had been so completely extinguished, the water so entirely absorbed, that a small aqueduct, constructed for the conveyance of them to the Durance, had become useless. But in this department these most remarkable results had, he said, been brought about by a simple communal guard, and the following citations from the *Annales Forestiéres* justify the statement :—

"M. Jourdan has commenced his works in the forest of Salignac; and in commencing them he has had to overcome a pretty keen opposition on the part of the inhabitants, as might have been anticipated. Until 1860 the greater part of the *barrages* erected in the forest of Salignac have been made by this guard alone; and he was obliged to repair many of those which the storms of rain or the malevolence of the people destroyed. From 1853 to 1861, the guard Jourdan has constructed and repaired by himself alone three hundred *barrages*, and the more is he to be commended that two-thirds of these are distant, upon an average, from 8 to 10 kilomètres from his place of residence."

M. Labussière, conservator of forests at Aix, who was honoured with a gold medal, decreed to him by the Central Society of Agriculture of the Puy-de-Dôme, for his beautiful works of *reboisement* in that department, commends in these terms to the functionaries placed under his orders the results obtained by Jourdan :—

"A communal guard of the Lower Alps has had the happy idea to establish *barrages* of *fascines* in the ravines formed in the clearings and fellings of the forest in proportion as they were exploited. Some of these ravines soon became veritable torrents, often dangerous, and they all caused damage more or less considerable to the lower-lying properties. This simple work, which required only some hours of work and a few *fascines* of almost no value, has given excellent results in the communal woods of Sistéron, Salignac, Entropierre, and Vilhose. I do not need to add that the *barrages* ought to begin at the origin of the ravines, and that they ought to be near to one another in proportion as the declivity is more steep and the soil more friable. Experience has shown that the best result is obtained by proceeding in the following manner : a first bed of *fascines* is laid on the ground across the ravine; this is covered with other *fascines* placed perpendicularly with the point looking toward the summit of the mountain ; and the structure is carried on in the same way to a height indicated by the condition of the localities. The first *fascines* laid in the direction of the ravine may be kept in their place and consolidated by large stones, or by turfs, if they can be found near.

"If such works of so easy execution could be constructed one after another, it would produce excellent results in a few years, if we may judge from those obtained by Jourdan, who had no motive for his undertaking but a desire to do good, and to turn his spirit of observation to account for the benefit of others."

The *Sociétié impériale et centrale d'agricultare de France* decreed to Jourdan a gold medal with the effigy of Olivier de Serres. "Is it not touching," says the report, "to see a whole city protected against the most terrible scourges by the intelligent hand of a forest warder, one of the most modest functionaries of the State.

"While savants were writing treatises, a simple guard was solving the problem. And to perpetuate the memory of this they have in the country given to the form of *barrage* just described the name of *Barrages Jourdan.*"

SECT. V.—*Department of l'Ardèche.*

All of the operations which have been now detailed are in the region of the Alps. In the region of the Cévennes, and the plateau of Central France, operations have been carried on extensively in Ardèche, Gard, Lozère, Herault, Puy-de-Dôme, Cantal, and the department of Hautes-Loire. And again selection becomes necessary.

In 1860 there was published a *Mémoire sur les Inondations des Rivières de l'Ardèche*, by M. de Mardigny. Mr Marsh has given the following statement of what is known in regard to that department, founded to some extent on the statements made by M. Mardigny :—"The river Ardéche, in the French department of that name, has a perennial current in a considerable part of its course, and therefore is not, technically speaking, a torrent ; but the peculiar character and violence of its floods is due to the action of the torrents which discharge themselves into it in its upper valley, and to the rapidity of the flow of the water of precipitation from the surface of a basin now almost bared of its once luxuriant woods."

He says in a foot-note,—" The original forests in which the basin of the Ardèche was rich have been rapidly disappearing for many years, and the terrific violence of the inundations which are now laying in waste is ascribed, by the ablest investigators, to that cause. In an article inserted in the *Annales Forestières* for 1843, quoted by Hohenstein, *Der Wald*, p. 177, it is said that about one-third of the area of the department had already become absolutely barren, in consequence of clearing, and that the destruction of the woods was still going on with great rapidity. New torrents were constantly forming, and they were estimated to have covered more than 70,000 acres of good land, or one-eighth of the surface of the department, with sand and gravel." And he goes on to say,—

"The floods of the Ardèche and other mountain streams are attended with greater immediate danger to life and property than those of rivers of less rapid flow, because their currents are more impetuous, and they rise more suddenly and with less previous warning. At the same time, their ravages are confined within narrower limits, the waters retire sooner to their accustomed channel, and the danger is more quickly over, than in the case of inundations of larger rivers. The Ardèche drains a basin of 600,238 acres, or a little less than nine hundred and thirty-eight square miles. Its remotest source is about seventy-five miles, in a straight line, from its junction with the Rhone, and springs at an elevation of four thousand feet above that point. At the lowest stage of the river, the bed of the Chassezac, its largest and longest tributary, is in many places completely dry on the surface—the water being sufficient only to supply the subterranean channels of infiltration—and the Ardèche itself is almost everywhere fordable, even below the mouth of the Chassezac. But in floods, the river has sometimes risen more than sixty feet at the Pont d'Arc, a natural arch of two hundred feet chord, which spans the stream below its junction with all its important affluents. At the height of the inundation of 1857, the quantity of water passing this point—after deducting thirty per cent. for material transported with the current and for irregularity of flow—was estimated at 8,845 cubic yards to the second; and between twelve o'clock at noon on the 10th of September of that year and ten o'clock the next morning, the water discharged through the passage in question amounted to more than 450,000,000 cubic yards. This quantity, distributed equally through the basin of the river, would cover its entire area to a depth of more than five inches.

"The Ardèche rises so suddenly that, in the inundation of 1846, the women who were washing in the bed of the river had not time to save their linen, and barely escaped with their lives, though they instantly fled upon hearing the roar of the approaching flood. Its waters and those of its affluents fall almost as rapidly, for in less than twenty-four hours after the rain has ceased in the Cévennes, where it rises, the Ardèche returns within its ordinary channel, even at its junction with the Rhone. In the flood of 1772, the waters at La Beaume de Ruoms, on the Beaume, a tributary of the Ardèche, rose thirty-five feet above low water, but the stream was again fordable on the evening of the same day. The inundation of 1827 was, in this respect, exceptional, for it continued three days, during which period the Ardèche poured into the Rhone 1,305,000,000 cubic yards of water.

"The Nile delivers into the sea 101,000 cubic feet or 3,741 cubic yards per second, on an average of the whole year. This is equal to 323,222,400 cubic yards per day. In a single day of flood, then, the Ardèche, a river too insignificant to be known except in the local topography of France, contributed to the Rhone once and a half, and for three consecutive days once and one third, as much as the average delivery of the Nile during the same periods, though the basin of the latter river probably contains 1,000,000 square miles of surface, or more than one thousand times as much as that of the former.

"The average annual precipitation in the basin of the Ardèche is not greater than in many other parts of Europe, but excessive quantities of rain frequently fall in that valley in the autumn. On the 9th of October 1827, there fell at Joyeuse, on the Beaume, no less than thirty-one inches between three o'clock in the morning and midnight. Such facts as this explain the extraordinary suddenness and violence of the floods of the Ardèche, and the basins of many other tributaries of the Rhone exhibit meteorological phenomena not less remarkable. The Rhone, therefore, is naturally subject to great and sudden inundations, and the same remark may be applied to most of the principal rivers of France, because the geographical character of all of them is approximately the same.

"The volume of water in the floods of most great rivers is determined by the degree in which the inundations of the different tributaries are coincident in time. Were all the affluents of the Lower Rhone to pour their highest annual floods into its channel at once, as the smaller tributaries of the Upper Rhone sometimes do—were a dozen Niles to empty themselves into its bed at the same moment—its waters would rise to a height and rush with an impetus that would sweep into the Mediterranean the entire population of its banks, and all the works that man has erected upon the plains which border it. But such a coincidence can never happen. The tributaries of this river run in very different directions, and some of them are swollen principally by the melting of the snows about their sources, others almost exclusively by heavy rains. When a damp south-east wind blows up the valley of the Ardèche, its moisture is condensed, and precipitated in a deluge upon the mountains which embosom the head-waters of that stream, thus producing a flood; while a neighbouring basin, the axis of which lies transversely or obliquely to that of the Ardèche, is not at all affected.

"It is easy to see that the damage occasioned by such floods as I have described must be almost incalculable, and it is by no means confined to the effects produced by overflow and the mechanical force of the superficial currents. In treating of the devastations of torrents, I have hitherto confined myself principally to the erosion of surface and the transportation of mineral matter to lower grounds by them. The general action of torrents, as thus far shown, tends to the ultimate elevation of their beds by the deposit of the earth, gravel, and stone conveyed by them; but until they have thus raised their outlets so as sensibly to diminish the inclination of their channels—and sometimes when extraordinary floods give the torrents momentum enough to sweep away the accumulations which they have themselves heaped up—the swift flow of their currents, aided by the abrasion of the rolling rocks and gravel, scoops their beds constantly deeper, and they consequently not only undermine their banks, but frequently sap the most solid foundations which the art of man can build for the support of bridges and hydraulic structures.

"In the inundation of 1857, the Ardèche destroyed a stone bridge near La Beaume which had been built about eighty years before. The resistance of the piers, which were erected on piles, the channel at that point being of gravel, produced an eddying current that washed away the bed of the river above them, and the foundation, thus deprived of lateral support, yielded to the weight of the bridge, and the piles and piers fell up-stream.

"By a curious law of compensation, the stream which, at floods, scoops out cavities in its bed, often fills them up again as soon as the diminished velocity of the current allows it to let fall the sand and gravel with which it is charged, so that when the waters return to their usual channel the bottom shows no signs of having been disturbed. In a flood of the Escontay, a tributary of the Rhone, in 1846, piles driven sixteen feet into its gravelly bed for the foundation of a pier were torn up and carried off, and yet, when the river had fallen to low-water mark, the bottom at that point appeared to have been raised higher than it was before the flood, by new deposits of sand and gravel, while the cut stones of the half-built pier were found burried to a great depth in the excavation which the water had first washed out. The gravel with which rivers thus restore the level of their beds is principally derived from the crushing of the rocks brought down by the mountain torrents, and the destructive effects of inundations are immensely diminished by this reduction of large stones to minute fragments. If the blocks hurled down from the cliffs were transported unbroken to the channels of large rivers, the mechanical force of their movement would be irresistible. They would overthrow the strongest barriers, spread themselves over a surface as wide as the flow of the waters, and convert the most smiling valleys into scenes of the wildest desolation."

M. Cézanne refers to the Ardèche as an illustration of the transition from torrents to rivers,—it may be described as a torrential river. And he refers to the work by M. de Mardigny as one in every way satisfactory, because the author, free from all foregone conclusions and theories, confines himself to description, provokes neither objection nor opposition. He says,— "The Ardèche is a great torrent, in which everything meets to produce a maximum of effect—a circular basin, converging affluents, denuded mountains, extraordinary rains. It is difficult to form any conception of the violence of the storms of rain which the wind from the south-east, the counter current to the mistral, precipitates on the amphitheatre of the Cévennes. M. Tardy de Montravel has received in his rain-gauge, in one day, 792 millemètres, about 32 inches, as much as falls in Paris in a year and a half. These tremendous rain storms occur only in September or October.

"The Ardèche is ordinarily a dry river-course, and the flood descends from the mountains and rushes along faster than a horse can gallop: the washer-women have to flee without thought of gathering up their linen; the delivery rises suddenly from 0 to from 7000 to 8000 cubic mètres, and next day the river is fordable: the deluge has flowed away.

"Sometimes the flood has been seen to rush across the Rhone presenting the appearance of a barrage, a wear, a broad wall of water, to break on the dike opposite the debouchure, and to spread itself over the plains on the left bank, and sometimes to cover the river with a continuous raft of trees torn from the mountains.

"The Ardèche alone creates in the Rhone at Avignon a sudden rise of more than 5 mètres, or nearly 17 feet, and in that country the sudden floods in the Rhone are called *coups de l'Ardèche*.

"If at any time a change of wind to the south-east, after having blown for a long time to the east, were to occur, and such a flood to occur when the Rhone and its affluents on the left bank were in flood, the river would probably exceed by many mètres the highest level known to have been reached by its waters.

"Such a thing might happen, and it has been asked,—What, in such an event, would be the effect of *reboisement* in the mountains? M. de Mardigny thinks that *reboisement* in the Cévennes would be of no effect; he seems to consider the Ardèche as one of those extreme and desperate cases, to which man, mastered by the elements, can only meet their fury with resignation."

Whether it be so remains to be seen.

In the department of l'Ardèche operations are being carried on in several périmetres, amongst others in these—Borée; Montpezat, La Champ, Raphaël and Laviolle, which conjointly comprise an area of 2,155 hectares.

SECT. VI.—*Department of Gard.*

In the department of Gard operations in some ten or more périmètres, one half of which, those of Pouteils, Malons, Concoules, Dourbies, and Bréau, cover an area of upwards of 4,000 hectares.

Of Pouteils, it was reported in 1869, that " this périmetre lay between 300 and 1400 metres of altitude, on the flanks of the Mount Lozère, the higher-lying region of the basin of the river Cèze, which is an affluent of the Rhone.

" The area of it is 741 hectares 93 ares, of which 654 hectares 93 ares belong to the commune of Pouteils, and 87 hectares to private proprietors.

" The soil, pertaining to the granitic formation, is deeply ravined by the different ramifications of the torrents which roll their waters into the Cèze.

" The restoration of these lands having been declared of public utility by a decree of date 13th January 1864, the works were begun in the spring of that year; and they have been prosecuted without interruption from that time onwards with the most hearty concurrence of the population.

" The State undertook directly the *reboisement* of the most elevated portions of the périmètre, measuring 451 hectares 87 ares. The different sections of the commune of Pouteils remained charged with the works on the remainder of the ground, but subventions were granted to them both by the Treasury and by the department.

" From 1864 to 1869 there had been subjected to *reboisement* or *regazonne-ment* 262 hectares 60 ares, viz:—

	H.	A.
" On communal lands at the expense of-the Treasury alone,	113	
" On communal lands at the expense of the commune, with subventions from the State and from the Department,	131	70
" On the lands of individual proprietors, with subventions from the State,	17	90
" Total,	262	60

" The half of the area might be considered as completely restored; it required further only works of maintenance. In the remainder of the extent, which corresponded to the higher-lying lands of the périmètre, the works had still to be made complete by new plantations.

" There had been employed 4990 kilogrammes of seed of resinous trees, 210 kilogrammes of seed of broad-leaved trees, (acacia, ash, chestnut, and oak,) and 1,500 kilogrammes of *fénasse*. There were planted, besides, in all, 81,000 young resinous and broad-leaved trees, and there were constructed 91 *barrages*, and on an extent of about 70 hectares there were pruned, with a view to recovering their vegetation, old trunks of chestnut trees.

" Exclusive of indemnities granted for deprivation of pasturage, which amounted to 1,221 francs, the whole expense of the works had been 22,045 francs 5 cents., of which 16,523 francs had been met by the Treasury, 3,600 by the department, and 1,922 by the sections of the communes and proprietors of the lands.

Sect. VII.—*Department of Lozère.*

The périmetre of Pouteils, though situated on Mount Lozère, is not situated in the department of Lozère, but in the department of Gard.

In Lozère, operations are being carried on in twelve or more périmètres, in seven of which—Saint-Bauzille, Lanuéjols, Badaroux, Balsiéges, Mende, Saint-Etienne-de-Valdonnez, and Chanac—there were comprised 5250 hectares.

Reference has been made in a preceding Part to what had been accomplished in the department of Lozère through the zealous co-operation of the prefect.

Speaking of the ravages committed in this department by torrents, and devastations occasioned by inundations, that honoured magistrate, at a meeting of the district *Société d'Agriculture*, held after the inundations of 1866, gave the following graphic account of the disastrous consequences which had come upon the department through the reckless destruction of forests. Speaking of these he said:—

"No, no, it is not God who has occasioned the evil which has come upon us, but men have done it in their improvidence, in their great desire to make the present minister to their enjoyment, without thinking of their children, without thinking of the future! When three years ago, on my arrival in La Lozère, the Municipal Council of Mende did me the honour to wait upon me; the first sentiment I expressed to them was that of my astonishment—my painful surprise—at not seeing more wood on the mountains, and on the steep declivities. And this was also the subject of my first address to this Agricultural Society.

"I brought before you the case of a land covered with sand—the poorest, the most miserable, the most insalubrious in France—the Landes; and I added that, thanks to *reboisement*, this department is to-day one of the richest, and the most salubrious in the empire. And, with a saddened feeling, I said to you, 'Take care; the rains and the snows are carrying off every day the lands of your mountains, strewing your valleys with the debris of rocks, raising up the beds of your rivers, and bringing about fearful and disastrous inundations. These lands, thus carried away, are lost forever, they go to the sea which engulfs them and never renders back what she engulfs.' Three years have not yet run their course, and see how this prophecy has been fulfilled! And do not persuade yourself that this is an accidental occurrence, which could neither be foreseen nor averted. How many warnings have you not received? These inundations are to a certain extent periodical; 1846, 1856, 1866, are three fatal years not likely soon to be forgotten! And note what a fearful progression may be seen in these disasters! In 1846 the Lot overflowed its banks and rose to a great height; but it did so slowly; it covered the valleys with mud and sand, it carried away the gathered crops, but it carried not away the land. The same was done in 1856, with this difference, that the rise of the river was less powerful. In 1866 the Lot, the Tarn, the affluents of the Allier and of the Ardèche, became furious torrents; within a few minutes they overflowed their banks, bearing down bridges, trees, mills, destroying houses, ploughing out for themselves new beds, and ravaging the soil so deeply that in many places they left only the hard rock. You have, with imprudent hands, overturned the barriers which retain the waters; you have cut down your woods, you have put your turf-covered banks under culture, and now you are astonished! Have you then no experience, and have times past taught you nothing?

"Glance over the Lot, from its source down to Barjac, and all along its course you will see nothing but stripped, bare, and naked rock; here and there some stumps of trees are there as if to attest that but lately there was there vegetable soil, and splendid woods and forests. Go from Mende to Florac, and over a stretch of 40 kilométres, and the same scene of desolation will be seen. Formerly all these sharp declivities were shaded by old pines and beeches and oaks; what a treasure! And to-day there is only the rock—the rock doomed forever to sterility, if you no not try to replant it with woods. Let a great storm come and our roads are covered with debris, and travelling is interrupted; and the State is compelled year by year to lay out for the restoration immense sums, which, but for this, had been more usefully employed in the improvement of our great highways. Communication is interrupted for many weeks, and agriculture suffers, because the farmer cannot go to the town to reach the markets for the disposal of his produce.

"Would you have a case still more striking? This summer I made a hurried tour through the south-eastern department. A mountain top caught my view. All the northern part of it was covered with a rich turf and trees; the southern portion, denuded, presented only a shapeless mass of rocks. The soil of it had disappeared. Low walls marked the division of the two properties. Alas! there was no need for this; it was sufficiently defined by the contrast of a rich vegetation with a soil of stones. I made inquiry, and I learned that the northern portion belonged to an intelligent proprietor, who had carefully conserved the turf and the trees, the heritage descending to him from his fathers. The southern part was a communal property. The inhabitants of the commune had partitioned it amongst themselves, and having cut down the trees, they had passed the ploughshare through the turf and sown it with rye. The snows and the rains had come, and the earth, little by little, had rolled down into the valley; ten years had sufficed to carry off all, and to leave there the rock alone. The communal property is now unproductive; the rest of the mountain crest has, on the contrary, acquired a considerable increase in value. The intelligent proprietor has enriched himself; the commune has impoverished itself. All the inhabitants have had was the delight to reap for some years a little barley. To-day they no longer reap anything, and they find themselves in misery.

"If I demand of history what she has to tell, I learn that for one hundred and fifty years the Gévandau was covered with forests, and the population rose to a hundred and fifty thousand souls. I do not know the result of any actual census, but what I do know is, that from 1850 to 1860 Lozère lost one thousand souls a-year. If one takes into account the increase of the population of France, which in the year 1800 reckoned only 18 millions of inhabitants, but counts to-day 36 millions, Lozère ought to have more than 300,000 souls! The earth being impoverished, the crops fail, the cattle become more sparse, the soil no longer supports man, and man expatriates himself; in this way the depopulation goes on increasing, agriculture fails to get labourers, extensive lands lie fallow, the cultivator cannot give to the land what it ought to produce, and thus the soil no more supports the man, and the man no longer makes the soil fertile; a vicious and fatal circle which infolds you, and the consequences of which, if you do not apply a remedy, will be to make this department a desert, which in a century, perhaps, having no more individual existence, will be divided amongst the neighbouring departments, and deleted from the map of France.

"God had made your country a country of woods, of pasturage, of cattle. To each country he has given its function; this was yours. You have changed it, you have uprooted the trees, you have ploughed up the turf, you have sought on these steep declivities to sow rye and wheat, you have run against the decrees of Providence and against the laws of common sense, and you have been sorely punished!

"In order to prove this, I have brought before you facts of the present and the teachings of history; permit me, in closing, to interrogate science; she also will give you instruction!

"The turf, the trees, which in your improvidence you have destroyed, retain by their roots the water of the storms. This water, of which one portion was absorbed, could only flow away slowly. The rivers could not enlarge and flow with the rapidity, with the violence, which makes of them to-day frightful torrents. The earth retained on our mountains was not carried away into the valleys; it did not raise the bed of our streams, it did not occasion their overflowing. Not only did the trees retain the water and absorb a great portion of it, but more than this, they caused it to penetrate to a certain depth into the soil. Their roots entering the rock lying under the vegetable soil, made as it were wells in which the water lost itself.

"To-day the torrents of rain have quickly carried away 20 or 30 centimètres in depth of the vegetable bed; underneath this they find the rock, and flowing over this as over a marble slab they carry away what earth remains. Would you have a proof of this? Run over the devastated cantons and everywhere you will see the fields, newly sown, cut into deep ravines; the neighbouring fields which have not been wrought, but which are covered by a dense turf, do not appear to have been ever touched by the storm. It seems as if God had desired to multiply proofs, for one cannot take a step on the mountain but they present themselves to the eye!

"All that was in wood, all that was in turf, has been preserved; all that was sown has been cut into ravines; and it is the earths detached from the mountains, it is the diluvial waters which nothing has retained, which have caused all our rivulets to overflow their banks, and has converted Lozère into ruins. Here even in Mendo you have a providential teaching. When the Lot inundated all the lower part of the town, threatening with death so many families, to whom it was impossible to us for eighteen hours to give succour—eighteen hours of agony!—I heard it said, 'What will the Merdançon do? Stopped by the rise of the Lot, it will change itself into a furious torrent, and it will carry away in the upper town the houses, as in the lower town the Lot bears down and overturns everything in its passage.' The Merdançon was an inoffensive rivulet, a few stones sufficed to make for it the semblance of a dike, it flowed slowly, it did not rise more than 30 centimètres, or 12 inches, it did no harm, it occasioned no disaster. Why? Because it flowed from a mountain the sides of which had been completely replanted with woods.

"See, gentlemen, the effects of *reboisement;* but these are not all. These trees, these turfs, which absorb, which retain the water in the storms, which prevent inundations and hinder the formation of torrents, these trees, these turfs, in the time of drought keep the springs from drying up. How many thousands of cattle have perished in the droughts which have fallen upon Lozère in 1864 and 1865? The more water, the more herbage!

"The farmers were ruined before because it did not rain, as to-day they are ruined because it has rained too much! And why? Because they have eradicated the woods and the turf which were their providence. They have killed—allow me the common expression—they have killed the goose which laid the golden eggs. It is true that, as some consolation, they say to themselves that they have cultivated a little wheat and a little rye on these precipitous slopes; but there they will soon find only misery, when the snows and new storms shall have carried away the little vegetable soil that remains."

Within three years after these appeals and statements had been made, the work of *reboisement,* as we have seen, was in full operation.

Sect. VIII.—*Departments of the Loire and of Haute Loire.*

The Ardèche has been shown to be a torrential river. The Loire presents to some extent the torrential character. The high mountains from which both it and the Allier descend—the sources of the Loire from an elevation of 1481 mètres, those of the Allier 1501 mètres, arresting in its passage the pluvial current from the west and from the north-west—receive very considerable rains, the produce of melted snows. The thaw, it has been remarked by M. de Coulaine, *ingénieur en chêf*, is sometimes brought about at Mende by a wind from the north-east; but this, it is stated by M. Cézanne, is originally a current of the west wind, which, caught in the valley of the Allier, having ascended it and passed the mountain, redescends into the valley of the Lozère, with a direction almost the reverse of that which it took from its point of departure; and thus is solved the apparent paradox; and the seeming exception, according to the popular misapplication of the expression, proves the rule.

The granitic slopes of the mountain are of course impermeable to water, and they are steep. Above Roanne, where the Loire becomes navigable for barks, the basin extends over an area of 6400 kilomètres, about the same area as those of the Eure and the Somme; but the Loire experiences floods of 7290 cubic mètres per second, which is a hundred times the magnitude of the greatest floods experienced by these rivers—a difference attributable to the greater abundance of rain, the more favourable ramifications of the *thalwegs*, the steepness of the slopes, and the absence of permeable soil.

By Cézanne has been brought forward the question,—Would the complete *boisement* of the basin of the Loire change the state of things in respect of floods? And he says,—" It is impossible, without having made a thorough and careful study of local circumstances, to answer the question, as some extravagant and enthusiastic advocates of forests desire to do; but what may be affirmed is, that in the Loire bringing down a very considerable quantity of sand, we have evidence that the higher-lying portion of the basin is subjected to an energetic action, whereby ground is cut up and washed away; and consequently there is reason to believe that the general *boisement* of the basin would to a marked degree modify the *régime* of the river. But as for the questions,—Would such *boisement* be a remunerative operation, or is it only possible that it might prove so? These are economic questions, the discussion of which has been often taken up, but the final settlement of which is still remote."

In the Haute Loire, operations of *reboisement* have been carried on in the périmètre of Megal, measuring upwards of 1440 hectares; and of Mézenc, measuring upwards of 1136 hectares and in several other hectares, measuring together 1026 hectares. And in the department of the Loire similar operations have been carried on in nine several périmètres, comprising in all an area of nearly 3000 hectares.

Sect. IX.—*Department of Hérault.*

In Hérault *reboisements et gazonnements obligatoires* have been carried on in some eight périmètres, four of which—Riols, Mons, Saint-Pons, and Saint-Julien—cover an area of nearly 4,800 hectares. Of the first of these it was reported in 1869:—

" The *périmètre obligatoire de Riols*, declared of public utility by imperial decree of the 13th January 1864, is situated entirely in the commune of Riols, canton of Saint-Pons. The mean altitude is 900 mètres. It comprises 1147 hectares 3 ares, of which 968 hectares 43 ares belong to the commune; 162 hectares 70 ares to sections of the commune; and 15 hectares 90 ares to different individuals.

" The périmètre, shut up by high mountains, has steep slopes, the denuded flanks of which are furrowed by numerous ravines.

" The mountains of Riol form two distinct groups: the chain of Sommail, with an altitude of from 900 to 1000 mètres; and that of Marcon, from 800 to about 900. The slopes vary from 50 to 75 in 100. In general they present a sad and desolate aspect; there are everywhere meagre pastures, cut up in all directions by ravines, where the eye meets not a single clump of trees, for the wood has by imperceptible degrees disappeared, destroyed by the hand of man, or laid waste by the teeth of sheep.

" The chain of Sommail is of granitic land; that of Marcon is calcareous.

" The first works undertaken in the périmètre date from the year 1864. At the beginning, the soil was prepared by digging holes, called *potets*, from 12 to 16 inches across, 12 inches deep, and from 3 feet to 3 feet 6 inches apart. Experience has led to some modification of these proceedings. Now, holes much longer, wider apart, and trenched to a greater depth, are made, which, permitting the roots to take more quickly possession of the soil, develope vegetation more in the places in which there is little depth of soil prevailing generally.

" These *potets*, when 6 or 7 feet apart, are about 40 inches square, and are dug to a depth of 18 or 20 inches.

" At the altitude at which the operations are carried on, resinous trees and the birch alone can withstand the rigorous temperature and the high winds which prevail on these unsheltered plateaux.

" Sowing or planting is the mode of operation adopted, according to the depth of earth. Sowings are preferable on the plateaux; but they do not succeed well on the elevated steep slopes, on account of the heat of summer, which affects injuriously the roots while they are too little developed to withstand the drought.

"Since 1864 all kinds of trees have been tried: amongst resinous trees—the silver fir, the larch, the Norway fir, the cedar, the Scotch fir, the Austrian pine, the Mugho or dwarf pine, and the Maritime pine; amongst broad-leaved trees—the oak, the beech, and the ailanthus. The silver fir and the larch have not succeeded well; the Norway fir and the Maritime pine on the heights tend to disappear, the ground being too dry, and the altitude too great; the cedar forms a thicket; but on the other hand, the pines withstand all extreme cold as well as heat, and snow as well as the wind, and thus the Austrian pine and the Corsican pine maintain their place.

"The beech takes well to the mountain, but grows slowly.

"The oak does not succeed on the plateaux, but grows strongly on the lower slopes.

"The ailanthus disappeared completely during the first winter.

"It may be well to add, that sowing broadcast short brooms has given good results.

"Thus it is now quite determined what kind of trees should be introduced into the périmètre of Riols.

"The law of 8th June 1864, which sanctioned the diversion from what was primarily set apart for *reboisement* of considerable areas for pasture grounds, has met the feelings of the inhabitants of the Sommail, proprietors of pretty large flocks. There has been nothing more to fear of dispute or opposition in this commune.

"From 1864 to 31st December 1868, there have been rewooded 415 hect.; and there have been expended on new works, and works of maintenance, 49,955 francs 28 cents., inclusive of 1445 francs provided by departmental conventions.

"The 415 hectares which have been rewooded from 1864 to 1868 may now be considered in a state of good keeping.

"The *reboisements* in the commune of Riols are of too recent a date for them to have been able as yet to exercise a marked influence on the *régime* of the water. The enclosures, and prohibitions of grubbing of box trees within the périmètre, have, however, prevented the increase and extention of ravines."

A corresponding report has been given of operations in the périmètre of Saint-Julien, in the canton of Olargues, in this department.

Similar are the operations carried on in the Puy-de-Dôme, and Cantal, both belonging to the region of the Cevennes and the central plateau; but these I pass to give information in regard to operations in the Pyrenees.

Sect. X.—*The Pyrenees*.

It has already been mentioned that it is only in the department of the Isère, the Drôme, and the Lower Alps, that we met with torrents like those of the High Alps. In the Lozère there are vallats somewhat resembling them; their representation in the Pyrenees are Gaves, which are torrential rivers or rapid water-courses in deep cuttings, losing themselves occasionally in subterranean canals.

Of the Pyrenees, the following account was given by Wild, some fifteen years ago, and relating to the everlasting mountain, it holds true, and will hold true, for a long time to come. In his volume entitled *The Pyrenees West and East*, he writes,—" Regarded in their largest extent, the Pyrenees may be said to extend from Cape Creux, on the Mediterranean, to the Gallician coast—a distance of about 670 miles ; but by the Pyrenean range is generally understood those mountains which divide France from Spain.

" Silius Italicus, whose voluminous writings throw light on the geographical history of various countries, says :—

"'*Pyrene celsa nimbosi verticis arce*
Divisor Celtis late prospectat Iberos
Atque æterna tenet magnis divortia terris.'

And the Pyrenees are still the barrier between those two countries.

" In a straight line these Pyrenees are about 280 miles long, 50 miles broad, and comprise an elevated area of about 1,100 square miles. The maximum height is nearly midway between the Atlantic and the Mediterranean, where the Maladetta attains an elevation of 11,424 feet, while several mountain peaks in the vicinity are but little below this elevation, and forty-five mountains are above 9,000 feet in height.

" The range is remarkable for its wall-like form indented by gaps, or 'ports' as they are called, which give passage between France and Spain. Through about fifty of these the principal traffic between the two countries is carried on, the intricacies of many of them being only known to the *contrabandistas* who abound in the Pyrenees. There are but five carriage roads in the chain, all lying to the extreme east or west. The 'ports' are generally higher than the Alpine passes, and present scenery of great grandeur. In consequence, however, of the Pyrenees being much more south than the Alps, and of their vicinity to the sea, the line of congelation is higher than it is in the Alps. Raymond fixes it at 8,600 ; Malte-Brun at 8,300 on the south side of the range, and 9,266 on the north side ; probably we shall not be far wrong if we assume 8,700 feet, or 1,300 above the line of perpetual snow in the Alps, as the Pyrenean altitude of perpetual congelation.

" Thus the grand glacial features which are characteristic of Alpine passes are frequently absent in the Pyrenees, when you are even on elevations which in the Alps are covered with ice and snow. But glaciers, snow-fields, and drifts, are not wanting in the high 'parts,' where the weather is generally so wild, and the path so bad, as to give rise to the proverbs— ' In the " part " where the wind rages the father waits not for the son, nor the son for the father ;' and ' He who has not been on the sea, or in the " part " during a storm, knows not the power of God.'

" A remarkable and very interesting feature in the Pyrenees, are the basins—' cerques,' or ' oules,' is their local name. They are situated in the transverse valleys lying between the buttresses of the principal range, and are generally surrounded on three sides by lofty walls of rock, opening into the valley by a narrow gullet. The scenery of these ' cerques ' is peculiar, possessing much sublimity with great pastoral beauty.

" The geology of the Pyrenees has not been as thoroughly investigated as is to be desired. Enough, however, has been done to inform us that the primitive rocks occupy but a very small portion of the chain. The arrangement of these differs very remarkably from that in the Alps, and elsewhere, where they burst out irregularly in the transition and secondary formation, whereas in the Pyrenees they run in bands, or zones, parallel to the chain. Thus, a very long granitic zone extends between Mont Perdu and the Maladetta, and other zones of the primitive formation may be traced to the east and west of those mountains.

"The secondary formation, or transition rocks, of which the greater mass of the mountain range is composed, consists of argillaceous schist, schistoze, and common grauwack, and limestone. These formations contain some minerals, principally iron ore, copper, and argentiferous lead. The iron ore is found in a white saccharine altered limestone, principally in the eastern portion of the range. The iron mines, in the valleys and gorges transverse to the Val d'Ussat, have been worked for centuries, and still employ a large number of miners; but the copper and lead mines are abandoned.

"On the other hand, the quantity of zinc ores, and especially calamine, yielded by the mines near Santander, within the last two or three years [previous to 1859], has been very great. I may also mention, that a remarkable deposit of rock salt, consisting of two vast masses, one of which measures 250 yards by 130 yards, exists on the side of the mountain of Cardona, and is still worked.

"French geologists formerly maintained that the Pyrenean range rose à *un seul jet;* but recent investigation shows that, notwithstanding the general unity and simplicity of its structure, six, if not seven, systems of dislocation, each chronologically distinct, may be made out.

"In great mountain chains, the lower elevations are commonly composed of secondary and transition formations, through which the granite pierces, and forms the highest mountain peaks. In the Pyrenean system, however, the case is different; for the highest peaks of the chain are composed of marine calcareous beds, the organic remains of which are pronounced by eminent geologists, including Sir Charles Lyell, to be equivalent to our chalk and green sand period. Recently, however, it has been discovered that the most modern of the Pyrenean rocks contain the same description of Eocine fossils as those found at Biaritz.

"Solemn thoughts fill the mind when we reflect that the proud peaks of the Marboré and Mont Perdu are studded with shells which once lived in the depths of the ocean. Looking wonderingly at them, we seem to hear the words,—'Where wast thou when I laid the foundation of the earth? and declare if thou hast understanding?'

"The dislocations in the Pyrenean system are intimately connected with the thermal springs; and as these form a prominent feature in the physical geography of these mountains, and possess high scientific importance, some account of the peculiar phenomena which they exhibit may not be unacceptable.

"Their number, as they 'spring through the veins of the mountains,' is extraordinary, no less than 253 being known; and there is a great and almost romantic interest in the fact, that they have for many centuries been ceaselessly pouring forth an almost unvarying quantity of water, for the most part of a high temperature, in some cases approaching ebulition. Remarkable, too, is the fact, that these waters, rising through vast earth and rock masses, undergo no change in their solid or gaseous composition. The same mineral water medicines, furnished in inexhaustible supplies centuries ago to our forefathers, still flows without change or stint."

There follow some disquisitions on thermal springs, which I omit as not so germane to the subject of *reboisement*. After these he goes on to say,—

"The Pyrenean valleys are much lower than the Alpine; few being more than 2,000 feet above the level of the sea, whereas those in the Alps are rarely less than double that height. Thus the mountains in the Pyrenees, when seen from the valleys, frequently assume a more imposing appearance than those in Switzerland of higher elevation.

" In the valleys and on the slopes of the lower mountains a great quantity of Lombardy poplars flourish; as we ascend, Spanish chestnuts, oak, hazel, mountain ash, alder, sycamore, and magnificent birch trees abound. Higher still, we come to the grand dark pine forests which form a prominent feature in the Western Pyrenees.

" The Pyrenean forests are classified as follows :—

" Imperial forests, . . . 129,440H.
" Communal forests, . . . 115,796H.
" Private forests, . . . 123,000H.

" Total, . 368,236H.

" There is, indeed, every reason to believe that the greater portion of the Western and Central Pyrenees were formerly covered by forests. In Bigorre many places were called *forum lignum*, and Roman writers allude frequently to the thickly wooded state of these mountains. As late as 1670 the royal forests were estimated to cover 174,300 hectares, of which, before the close of that century, 51,300 hectares were destroyed by fire. Communal forests are those, however, which have suffered most from reckless cutting and general mismanagement. In the early part of this century there were 31 saw-mills in the commune of St. Gaudens, at which trees were cut, abstracted, according to the Government report, from the royal forests. These mills are now suppressed.

" The Pyrenean pines are a variety of the *pinus sylvestris*, frequently attaining a great size, though not so thick in the bole as the graceful stone pine. This tree never forms forests in the Pyrenees, and is only met with in isolated groups. The peasants have a reverence for the stone pine, or rather for the kernel. When this is ripe and split, the cotyledons roughly resemble a hand, which they call '*la main de Dieu*,' and believe that by swallowing the kernels in odd numbers, as one, three, five, fevers and other maladies are cured.

" The streams—not turbid like those in Switzerland, but clear and bright— gush from every hollow, and water every valley, and impart an exquisitely bright verdure to the lower lands, nourishing at the same time an almost endless variety of lovely flowers. These are not, however, confined to the valleys, for, like all mountainous districts, the flora of the Pyrenees present an epitome of the vegetation from the equator to the poles; and botanists may like to be informed that in the valleys around the Cangou, and on that mountain, a very large proportion of the flora of the Pyrenees may be found. Two botanists collected in this district, during three days in June, 5,500 specimens."

Entering the Pyrenees by the Val-d'Ossau, about ten milles from Pau, he thus describes the scene :—" Adieu, now, to level roads, for we are on the spurs of the Pyrenees, ascending fast, as you may see by the rapidity with which the Gave-de-Gabas flows past from the rocky mountain, far above where the streams are born. Oh, the beauty of the Pyrenean rivers! Unpolluted by alluvial soil, they retain, throughout their bounding course, crystalline purity, reflecting as they flow varied hues from sky, scar, and wood, studded with moss-clad rocks, and fringed by lovely flowers. The road runs parallel to the brawling stream, retreating occasionally from the bank into the recesses of magnificent birch and chestnut woods.

"Few places are more singularly situated than Eaux-Bonnes. Viewed from a distance, you are puzzled to understand how the houses can find standing room in the wedge-like ravine containing them, and your surprise will not be lessened when you reach the smart little town. Fancy a section of a bustling Paris street, peopled by a curious mixture of gaily-dressed women, black-robed priests, prosaic bourgeois, Spanish and French peasants —the former wrapped in capacious brown cloaks, the latter wearing the picturesque berret,—cavalcades dashing to and fro, lumbering charrettes, and big oxen, and you have Eaux-Bonnes.

"Once housed, I set out for a ramble, unheeding the numerous offers from guides to conduct me to the Cascades. Indeed you cannot go wrong, for walks, zig-zagging up the mountains, through the woods, lead to various points of view. The most picturesque fall is the Valantin, which sweeps down amidst great rocks in a very striking manner. But the remarkable features of the walks around Eaux-Bonnes are the mountain forms,—particularly that of the Pic de Ger,—the dark pines and the patriarchal beeches. The huge roots of these trees assume the most fantastic shapes vying with the branches in length and thickness; you might imagine that the trees had been half torn from the earth by titanic force, and that the roots were writhing in agony.

"About half-a-mile from Les Eaux-chaudes, the Gave is crossed by Le Pont d'Enfer, an undeserved name, as there is nothing infernal about the structure. It leads, however, to wonderful scenery; a short way beyond, the Pic-du-Midi Ossau appears with its twin summits—a magnificent object towering over a crowd of mountains. Cascades stream down the precipices; and on passing the hamlet of Goust you plunge into a dark pine forest, which continues to Gabas. This is the last village in France, scarcely meriting that name, and consists of but half-a-dozen houses, whose inhabitants live by the traffic carried on between France and Spain. Nearly 20,000 mules pass the frontier annually."

The traveller describes his journey from Louvie to Lestelle, as made partly across a plain covered with maize, by a road frequently bordered by vines hanging in festoons from apple to cherry trees, and entering Lestelle, charmingly situated at the entrance to the valley of Lourdes, by a bold single arch-bridge spanning the Gave de Pau, here a soft blue stream, the crown of the bridge mantled with ivy hanging in long pendants below the arch—and the entire structure, with its back-ground of wooded hills, being highly picturesque. "Shortly after leaving Lestelle (says he) we enter the department of Les Hautes Pyrenees, and are again in mountain land; the valley now contracts, the hills are higher, and we see on a precipitous rock the old castle of Lourdes. Around this war raged long and fiercely. The Saracens, driven from the plains of Poictiers, took shelter beneath its walls from the victorious sword of Charles Martel; and our own history records how long and bravely English soldiers struggled to hold this, our last possession in the south of France.

"Beyond Lourde the scenery becomes barren and mountainous, which, however unpleasing, has the effect of heightening by contrast the exquisite beauty of the valley of Argelez, declared to be, and justly, the paradise of the Pyrenees; and if a combination of swelling hills, crowned by forest-clad mountains, clear flowing waters, deep green pasture, varied crops, orchards, picturesque villages, and a great number of churches and ruined castles, can be said to constitute an earthly paradise, the Val d'Argelez has these in perfection.

"At Pierrefitte the road bifurcates; that to the right leads to the Cauterets, and the left to Luz. The Pic-du-Midi-de-Viscos, a bold mountain, rising 7030 feet above the sea, divides the two gorges through which the roads are carried. The entire distance from Pierrofitte to Cauterets, five miles, presents a succession of wild mountain scenery, which, thanks to the road engineer, you are enabled to see to great advantage as you journey along. Beneath beetling buttresses glowing with lichens,—over the foaming Gave, —now surmounting seemingly impassable rocks, and then plunging deep into the gorge—your wonder increases as you proceed, until a turn of the defile discloses Cauterets."

It is a peculiar feature of nearly all the Pyrenean brunnens to be nearly buried in ravines. Cauterets is overhung by mountains which almost meet, leaving only a small triangular-shaped piece of ground on which the houses are built. The vicinity of Cauterets abounds with subjects for the landscape painter; but more to our purpose is the account given of the scenery enjoyed on an excursion to the Lac de Gaube, said to be one of the most interesting excursions to be made in the Pyrenees. Leaving Cauterets, the path, after a little way, leads to a point where the mountains seem to close, and the path winds up the defile of the Marcadaou, among huge boulders by the side of the foaming Gave. One spring after another is passed; that of Le Bois is the highest. "And here (says our traveller) I came upon a group of Spaniards, wrapped in their mantas; five stalwart fellows, with huge legs and feet cased in rough hempen sandals. They were drinking the water with much gravity; presenting a great contrast in this respect to the French, who gulp the nasty stuff amidst music and laughter.

"Now, however, you bid farewell to the springs and their votaries, and the scenery changes. The trees relieve the wrinkled face of the granite precipices; the Gave plunges down the gorge in a series of cascades; one, the Cerizet, is of great beauty; and the mountains on either side tower to a prodigious height, crowned by peaks. Higher still, you enter a pine forest, which clothes the summit of the lofty mountains—every ledge is fringed with pines, and only where the rocks are actually vertical are they bare. I rode slowly through the forest, being animated frequently by the exquisite views appearing between the pines. These, steeped frequently in the glowing prismatic hues of minature rainbows, formed by the water-falls—the underwood, matted by lovely creepers, shaded by even lovelier flowers—the trees, sturdier and more varied as the elevation increases—occasional glimpses seen through their branches of the peaks far above—such are the features of the ride to the Pont d'Espagne, six miles from Cauterets.

"This bridge, leading to the Marcadaou Pass into Spain, is a frail-looking structure of rough pines, thrown across a deep gulley, down which thunder the waters from the Lac de Gaube, and the snows and glaciers of the Marcadaou. The torrents, leaping together from the precipice, meet in mid-air, and plunge roaring and foaming down the gorge. Compared to the falls in Switzerland, these in the Pyrenees are diminutive, but the setting of the Pyrenean cataracts is, in my opinion, more picturesque. The rocks amidst which the water falls are invariably massive, and the vegetation displays a luxuriance unknown in more northern Helvetia. The Pont d'Espagne water-fall has, moreover, the grand features of magnificent mountain scenery—pine forests on the one hand, on the other bare precipices, above which you will probably see eagles wheeling in great circles."

Beyond this there was yet another water-fall. "From a small grassy plateau, you see the Gave rushing out of the Lac de Gaube, descend in a bold unbroken heap down a lofty precipice; but though the fall is unquestionably fine, it does not possess the interest of the double fall of the Pont d'Espagne, which has the advantage of a far greater abundance of water.

"The path from this point to the Lac de Gaube, above three miles, is wretched, being carried over fallen trees and great roots, among huge rocks, and frequently through swampy ground. At length, after a long climb, I emerged on a plateau, and saw a tiny sheet of turquoise-hued water,—

"'A lofty precipice in front,
A silent tarn below.'

The Lac de Gaube, though only two miles and a half in circumference, is yet the largest lake in the Pyrenees. What it lacks in extent is, however, in some measure compensated for by depth, for the sounding line shows that it is 425 feet deep in the centre. The mountains around the lake are bare, except where seamed by lines of straggling pines torn by the avalanches which plunge down in the spring. The centre of the picture is entirely filled by the noble Vignemale, 11,000 feet high robed with snow and streaked by glaciers; mists were curling up and wreathing the head of this grand mountain while I gazed upon it—now settling like a pall upon its crest, and now rent by blasts, disclosing the snowy heights and blue glaciers. The sublimity of the scene is greatly heightened by the absence of all cultivation.

"Long before the Lac de Gaube is reached you exchange the glowing warmth of Cauterets for a chilly temperature, and at the elevation of the lake, nearly 6000 feet above the sea, vegetation is confined to stunted pines and mountain flora."

From Cauterets Mr Weld proceeded to Luz. Writing of the journey from Pierrefitte, he says,—"On reaching Pierrefitte the road turns to the right, and you immediately enter the grand gorge through which the Gave-de-Pau descends from the mountains above Gavarnie. This defile is, if possible, finer than that between Pierrefitte and Cauterets, the mountain walls are closer and higher, frequently appearing to overlap each other; the woods are darker, and the torrent, which you cross over single-arched bridges no less than seven times, makes perpetual music, chaffing over its rocky bed. The present road is carried along a succession of shelves, overhanging the Gave, at a great height; very much lower, however, than the old road, the dizzy elevation of which may be seen by the remains of the Pont d'Enfers, which hangs 300 feet above the present structure. On our way up the gorge we met large flocks of goats, and droves of gaily caparisoned mules, which had left Spain but two days before. Picturesque animals are these goats and mules; they are in such excellent keeping with the scenery; and all along the border-land between France and Spain you meet them, often in places where you wonder they find foot-hold.

"A short distance from Luz the mountain walls recede, agriculture reappears, and in the midst of a small triangular plain stands the small town. It would be difficult to find a lovlier, or apparently more desirable spot for a residence, and yet frequent floods during winter compel the inhabitants to remove to higher habitations, consisting of mere huts erected to meet immediate requirements.

"Pleasant rambles lie around Luz. Three valleys invite you to wander—one leading to Pierrefitte, another in which St Sauveur is situated, and the third opening to Gavarnie. The last two are watered by rivers which meet at Luz and flow down the Pierrefitte gorge.

"Luz was formerly the chief town of the district, comprehending the adjacent mountains, and the three valleys from which they arise. The district formed a small republic. Laws were enacted, and the registers were kept by tallies, called *totchoux*, meaning cut sticks. This custom being unknown to an official who was sent from Paris to Luz at the close of the last century, on the part of the Government, he desired that the registers of the commune might be brought to him, and was not a little surprised to hear that a man was waiting outside his house with the registers in question, in the form of two waggon-loads of *totchoux*. Primitive simplicity! And, although many governments have ruled Luz since the talley days, the people of her valleys continue rude and simple."

From Luz he took a morning stroll before breakfast to St Sauveur. "The situation of this place (says he) is very remarkable: the little town of one street, standing upon a shelf of slaty limestone, overhanging the blue Gave, and commanding views of the valley of Barèges, or Lavedan, the entrance of the Val Bastan, and the Pas des Echelles, leading to Gavarnie." In the afternoon he ascended the Pic de Bergous, a Pic ascended by many visitors, which, though rising 6916 feet above the level of the sea, is but a pigmy among the giant summits which form the crest between France and Spain. It is (says Weld) to the Pyrenees what the Rigi Kulm is to Switzerland. "The lower slopes of the Pic are cultivated; for in this southern clime, elevations which in more northern latitudes would be clothed with heath, yield crops of golden corn. Above this cultivated zone the path winds among a great variety of trees, and above them it zig-zags up the cone of the peak. Herdsmen's huts, at this elevation, dot the mountain sides, fragile structures, which look as if a storm blast would uproot, and send them reeling down the steep. The climb near the summit is rather tough, but my pony made light work of it, and in about three hours from the time that I had left Luz he was cropping the herbage on the top of the Pic. Not a cloud obscured the panoramic view, which embraces a multitude of mountain masses. The Brèche de Roland is seen distinctly, appearing like a tiny notch in a mighty wall. To the left rose the snowy summits of the Marboré, Tremouse, and Mont Perdu; on the right the Vignemale, streaked with glaciers; and to the north-east the grand rugged Pic du Midi De Bigorre. These are the giants towering over a host of cones and pinnacles, furrowed and riven by winter storms; and the picture is filled up by dark dells, purple glens, green valleys, and gleaming streams, winding through pastures, corn-fields, and woods, which at this elevation seemed like a rich mosaic."

From Luz several other interesting excursions were made, the natural scenery and interesting incidents of which are all described with graphic power. In one of these, an excursion to the Brèche de Roland, he passed through the Cerque de Tremouse, his description of which I require to site to give to my readers a definite idea of the *cerques* or *oules*, or basins, of the Pyrenees, to which reference has been made; and I shrink, as I would from an act of vivisection, from attempting to extract it from the setting in which it occurs, and yet the whole of this it is impossible for me to give. In a preliminary statement is given the following, as a brief sketch of the leading features of the Brèche :—" On the west flank of the gigantic Mont Perdu rises Mont Marboré, consisting of a series of colossal steps or ledges, from the highest of which a huge stone wall stretches to the west, from 400 to 600 feet high, in most places absolutely vertical. This vast natural wall forms the crest of the Pyrenees at this part of the chain, and divides France from Spain. In the middle of the natural barrier is a gap, which seen from a distance appears a tiny indentation, but which is in reality a magnificent and colossal portal, 134 feet wide and 370 feet high.

"Near Gèdre, which is about half-way to Gavarnie, the mouth of the Val d'Héas is seen, one of the wildest and the most savage of the valleys of the Pyrenees. Close to Gèdre a grand view of the Brèche is obtained, making you wonder from its locality, high amidst the eternal snow-covered Tours de Marboré, how you are to reach it. At Gèdre, the Marboré disappears; but there is an almost over-abundance of grand scenery in the mountains towering to the right and left of that elevation, while there gorges are noisy with foaming cascades which swell the torrent. Close to these cascades—so close that they seem on the point of being swept away—are mills, not much larger than sentry-boxes, one above the other. These mills are of very primitive machinery, closely resembling that of the old hand-mills, but they grind the corn, and what more could the best mill in Europe do?

"Beyond Gèdre, you come upon a singularly grand and savage scene, called the Peyrada, or Chaos. It consists of an *éboulement*, or slip, of vast masses of gneiss, which have fallen from the precipitous sides of Mount Coumélie, and so vast and great is the ruin that you would suppose an entire mountain had been shivered to supply the blocks which lie around in grand confusion. The path winds as if it were perplexed how to find an issue from the rocky labyrinth; and the blocks are so huge that my herculean guide seemed a mere pigmy among them.

"The mountains increase in majesty as Gavarnie is approached; the Vignemale, with its glaciers, to the west, and the Pimené to the east, ranging among the loftiest. The morning continued highly favourable for our expedition; the mountain summits stood cloudlessly out against the deep-blue sky, crowned by myriads of soaring peaks, and pinnacles frosted with glittering snow. The path, about half-way between Gavarnie and the cerque, is carried over the torrent by two terribly narrow planks, without any manner of railing. Over this frail bridge, not three feet wide, my guide, greatly to my astonishment, rode his pony; and as my steed manifested no asinine disinclination to follow, but, on the contrary, evidently regarded the proceeding as nothing extraordinary, I slackened the bridle, pressed my knees a little closer to the saddle, and committed myself to my fate. The torrent rushed and roared some twenty feet beneath, but my pony was proof against these things; and what would have tried the nerves of many pedestrians was so familiar to him that he passed steadily over the narrow causeway as if it had been a broad highway.

"The passage of the torrent issuing from the cerque was the last feat of our horses; for after a brisk canter we dismounted in the arena of the amphitheatre, and turned the animals loose to graze.

"To render the first impression of the cerque, or oule, more impressive, a small projecting wall of rock masks the entry to the gigantic amphitheatre. This passed, the end of the world seems gained, for a vast barrier of rocks rises semicircularly before you to the height of between 1000 and 2000 feet. This gigantic wall is divided by three or four steps or ledges, each supporting a glacier from which stream cascades. That to the left, as you face the cerque, is 1266 feet high, and has the reputation of being the loftiest water-fall in Europe. The summit of this wonderful amphitheatre is covered by perpetual ice and snow, resting on the crests of the Cylindre, 10,500 feet high. The base of this fine mountain is embedded in a huge glacier, which gives birth to the high fall. Adjoining the Cylindre rises the Tours de Marboré, forming gigantic spurs of Mont Perdu. Stunted lichens alone vary the ruggedness of the vast semicircle of rocks, and the only sound breaking the stillness is that of the streaming cascades.

"The floor of the cirque consists of chaotic masses of debris. Immediately under the base of the precipice are large heaps of snow, beneath which the waters of the cascades flow, like the torrents spanned by the Alpine snow-bridges.

"You are unable to take in the sublime spectacle at once, so overpowering are its features; and gazing at the walls of the huge cerque, seamed by the rushing cascades, you fancy they are about to fall and crush you beneath their ruins.

"Within a few yards of the last water-fall on your right hand, the ascent to the Brèche is made. Without a guide the precise spot would be exceedingly difficult to find; and from the forbidding nature of the precipice, few would be bold enough to make the essay unadvised. It is literally a natural rock ladder, and is the only spot throughout the wide sweep of the cirque affording a means of ascent. The rugged strata, here nearly vertical, afford slight foot and hand hold; but there are places where the precipices are smooth, and you are puzzled to find a coigne of vantage. Here a steady head is necessary, as occasionally you have nothing between you and the bottom of the precipice but a thin shelf of rock on which you are standing.

"As we ascended new wonders were revealed,—precipices, cascades, and glaciers, alternating with wreaths of snow.

"The top of the great water-fall was still above us; and you have a very good idea of the altitude of the cascade, when, after more than an hour's ascent, you are still beneath the level of the glacier whence it is supplied. About two hours were occupied in surmounting the first series of precipices, and then we left the high mountain pastures, called by the peasants Malhada de Serrades, where goats pick up a scanty subsistence, and entered the snow-fields. Our course now lay through a very steep gulley filled with snow; up this we scrambled, taking advantage of the hardness of the snow to make it our path. Above us rose tremendous precipices, terminating in jagged peaks, on which my guide, with his practised eye, discovered numerous izzards. I saw them extremely well through my telescope, balanced like aerial creatures on the giddy heights, one of their number evidently acting as sentinel. Their attitudes were very graceful, all being ready at a moment's warning from their watchful leader to bound from crag to crag, or descend precipices untrodden by the foot of man.

"We now fairly lost sight of the cerque, and were in the midst of snow and glaciers, at a steep incline of about 40°. The climbing of this slope was most fatiguing to me, as the frozen snow was very slippery, and I retrograded nearly as often as I advanced. This part of the ascent occupied about an hour. My guide now turned to the left, for the purpose of crossing a glacier, so highly inclined that it cannot be ascended in a direct line. The passage of this glacier, beyond which lies the Brèche, is by far the most dangerous part of the undertaking. At the place where we encountered the ice the breadth of the glacier may be about 400 yards, but throughout, the inclination of the smooth polished ice is such that a false step might prove fatal; for beneath are grim precipices."

Here our traveller had a fall, the effects of which might have been serious. But, says he, " bracing my nerves, I resumed my slippery walk, taking care to hold my guide's hand, and resting occasionally. During one of these pauses a dull sound fell on my ear, and looking in the direction from whence the noise proceeded, I saw a grand snow-fall streaming from one of the ledges of the Marboré. Down it plunged with increasing roar, as the white mass loosened and gathered the snow in its course; but, before reaching the ledge below, a *tourmente*, or sudden gust of wind, caught the snow-fall and sent the scattered fragments whirling high in the air. The effect was extremely grand. This phenomenon is called in the Pyrenees *La Lid de vent*, in contradistinction to *La Lid de terre*; the snow in the first instance being, as I have described, borne upwards, sometimes whelming unfortunate mountaineers; while in the other case the snow descends, like the Swiss avalanche, into the valleys.

"At length I had the inexpressible satisfaction of achieving the passage of this formidable glacier. The rest of the climb was comparatively easy, though the steepness of the ascent, and the slippery nature of the footing, were trying enough. But all sense of fatigue forsook me when the huge portal—the tiny notch as seen from the valley near Gèdre--yawned in all its stern magnificence before me. The spectacle was a reward for all my toil; and I felt that I would have willingly endured even greater fatigue to make acquaintance with such a scene as now met my astonished gaze.

"Eager to attain the limit of my undertaking I hastened onwards, and with beating heart soon stood within the jaws of the mighty portal, through which roared the rushing wind. A step more and I was in Spain. Smooth glaciers slope away on each side of the wall; but opposite the Brèche, the action of the sun and force of the wind, here rarely at rest, through the great rock-rent, have tortured the ice and frozen snow into wierd forms, leaving the rock entirely bare.

"A wild world of barren mountains appears to the south; these in the foreground being covered with snow, the more distant looming hazily over the plains of Aragon. With a powerful telescope, Saragosa, it is said, may be seen if the atmosphere be clear; but although my glass was good, and the weather favourable, I could not discern it.

"Towards France the scene is softer. Mountains are there too, sky-piled; but also forests, the homes of wolves and bears, emerald vales, silver streams, and gleaming lakes. But how hope to portray the mighty phenomena of mountains and rocky pinnacles,—

'Dark, heaving, boundless, endless, and sublime;
The image of eternity,—the throne of the Invisible!'

"The wall, however, here about 600 feet high and 800 feet thick, is the great feature of the scene. Besides the Brèche de Roland, there is another opening in the wall to the west, called the Fausse Brèche. The precipices and glaciers between this and the Taillon, a lofty mountain which rises west of the false Brèche, are very grand; but the proportions of the gap are much inferior to those of the grand Brèche.

"The walls on either side of this Brèche are rendered still more imposing by being dominated by the Mareboré, the towers of which seem like a gigantic citadel protecting the approaches; and the similitude to a huge fortification is increased by the circumstance, that at each extremity of the wall, and close to the Brèche, is a hole which fancy suggests might have been pierced for titanic cannon."

Such are the High Pyrenees, and such, at this point, is the boundary between France and Spain.

The Brèche has had associated with it the name of Roland the brave Paladin, who, according to the legend, mounted on his war-horse in hot pursuit of the Moors, clove with one blow of his trusty sword a passage through this mighty wall.

It may be these pages will come under the eye of some who may be ready to say, as did the disciples of our Lord that had indignation within themselves,—Wherefore this waste; pages filled with pictures, and these not pictures by the author, but pictures by another? I read in my Bible, a soft answer turneth away wrath; and I reply, that this volume has been prepared, primarily, for readers in our colonial territories, most of whom live far from cities, and libraries, and bookseller's shops, to whom it is desirable that some idea of the country in which the works under consideration are being executed; and this I could do only by giving in full the graphic sketches I wished to cite. The physical geography of the Pyrecens is very different from that of the Alps and the Cévennes, and I deemed it proper that this should be shown. And being prevented by occurences, over which I had no control, from availing myself of the facilities for my visiting and seeing for myself the works of *reboisement*, and the localities in which they are carried on, referred to in the preface, I could only do this in the words of another.

In the High Pyrenees are several périmètres in which are carried on works of *reboisement*. One of these is at Lourdes, another is at Baréges, the famous watering-place, in the region now described. And on the day after his ascent of the Brèche, Mr Weld made an excursion to Baréges, which, says he, if not the most picturesque, is by far the most celebrated of the Pyrenean brunnens.

The road lies through the Val Bastan, continually ascending by the side of the Gave of that name, which is one of the most riotous and desolating torrents in the Pyrenees. Even in summer the Gave de Bastan is a very noisy watercourse; though you would not imagine, from the variety of lovely flowers gemming its banks, that the valley is yearly devastated by the floods which pour down from the surrounding mountains after heavy rains. About a couple of miles from Luz the valley contracts, and the vegetation is confined to shrubs, among which the common box is very conspicuous. This shrub is extremely common throughout the Pyrenees, flourishing on the ledges of precipices, where it might be thought impossible for any plant even to live. In such localities it never grows beyond the dimensions of a shrub, but there are places in the Pyrenees in which in attains to those of a tree. Of Baréges, he writes:—"Nothing but dire necessity would tempt you to stay at Baréges more than a few hours; for, independently of its situation, which has scarcely a redeeming feature, almost every one you meet is crippled, wounded, or in other respects diseased in body, and unlovely in appearance.

"The ground on which the permanent houses of Baréges stand is so limited in extent, that they are necessarily few in number, and although erected in the most eligible locality, are perpetually subject to the risk of being overwhelmed by *éboulements* from the mountain which rises precipitously immediately behind them, or swept away by the torrent before them. Buttresses of great strength prop these buildings, but even with this protection they are occasionally seriously injured, and a few years ago were nearly annihilated by the bursting of the Lac d'Oncet, on the slopes of the Pic du Midi.

"The mountains impending over Baréges are composed almost entirely of clay slate, coated by vast alluvial deposits, which frequently descend in the form of mud avalanches. This soil is highly unfavourable to vegetation. Even the hardy pine cannot thrive at Baréges, and a few straggling trees above the town are so stunted that they look, on the scared face of nature, like the result of an abortive attempt to grow a beard on that of man.

"Baréges in summer consists of numerous temporary wood tenements, which far out-number the permanent houses, and afford accommodation to the visitors. Many of them are devoted to purposes of trade, and the name and business of the shop-keepers are announced on stripes of red cloth hung over the door. The articles sold are for the most part exceedingly trashy; but the traders doubtless know their customers' wants, and the depth of their purses. I asked the price of 'barege,' not, by the way, made here, but at Baguères, and found it to be actually dearer than you can buy it in London; however, the shopkeeper was quite willing to bargain. These wood structures are set up in the beginning of April, care being taken to plant them at a respectful distance from the torrent; and at the end of the season they are taken down, and stowed away until the following spring. Thus the population of Baréges in winter, when the snow is fifteen feet deep, does not amount to above more than fifty persons, whereas in the height of the season it frequently exceeds 2000."

Of the works of *reboisement* the following is the account given in 1869:—
"The valley of Baréges is, strictly speaking, only a narrow gorge enclosed between two links of a chain of high mountains, with abrupt slopes, and tooth-like crests. It takes its departure from the Col de Tourmalet, to issue on the valley of Luz, following in so doing, from east to west, a direction almost parallel to the central chain of the Pyrenees.

"The altitude of the Col is 2122 metres, that of Luz 710 metres; there is reckoned between these points a distance of 17,730 mètres; the average slope of the *thalweg* may then be given as nearly 8 in the 100.

"The torrent of the Bastan rolls its waters at the bottom of the valley on a bed of granitic rock, torn from the flanks of the mountain. On the left bank an imperial road goes up to Baréges, whence goes a carriage road, of recent construction, which leads by the Col de Tourmalet to Bagnères-de-Bigorre—the last is only passable during four or five months of the year.

"Five communes—Esterre, Viella, Viey, Sers, and Betpouey—compose the valley of Baréges, and constitute what is called the Vic-de-Labatsus, the vast undivided pasture lands of which are exploited in part by the inhabitants, in part by strangers.

"The general aspect of the country is that of all mountain lands from which the improvidence of man has caused the forests to disappear—it is bare and sad. From whatever side it is contemplated, there meets the eye scarcely anything but clean shaven mountains, the flanks of which are furrowed by numerous ravines, which go on digging themselves year by year; in winter the snow is heaped up, and holds the country exposed to formidable avalanches; in summer every storm of rain transforms them into torrents, the dejection of which collect in heaps at the foot of the valley, lay waste the meadows, and obstruct the roads.

"The spot most seriously threatened is the burgh town of Baréges, a hamlet of the commune of Betpouey, situated 6500 mètres from Luz, at an elevation of 1232 mètres above the level of the sea.

"It is generally known that Baréges has within its bounds sulphurous springs, which are justly renowned, and to which come numerous patients in quest of a remedy for their sufferings. The valley people have erected there baths of a monumental appearance; the Minister of War has caused to be erected an extensive and beautiful military hospital; and a civil hospice has recently been erected.

"These are the only important erections which can be spoken of. At Baréges the dwelling-houses have nothing of the comfort and elegance seen at most of the fashionable hot-springs. Hotels and private houses have a poor and pitiful aspect, which need not excite surprise, if one considers for a moment that the shock of an avalanche may destroy from the foundation, and lay in heaps, an edifice erected at great expense. Baréges, moreover, is during the winter inhabited only by some forest-guards, and a few individuals left in charge of the public and private buildings. The thermal season lasts from the 1st of June to the 30th September; when that is over, all the tradesmen, and others, who depend on visitors and their requirements for a livelihood, hasten to close up carefully their dwellings, barricade the doors and windows with planks and beams, and make off for the plains. And one peculiarity is to be noted—all the shops erected in the vicinity of the military hospital are wooden erections, set up in spring and taken down in autumn, lest an avalanche should come and carry them away.

"It appears as if every moment were spent in Baréges in the consideration of means of protecting the place against the action of the snow and of the waters, and thus does it appear to have been long.

"Thus in 1594 the Conseils of the valley interdicted the felling of trees there.

"On the 6th May 1732 a resolution of the Council of State made it expressly forbidden, under pain of corporal punishment, to cut or lay waste the trees and woods which surround the hamlet of Baréges, and protected it against ravines.

"A decree of the *30th Prairial an XII.* renewed these prohibitions, and charged the prefect to propose to the Government any measures which he might believe would be useful to prevent the formation of ravines and avalanches.

"On the 22nd February 1815, a resolution of the prefect of the High Pyrenees, approved by the Minister of the Interior, determined the measures to be taken to prove infractions of the decree, and to put an end to them.

"In 1839 M. de Verdal, captain of military engineers at Lourdes, proposed to construct in the ravine of Theil—that which threatened, more particularly, the hospital and the baths—a system of stone dikes.

"About the same time, Major Itiet, of the 5th Regiment of *Chasseurs à Cheval*, devoted his leisure to the study of the same subject, and submitted **a measure deserving of consideration**, in which he gave an exposition of the means of securing the height of Baréges by the help of stoccndoes of stone, or dry stone dikes consolidated by being filled up behind with earth.

"In fine, in 1843, the Departments of War, of the Interior, and of Agriculture and Commerce, brought about the meeting of a special commission, called *Commission des Avalanches*, charged to consider the localities, and formulate such propositions as might relate to the matter. Unhappily, projects, resolutions, and regulations were passed, but never applied, and they remained in the state of a dead letter.

"In 1859, the Emperor was sojourning at the waters of Saint-Sauveur. He visited Baréges; with his own eyes he took the measure of the danger which threatened this important thermal station, and determined to exercise it.

"The effect of this resolution was soon felt. On 31st August 1859 there met, summoned by the Emperor, a commission composed of the prefect of the department, of the *ingénieurs en chêf des ponts et chaussées* and of mines, of the commandant of military engineers, of the syndic of the valley, of the sub-prefect of Argelès, and of the inspector of forests. It was perceived, after a thorough discussion, that *reboisement* was the only means of combating the evil, save indispensable artificial works to be executed under the corps of military engineers, to guarantee the hospital from the attacks of the ravine Theil.

"The work was begun in 1860 as soon as the ground could be reached; and on the 22nd August the same year was prepared a detailed project— prepared with a view, on the one hand, of replanting the mountain of Lacgrand; and, on the other, of arresting the dejections of the Rielut, a ravine, the effects of which will be afterwards described.

"Finally, on the 11th May were commenced the sowings, and before the heat of summer had come to interrupt them, they had been carried over an area of 47 hectares.

"But it was necessary to regulate the execution of the works, both as regards the law and regulations, and as regards the interest of the commune and of the individual holders of property along the banks. The requisite formalities were duly observed; and an imperial decree of the 21st February 1863 declared of public utility *reboisement* to be effected on 280 hectares of land situated on the territory of the commune of Sers and Betpouey.

"It remains to considered what has been done since, against what difficulties it has been necessary to contend, and to what extent it has been found practicable to surmount them.

"I.—The imperial decree of 1863 ruled the area to be rewooded as 280 hectares 38 ares, but it was in point of fact 302 hectares 32 ares, of which 232 hectares 87 ares belonged to the canton Lacgrand, and 64 hectares 45 ares to the canton Ayré.

"It was in the canton Lacgrand that the first works were undertaken; and therefore attention will first be given to these districts.

"The mountain Lacgrand is situated in front of Baréges, on the right bank of the Bastan, with a southern exposure. The crest of it terminates in the peak Capes at an altitude of 2400 mètres. Three principal ravines furrow the slope : the ravine Midaou, which debouches below Baréges; the ravine Theil, which menaces the central portion; and the ravine Aygunave, coming down a little above the town. All these bring down avalanches: the ravine Theil is the most formidable in this respect.

" The entire length of this is about 1200 mètres, the fall varies from 5 to 6 in 10; it is surmounted by a vast funnel, with precipitous walls, which crowns the peak Midaou. Through the configuration of the valley the north-west wind rushes into it with violence, heaps up the snow in a considerable quantity in the funnel which terminates the ravine, and facilitates thus the formation of avalanches sufficiently powerful to expose to the most serious danger the hospital, the baths, and the neighbouring erections.

" It is the military engineers who have charge of the works to be executed in the bed of the torrent, with a view to the creation of an immediate and permanent obstacle to the descent of the snow, in anticipation that the *reboisement* will produce the results which we have a right to expect. For the *barrages* constructed with hurdles, &c.,—first tried, and broken down in the first winter—there have been substituted high and strong stoccadoes in solid masonry.

" With regard to the forest agents, they, as has just been stated, put hand to work in the month of May 1861, and they sowed in the spring and in the autumn 140 hectares of ground with seeds of the larch, the Norway fir, the Corsican pine, and the Austrian pine.

" The year 1862 was devoted entirely to works of maintenance.

" The new works resumed in 1863 consisted of sowing acorns and seeds of the Austrian pine on 38 hectares and 54 ares, so that at the close of the year they had operated on 178 hectares 54 ares—that is to say, on the whole of the portion of the périmètre capable of cultivation, the remainder being only an agglomeration of bare rocks, entirely stripped of vegetable earth.

" In 1864, 1865, 1866, the reclothing of the ground, consisting in sowings and plantings of the different kinds, were carried on on the Mountain Lacgrand. Nothing was done in 1867, 1868, 1869, the available resources having to be spent on other spots.

." The area of the ground belonging to the canton Ayré, if we deduct the denuded rocks, is reduced to about 50 hectares, which overhang the orifice of the ravine Riculet, with a northern exposure, in front of the communal wood of the Trouguet, the only block of forest which exists in the country. The mean altitude of this canton is 1800 mètres; as is the case with Lacgrand, schist and chalk constitute the mineral basis of the soil, with this difference, that the northern slope is characterised by blocks of granite, sometimes scattered, sometimes lying together in considerable masses.

" In 1862 and 1863 they had sown the whole canton with seeds of Norway fir, the Austrian pine, the Corsican pine, the Mugho or dwarf pine, the alder, and the birch.

" In 1866 the work was resumed in the form of works of maintenance, and this on a scale which showed the insuccess of the first experiment; they consisted of sowings of the Mugho or dwarf pine, and in the planting of 274,000 plants of the Austrian pine, the Mugho, the beech, the birch, and the acacia. Some replanting remained to be accomplished in 1867, 1868, and 1869. These works consisted almost entirely of plantings of the Mugho.

" If now it be asked, what have been the results definitely attained, it will be found :—

" On the mountain Lacgrand 81 hectares have been completely rewooded in seven distinct blocks, viz :—

" 30 Hectares of oak mixed with larch and Austrian pine.
" 9 Hectares of oak and Austrian pine.
" 5 Hectares of Scotch fir, beech, and oak.
" 14 Hectares of Mugho, larch, Norway fir, and Austrian pine.
" 13 Hectares of Mugho, larch, Austrian pine, and Corsican pine.
" 4 Hectares of Austrian pine and Mugho.
" 5 Hectares of Norway fir.

" The remainder of the ground capable of cultivation shows no complete block, but scattered clumps and trees of different kinds, which may be estimated to cover about a fifth part of the whole of the area, which may be 19 hectares 54 ares.

" The portion wooded may be, to what remains still bare, as 100 to 78 ; or, in simple terms, four-ninths of the périmètre remains to be rewooded.

" Passing from the mountain of Laegrand to that of Ayré, it may be stated, that in this canton 40 hectares have been completely rewooded, viz. :
" 10 Hectares of Corsican pine, in a promising condition.
" 15 Hectares of Mugho.
" 5 Hectares of mixed trees—beech, birch, Austrian pine, larch, and Mugho.

Now, as the total area—with the deduction of works—is reduced to about 80 hectares, it may be admitted that the wooded portions are, to the portions still bare, as 4 to 5.

" Though still incomplete, these results are remarkable. They show that the agents entrusted with the direction of the works have not rested satisfied with the accomplishment of any thing short of their task. They have surmounted, indeed, difficulties by no means inconsiderable, which it may be well to pass in review.

" From the first there was the altitude of the périmètres, the enormous quantities of snow which covered them for six months in the year, the extreme cold which prevails there in winter, and the heats which suddenly succeed this. Then, in the commencement of the enterprise, the whole staff of agents, brigadiers, guards, workmen, &c., had to be formed. And, in fine, there had to be resolved two questions of the greatest importance : the selection of the kind of trees, and the mode of restocking the waste lands with these.

" There was the selection of kinds of trees : they could not confine themselves to those which almost exclusively grow in the country—the beech and the silver fir. Both the one and other of these adapted themselves badly, during the first years of growth, to the slopes, completely bare or subjected to late frosts, which the sun dried up in July and August, to say nothing of the summer preceding, and the check to vegetation which would thus be given. It was a matter of necessity to proceed by way of trials—like groping in the dark.

" Then, should they sow or plant? At first, the plans were very defective here. Then, from theoretic considerations, the arguments in favour of sowing greatly preponderated, and a great many good agents deprecated planting as being more expensive and less likely to prove efficient than sowing.

" Without taking up this question in the abstract, it may be enough to affirm that, in so far as Baréges is concerned, experience has decided in favour of planting. This is a necessary consequence of the climatal condition of the country. In point of fact, towards the end of October, or in the course of November, the snow invades the périmètres, and does not disappear in general till the end of April, or in the course of May. From the time that the soil is thus made accessible to atmospheric agencies, vegetation takes on the character of activity which is peculiar to mountainous regions. Let seeds of good quality be then sown, they will spring up speedily, satisfactorily, and in such a way as to give rise to the greatest expectations. But then comes a hot and dry summer (and this is what is generally the case in the valley of Baréges); the vegetable layer of soil, wanting compactness and depth, dries up under the action of the solar rays; the young rootlets which creep along near the surface of the soil infallibly perish; and it may happen that, with the exception that in streaks of deeper earth in which moisture has been retained, before the end of autumn there remains scarcely a trace of sowings executed with all imaginable care. To this the oak is an exception, but it is such an exception as may be said to confirm the rule, as it is known to every one that the acorn committed to the earth developes its radicle in the form of a long taproot, which buries itself to a sufficient depth in the soil, not to be affected by drought to the same degree as young resinous trees.

"To the instruction yielded by past years may be superadded that yielded still, every day; for the work of *reboisement* is being now carried on in the périmètre of Sers and Betpouey, contiguous to that of Baréges. There also, some partial losses, now repaired, have taught the agents the best course to follow in order that from this time forward it may be easy to complete, with certain prospects of success, those portions of the périmètre of Baréges which are still in the state of gaps in the work.

"The rules proved by experience to be necessary to follow are these:—

" 1.—Excepting for the oak, to give the preference to planting, and only to sow in exceptional circumstances.

" 2.—To give up planting or sowing in spring, the season being too far advanced before the périmètres become accessible.

" 3.—In the more elevated parts, to prefer the Mugho or dwarf pine to all other trees. It can withstand much—it developes itself slowly but surely—while the heat of the summer causes the Norway pine to perish; as for the larches, generally long and slender, the weight of the snow lays them and deforms them. In lower-lying parts, to mingle the Mugho with the Austrian pine, and with the Scotch fir to the south. In fine, to employ the oak at the base of the périmètre of Lacgrand.

" 4.—To employ as much as possible trees of only three years' growth, or even of two years' age; at a greater age their taking root is very uncertain.

" 5.—In fine, so to arrange always that the plants, taken from the central nursery at Luz, shall all be planted within forty-eight hours *at most* after their extraction. This precaution is essential in gorges where prevail, especially in autumn, south winds, the effects of which are hurtful to the young plants.

"By proceeding thus the plantations already created will be completed without fail. This will now be done if it do not prove needful before all to push on actively the works of Sers and Betpouey, in such a way as to counteract the attempts made to put a stop to the execution of them.

"The question may be raised, when once the *reboisement* of Baréges shall have been happily completed, may we reckon for the future on a fine stock of trees? This question it would be rash to answer definitely at present. It may legitimately be doubted whether at such altitudes vegetation will ever show a lofty growth; but in any case the object aimed at will be constantly attained: an end will be put to the scourges of ravines; the occurrence of avalanches from the quarters rewooded will be stopped; and, in fine, Baréges will be delivered from the dangers which threaten it.

"At the present moment, guarding against optimist estimates which may not be fully warranted, it is difficult to determine exactly the results obtained, or to determine to what extent the first works, and to what extent the works of the military engineering corps, have contributed to securing these results. As a matter of fact, while the last two winters have witnessed the fall of a considerable quantity of snow, no serious accident has occurred. Instead of formidable avalanches, the ravine Theil has only given some successive smooth snow-slips, which have slid into the bed of the Basan, and have come to lie on the north façade of the military hospital without occasioning the least damage. One of these snow-slips, more considerable than the others, temporarily obstructed the bed of the torrent, the waters of which cleared for themselves a passage across the only street in Baréges, inundating some cellars, and some rooms on the ground-floor, but only occasioning altogether trifling damages.

"A fact to be noted is the action on the régime of the waters exercised by the prohibition of pasturage on the périmètres. From the time that the flocks ceased to come on these there has been developed naturally a mass of herbaceous and of ligneous plants, which constitute a true natural *gazonnement*. These plants supply to the soil a precious covering; they arrest a portion of the snow which the winds blow towards the ravine; their roots envelope the vegetable soil in a powerful net-work which keeps it in its place; in fine, at the time of the melting of the snow, and at the time of great storms of rain, they divide the waters, and by so doing retard their flow. Perhaps we ought to see, in this prohibition of *parcours*, one of the causes why Baréges had no avalanches in 1867 and 1868. In any case, it would be a grave error to cramp the vegetation by uprooting, and by inconsiderate clearings. Such like operations should be limited to suppressing plants, the shade or the immediate contact of which might be adverse to the growth of the young plants.

"II.—There is to the south, in front of Baréges, a small communal forest stocked with beeches and silver fir, which is situated on the base of the mountain of Ayré. It is at the summit of this wood that the ravine Riculet takes its birth, and in about a straight line it makes for the imperial road, near the entrance to the town.

"The formation of this ravine is of recent date; according to the local tradition, the origin of it does not date further back than sixty years ago at most; it is, moreover, within half-a-century that the Riculet must have taken such a development as to occasion disquietude.

"It is composed of an open abyss, in the form of a funnel, in magnitude about 100 mètres at its orifice, the precipitous walls of which show exposed the rocks which constitute the mineral basis of the mountain. On leaving this funnel, the waters, in making for themselves a way, have opened up a large and deep bed of exceedingly steep fall, cut up in different directions by precipitous perpendicular falls. The slopes of this ravine have a height exceeding 25 mètres in some places; they are formed of transported earth successively subdivided, in which are lodged enormous blocks of granite. In summer, when the drought is continuous, these earths acquire a consistence and hardness which gives to them the appearance of a cement. But when there comes a storm of rain, or continuous rain, which softens the earth, then follows a phenomenon of which it is difficult to form a correct idea unless it have been witnessed. At a definite moment the transported earth becomes semi-fluid, suddenly puts itself in motion, and changed into a thick, blackish mud, descends in a compact mass towards the bottom of the valley, carrying off in its course the blocks which it held suspended. These, striking against each other, roll on with a dull, continuous sound, somewhat like that of thunder—a sound familiar to the inhabitants of Baréges, who, hearing it from afar, hasten to see pass what they call the *Barranque.* In an instant the imperial road—flooded, covered with mud and débris, and stopped up with large blocks of granite—becomes impassable for carriages and for beasts of burden, and sometimes for foot passengers. And at the same time the adjacent meadows are covered with sand and stone from the one end to the other.

"For some years past—thanks to the works executed by the Forest Administration in the bed of the ravine—the dejections of the Rieulet have become reduced to a very trifling amount; and all devastation, even the most trifling, would have ceased entirely if there did not exist, between the road and the limit of the forest soil, some hectares of land which it was considered should be left at the free disposal of the municipal authorities. It is from this come, under the impulse of waters, the few blocks which still land on the imperial road.

" On the *Administration des Ponts et Chaussées* devolves the charge of clearing the way. A few hours of a few workmen will from this time forward suffice for this work. The average annual expense, which was 3500 francs, has been reduced to 700 francs, a sum destined to be still further decreased.

" It is by means of dry-stone *barrages* that the Rieulet has been bridled. The first of those erected at the head of the ravine were not of strength sufficient to withstand for a lengthened time the action of the waters; undermined at the base and at the sides, this ended in their fall, and they have not been re-erected, because it would have been necessary every year to renew the underworks, and because, moreover, there have subsequently been erected others sufficient to maintain their position against all that may occur.

" In 1861, 1862, and 1863, a beginning was made to open at the summit of the Rieulet large and deep cuttings, to stop up the entrance of the ravine, and turn away the waters of the rainfall.

" At the same time, they consolidated the ground by means of turfed terraces, and wooden stakes and hurdles; they lowered the sides of the ravine in the more abrubt slopes; they tried, moreover, to plant there cuttings of willow and of alder, but all imaginable care has failed to cause these cutting to sprout in a soil which becomes desiccated and hardened to a great depth.

" It was in 1861 that the erection of *barrages* was commenced. Undertaken and continued without interruption from that time onward, there are now of these fifteen, exclusive of those the abandonment of which has been determined. The total cubic contents of them is about 4300 mètres,

" As all of these are similar, and differ only in dimensions, it will suffice to describe briefly the strongest of them, that known as *barrage* No. 4.

" Begun in 1867, it was finished in 1869.

"Its thickness is 4 mètres, its height 14 mètres, its length at the middle 28 metres—in all 1568 cubic mètres.

"Encased, in the base and the two sides, in the calcareous rocks, in which by mining there was dug a deep groove, it is protected against undermining, so much to be feared for works of this kind. It is composed exclusively of large blocks of granite, laid in successive courses without any mortar, but dressed and put in their places with minute care. Its form, slightly convex, gives it more solidity to withstand the shocks and the pressure of the earth coming from above. In fine, they have dug or left large open sluices for the flow of the water.

"A similar *barrage*, known as No. 5, has been begun in 1869; it may be completed next year [1871]; and it will complete the system of defence for the imperial road and neighbouring properties.

"When the Rieulet is in movement, the mass of its dejection is divided and retained by the *barrages*; the water and the sand alone escape by the open sluices; the solid materials heap themselves up behind the walls, and these give birth to extensive horizontal platforms on which are stopped the blocks detached from the summit of the ravine, which formerly came down without hindrance to the imperial road like veritable projectiles.

"At the commencement of the works nobody in the country believed they would be successful. To undertake to restrain the Rieulet was, they would say, an impossibility. To-day the most incredulous are constrained to yield to the force of evidence.

"III.—To complete the description of the périmètre of Barèges it is proper that mention should be made of the opening of a forest road, which, setting out from the civil hospice, goes up by a series of *lacets*, of a pretty gentle slope, to the base of the peak of Ayrè. It is of a total length of 11,782 mètres; sustaining walls line it where the mobility of the slope might create fears, and the roadway is in such a state of solidity that the expense necessary for its maintenance will be very small.

"The formation of this road was indispensable that they might come with beasts of burden to the origin of the ravine Rieulet, or to the extremity of the canton Ayrè, for works connected with the périmètre of Barèges as well as for those which were connected with the périmètres of Sers and Betpouey. Undertaken in its lower part in the course of the year 1862, it reached in the year following a development of 8100 mètres; repaired and consolidated since then, it was finished in 1868.

"Apart from its utility in regard to the works, the *route d'Ayrè* is a real benefit to the bathers who frequent Barèges, a benefit greatly appreciated by them and by the whole of the intelligent portion of the population. There is not, indeed, in the locality other place of promenade than a level alley along the front of the houses of the town, devoid both of shade and shelter. Now the forest road stretches for nearly 8 kilomètres across the communal forest, sheltered from wind and sun in running through the midst of massive blocks of beech co-mingled with silver pine, which give a smiling aspect to the road, which being moreover frequented by a great many of the bathers, the presence of these gives to it a character of great animation. The more vigorous get on horseback to the peak d'Ayrè whence they see the valley of Luz, that of Lians and the glaciers which crown the Néouvielle. Those who cannot undertake so long an ascent, content themselves with making shorter tours, and find, if necessary, a sheltered resting-place in the little nursery adjoining the road, where they are hospitably welcomed.

"IV. Having thus enumerated all the works accomplished up to this time in the périmètre of Baréges, it only remains to be stated how the works were received at first, what opposition was made to them, what complete change his since been produced in the spirit of the people, and what have been the causes of this change.

"At first the works encountered in the Administrative inquests and in the Municipal Councils an absolute and determined opposition, which showed itself first in silent or avowed combinations, then by daily acts of malevolence, and, in fine, in a hostile demonstration made at Viella, quite near to Baréges, in the form of an assemblage who stopped by main force the works begun shortly before.

"In consequence of these doings the authorities issued the revocation of a mayor who had been connected with them.

"Subsequently, evil-disposed men, who kept unknown, destroyed altogether the nursery at Baréges, burned the barracks which had just been erected there, and smashed with stones the door of the house inhabited by the forest brigadier.

"In fine, on the night of the 10th March 1867, fire was set to the sowings of oak on Lacgrand at fourteen different points at once.

"On this last proceeding the judicial authorities took action. The chief of the Court of Lourdes hastened promptly to the spots, with the forest agents, and caused to be arrested some inhabitants against whom there was grave presumptive evidence.

"In default of positive proofs, the prosecutions had to be abandoned; but the firmness displayed by the imperial prosecutor had sufficed to inspire the inhabitants with respect for the law and the works. Besides, it must be acknowledged, these acts of vandalism had raised in the country a feeling of disapprobation of such a nature as to bring the malevolent to reflection. Meanwhile, the employment at Baréges of an auxiliary brigade had the effect of keeping them in order; and no second attempt of the same kind has been made since.

"But if in these circumstances it was necessary to give proof of energy, and in like circumstances it may be proper to do likewise, it is proper that rigour should only be used in cases of absolute necessity.

"Thoroughly imbued with this idea, the agent charged with the direction of the works has endeavoured to maintain, in his relations with the inhabitants of the valley, the kindest consideration. All their objections and complaints have been listened to with attention, and calmly discussed; and none of them which could be met with compliance, without serious inconvenience, have been withstood. The forest guards have been invitingly counselled to avoid all irritating remark or proposal, and all irritating discussion. As the local brigadier, naturally active and moreover devoted in the prosecution of the enterprise, could not, face to face with the population, give up an aggressive style of procedure, his removal to a distance has been applied for and granted.

"By degrees, pacification of the people has come about; all physical violence has ceased; there no longer occurs either open resistance or threatenings, or more slight offences; the inhabitants look with satisfaction on the works which are being prosecuted. The work of *reboisement* counts now but a limited number of opponents in the country, and these, doubtless, will finish by rallying themselves with the supporters of the measure, when enlightened by experience of the benefits it brings."

It may be more than one of my readers may now be almost willing to thank me for having brought them to the scene of these operations by the route I have done, in the pleasant company of such a tourist as Mr Weld. But our errand accomplished, it would be pleasant, but apart from our purpose, to accompany him further; and, thanking him for all his details, here he and we must part. I found myself just now heaving a sigh at the thought that the description of the country seen by him, in his descent to Toulouse, must not be touched. Nor his details of what was seen by him in a journey thence through the department of Ariége, and the Eastern Provinces. May we never meet a less pleasant *compagnon de voyage!* We may journey long and not meet with a better, sharing not our passion, but not less interested than we in the natural scenery to which it leads us.

In the department of the Lower Pyrenees, operations are carried on in three périmètres; and in the Western Pyrenees, in the périmètre of La Tet, covering an area of nearly 5000 hectares.

SECT. XI.—*Department of l'Aude.*

In the region of the Pyrenees are comprised, not only the departments of the High and the Low and the Western Pyrenees, but also the departments of Aude and Ariége, in which are carried on works of *reboisement.* Of two périmètres in the former—those of l'Argent-Double, comprising 2842 hect., and of the Rialesse, comprising 1080 hect.—in monographs are given special reports. Again selecting one, I take at hap-hazard the monograph on the former :—

"This périmètre lies in three communes of the canton of Peyriac-Minervois, arrondissement of Carcassonne : they are those of Caunes, Citou, and Lespinassière.

"It was decreed of public utility on the 22nd April 1863.

"L'Argent-Double takes its rise in the territory of Lespinassière, at the bottom of a valley closely confined, running east and west, and the crests of which constitutes the boundaries of the departments of Aude and of Herault.

"From its origin to the end of the périmètres, this mountain stream receives a great many affluents, of extemely irregular delivery, which take their rise in the three communes of Lespinassière, Citou, and Caunes. Besides, it receives, on the territory of Caunes, a considerable affluent called the *Ruisseau du Cros,* which takes its rise in the commune of Félines-d'Hautpool (Herault), and flows along and across the périmètre before its junction with l'Argent-Double. Nothing has been done in the commune of Félines-d'Hautpool to regulate the flow of this mountain stream. The mountains in which it takes its rise are all of them in a bare state, or a state of grievous destruction. It would be desirable that *reboisements* were executed on the territory of this commune, to secure in time coming the advantages expected from the works being executed in the Aude.

"The actual périmètre comprises the steepest slopes, and the most devoid of soil, of the valleys of l'Argent-Double, and of those of the affluents situated in the department of l'Aude.

"The highest altitudes are of 934 and 1022 mètres, in the commune of Lespinassière; and the lowest, of 413 and 754 mètres, in the commune of Caunes.

"The climate is dry and hot, the valley being a southern rent of the Montagne-Noire. In general, there is no rain in summer, but when storms of rain come. When these are violent, the havock caused by the l'Argent-Double is considerable. Sometimes there is no rain for many months, and all vegetation is stopped in its growth, if it be not altogether destroyed.

"This has been the case in 1869, in which there has not been a rain-fall in storms of rain, or otherwise, penetrating the cultivated soil to more than three centimètres, or an inch and a quarter, since the month of April. This long continued drought has given a terrible blow to the replanting of years preceding, and is being hurtful also to the works of the autumn of 1869, for it does not allow of the works being carried on economically, nor of planting being employed there when this mode of operation is admissible.

"The prolongation of these droughts in summer, and the severity of the frosts in winter, cause that in this périmètre sowings be practised generally, and plantings only exceptionally.

"The soil is very variable. At the highest summits of the mountain in the communes of Citou and Lespinassière, it consists almost entirely of micaceous schists, easily disintegrated, and still more easily flooded into ravines; the earthy residua of this kind of rock are dry and much under the influence of the smallest thread of water. As fast as the ravines thus created go on lengthening, the carrying off the hills goes on extending with rapidity, and earth, stones, and every thing are swept away into the tide of the principal streams, and often thrown out upon the cultivated grounds situated on their borders.

"In one part of the territory of Citou, and in the greater part of that of Caunes, the soil is argillaceous-limestone, or calcareous clay of little depth, tufted with innumerable calcarious shelves. There the floods have scarcely any debris to sweep away: the work of destruction has been almost completed.

"This périmètre comprises private lands, communal lands, and dominal lands. These have been derived from the old Benedictine convent of Caunes, and amount to 420 hectares, 306 hectares 76 ares of which only are comprised in the périmètre.

"The division of the lands, according to proprietorship, may be summed up thus:—

"Caunes,	Dominal lands,	306H.	76A.
	Communal do.,	178H.	06A.
	Private do.,	371H.	14A.
"Citou,	Communal lands,	360H.	89A.
	Private do.,	783H.	88A.
"Lespinassière,	Communal lands,	657H.	03A.
	Private do.,	184H.	26A.
	Total,	2842H.	02A.

"*Reboisement* has been carried out only on the lands belonging to the State and to communes. The commune of Lespinassière erected the works by aid of a subvention of 60 per cent. guaranteed by the State; the two other communes have left the expense of the execution of the work to be met entirely by the State.

"*Gazonnement* has been employed on no part, as neither the land nor the climate permitted of this.

"During the preparatory inquests, and at the beginning of the operation, the Administration was sustained in the enterprise by the whole of the more enlightened part of the population, and also by that portion of the lower classes who derived no personal benefit from the communal property; but it was violently opposed by the proprietors of flocks, who constituted the minority of the population. The disposition of these began to improve from the time that indemnities for the deprivation of pasturage were granted to the communes; and everybody, even those who were once the most opposed, have become convinced of the utility of *reboisement*.

"The efficacy of the *barrages* has been generally acknowledged, and many private proprietors have constructed them at their own expense on those parts of their properties which they wished to retain for culture.

Notwithstanding these facts, it is the case, that few of the inhabitants consent to carry out the work of *reboisements* and *barrages* on the lands comprised within the périmètre. They wish to retain their lands as pasturage, as the communal lands escape from this.

"To complete, then, the work on the périmètre, it is absolutely necessary to acquire by degrees the required lands belonging to private proprietors, and to continue the grants of indemnities for deprivation of pasturage.

"These indemnities are based on an average of 4 francs per hectare received in Citou and Lespinassière, and of 2 fr. 70 cents. at Caunes.

"The work done up to the 31st Dec. 1868 may be reported thus, over the whole extent of the périmètre:—

"*Reboisement* of 224H. 11-50A. dominal, 199H. 25A. communal, and 2H. 80.60A. of private lands—in all, 426H. 17.10A.

"Pruning of plants scattered over the communal and dominal lands.
"Preparation of 5000 layers of beech.
"Construction of 932 rustic *barrages* of dry stone.
"Construction of 175 mètres of enclosing walls.
"Opening up of 5330 mètres of path.

"The *reboisements* were made principally by sowings. The kinds of trees employed up to this time were, according to the altitude and the soil, the green oak, the hard oak, the chestnut, the Aleppo pine, the Mugho, and the Atlas cedar.

"When plantations could be made, use was made of the hard oak, the chestnut, the ailanthus, the ash, the Mugho, and the beech. Of the Mugho employed in 1867 in the commune of Lespinassière, to the number of 50,000, almost all were uprooted by a continuous frost, which lasted from the month of December 1867, to the 1st April 1868. These hard frosts, followed in the month of April by heat, which was considerable, prevented the principal plantations being made in the spring, which is a serious matter, on account of the facility with which the schistoze soils are heaved up. When it does not freeze much in the months of February and March planting can be effected; but the grants not being available at this season, nor the seed to hand, it is necessary to renew the whole in autumn.

This course of procedure has been followed since the commencement in 1864, and has not given very bad results, notwithstanding the deplorable atmospheric circumstances to which the young plants had to submit. The excessive heat of summer is the cause of the principal failures in success. The year 1869 may be cited as having been particularly disastrous. There had not fallen a single penetrating shower in the territory of Cannes and over half of Citou from the month of April to this time (20th November). A considerable dessication was the result, but as each *potet* sown has still living plants there is no occasion to renew the greater part of these sowings. A dominal lot in Cannes, however, replanted in 1865 with green oak, has been burned up to the roots by the drought, on from 20 to 22 hectares. In deducting these from the 426 hectares in all, it may be considered that there has succeeded perfectly and is now in good condition about 400 hectares, for there is little probability that they will have to withstand now droughts like that of this year, which they have stood pretty well. Certain sowings suffered in 1866 from very violent storms of hail, but they have been completely restored

" Since 1868 this périmètre has had a nursery of which 50 ares only have at this time been sown. This will suffice for the wants of the district, as sowings are almost always the only mode practicable in *reboisement* here.

" The expenses have been pretty high, on account of the nature of the soil, which is very stony. Every *potet* had moreover to be wrought to a considerable depth, that the roots of the young plants might reach as soon as possible a depth such as would protect them from drought.

" This method of preparing the soil is indispensable with this view; unfortunately it contributes to the striping of the plants at the roots in hard frosts. To combat as much as possible this scourge the root of the young plant is surrounded with stones of a greater or less size.

" The total expense incurred to the 31st December 1868, inclusive of the value of all that has been supplied, amounts to 43,264 fr. 28 ct., which has arisen thus—including the priming and preparation of layers in *reboisement* :—

Works of *reboisement*,	Fr. 41,937 13
Construction of *barrages*,	418 45
„ of paths,	735 70
„ of walls,	123
	Fr. 43,264 28

" These sums may be divided according to the ownership of the property on which they have been spent, thus :—

" The whole of the works executed on dominal lands have cost	Fr. 25,825
" Works on communal lands executed at the expense of the State,	14,089 03
" Amount of subvention of all kinds granted to the Commune of Lespinassières,	3,157 50
" Subventions granted to private proprietors,	39 75
" *Reboisement* effected on private property at the cost of the State,	155
" Total as above,	Fr. 43,264 28

" The re-planting effected in the périmètre of l'Argent-Double have not yet been able to exercise a considerable influence on the régime of this affluent of the Aude, for they do not occupy more than a seventh part of the area. Nevertheless it has been ascertained, after the storms of rain which came on in 1865 and 1866, that the *barrages* had retained great quantities of earth and stones, and that the *potets* dug on the flanks of precipitous mountains had also retained in like manner a considerable proportion of materials, acting individually as a multitude of small *barrages*."

Besides these and similar operations carried on extensively in the departments named, similar operations have been carried on in different localities in the region of the Vosges and of Mount Jura; but enough, if not more than enough, has been brought forward to enable the student of Hydrology—or the student of Forest-Science—to compare the present with the past.

CHAP. VI.—LOCAL FEELING AND PUBLIC OPINION IN REGARD TO REBOISEMENT.

Again and again in the preceding pages have indications of a strong local feeling against the prosecution of the enterprise appeared, with indications of these having subsided, and in some cases,—and these not few—given place to feelings of satisfaction and of a disposition to help forward the work. This appears to have been the case everywhere in so far as those who were more immediately affected by the operations—landed proprietors, and the communal population—have made themselves heard.

It may be attributed, perhaps, in part to prejudice, but apart from this, (1) to the operations interfering with long-accustomed usages, and even with the livelihood of many who saw and knew and felt their immediate effects, but did not see, and could not foresee, the benefits which were to follow; (2) to the sincere desire of the Government to avoid all private wrong in seeking to secure a public good, and their making this manifest in their legislation in regard to the matter; (3) to the Administration and their agents endeavouring, and that successfully, to act with the meekness of wisdom, fixing not their attention on the passionate expression of opposition, but endeavouring to find out the irritating grievance, and acting in a conciliating spirit, and taking proper measures to get this redressed; and (4) to the adaptation of the means employed to effect the end desired.

I consider it of some importance that the successive phases of local feeling and of public opinion in the localities in which these operations were carried on should come under the attention of legislators, and of practical men, who may be led to adopt similar works of *reboisement* and *gazonnement*, or modifications of them required by local circumstances, as means of counteracting similar evils elsewhere. Opposition in any such case may be anticipated, and it may be that the opposition will not be without good cause; but if the means be equally adapted to accomplish the end designed, and that end equally desirable, if the same meekness and forbearance be maintained by those entrusted with the direction and prosecution of the enterprise, and the same enlightened legislation characterise the enactments upon which the enterprise is based and by which it is regulated, the same or a similar issue of the opposition may be anticipated. If any of these be lacking it may be otherwise; and I know not which of them is of most importance. My aim and desire has been limited to showing what has been done in France, and in what circumstances and with what results this was done, and to supply, in so far as this can thus be done, information which may be utilised by those who may be called upon, or may feel themselves otherwise moved, to endeavour to counteract similar evils now or at any time hereafter.

Local feeling has been frequently referred to. Public opinion in connection with local feeling may be learned from the record of deliberative councils in the districts in which the operations were being carried on, and these show it to have been in accordance with what I have stated in regard to local feeling.

It was in the High Alps that these works of *reboisement* and *gazonnement* were most urgently called for; it is in that Department and those immediately adjacent that the works have been most extensively carried on; and the populations there have not been reticent of their opinions, nor have they failed to secure that these should be heard.

M. Cezanne, writing at the close of the first decennial period of these operations, and after the interruption which had been caused by the war, says,—" It is interesting to read now, in the proceedings of the Conseil Général of the High Alps, the reflection of the different states of feeling and public opinion in the Department. In the Session of 1860 the prefect announced the law *pour la mise en valeur* or improvement of communal lands; and for the *reboisement* of the mountains *the Conseil Général voted a subvention of 500 francs.*

"In Session of 1861 a report was given of what had been done. *The Conseil voted 1200 francs.*

"In the Session of 1862 it appeared that in regard to *twenty-five* proposals there had been fulfilled the legal formalities required. These embraced 60,000 hectares, on 6800 of which *reboisement* was to be begun immediately; on 790 *gazonnement* was to be carried out; and 13,533 to be put *in defends*, or conserved by the temporary prohibition of pasturing and passage. *The Conseil voted a subvention of 3000 francs.*"

"But the mischief had begun to manifest itself. The agents of the Waters and Forests and the Engineers had co-operated with zeal. The reports read to the Conseil gave evidence of a lively faith, but the Conseil itself received these communications coldly, and it was felt that the opposition, thus far kept down, would not be long in bursting forth in flame. The prefect replied to the objections that it was impossible to reduce them to words, but that it was felt that they were in the wind. . . They had, it appeared, pictured to themselves the forest agents as *ogres ready some day to devour both shepherd and sheep*. . . The nature of the opposition showed itself clearly; it was the mountain *versus* the plain. . . . Let an example suffice. The proposed extinction of the torrents of Sapet and of Devezet was submitted to discussion in these communes: Ancelle and Saint-Leger on the mountain, and Labatie-Neuve on the plain. The prefect reports on this matter in these terms: "Called to give their counsel, the representatives of the commune of Ancelle formally announced that their vote was against the proposal. The Chief of the *Commission des Reboisement*, M. Costa de Bastelica, who took part officially at the séderunt at my desire, in order that we might be in the best possible position for supplying the information which might be needed, astonished at a refusal for which a motive could scarcely be imagined, asked permission to speak, and called attention to the circumstance that the measure affected Ancelle scarcely at all, but the Commission, on the other hand, has had in view to preserve a whole valley. Nothing was done: the vote was negative. The representatives of the commune of Saint-Leger were next called, and gave the same negative vote. 'But,' said some to them, ' you are only consulted on account of an interest which is very indirect, through your interest in a portion of the land taken as a whole. Would you please explain what is the ground of your opposition to the execution of a project the utility of which it is impossible to dispute?' Reply: 'The torrent passes far below our place, and cannot do us any harm.' 'But this little hill of undivided territory has been already subjected to the forest régime?' Reply: 'Yes, but tho Forest Administration allow us the run of it; this will be prohibited tomorrow if sowing is carried out on it or on the felled wood.' It was now the turn of the delegates of Labatie-Neuve. These praised highly the project, and protested energetically against what they called the selfishness of their neighbours. They showed over and over again, from different aspects of the subject, what there would be unjust and cruel in leaving much longer whole communes situated on the lower parts of the torrents exposed to disasters every day, when in reality it would occasion no appreciable damage to Ancelle and to Saint-Leger. The efforts of the delegates from Labatie were powerless to obtain the least concession. They were two against four, and a majority was obtained only by support lent by the Counsellor-General, the Counsellor of the Arrondissement, and the other permanent members of the Commission."

The quotation is given from the formal report on *Comptes rendus du Conseil Général des Hautes-Alps, Session 1862*, and M. Cézanne remarks— "This little life-portraiture is full of instruction. Who now will question that the intervention of the State, so strongly urged by Surell, was necessary?"

In the Session of 1863 twenty-six new proposals were submitted. The Commission of the General Conseil opposed to them the mournful complaints of the communes. The prefect, pre-occupied doubtless by the political state of the country and approaching elections, showed himself much less firm than he did the year preceding. The Conseil voted, however, again 3000 francs, but demanded at the same time the revision of the law, and the stipulation of a previous payment of indemnity for the communes deprived of their pastures.

In the Session of 1864 the reports of the prefect maintained a prudential silence on the subject of *reboisement*. It spoke of everything but this, the one most important matter affecting the district. The Commission of the Conseil was less reserved. The Commission of the Conseil confirmed the unanimous protests of the peasants. There had been violent outbreaks in the environs of Embrun, the intervention of the military had been deemed necessary, and from the plain had been seen the mountain flashing with light reflected by the bayonets. The Conseil demanded as extensive an application as possible of the law of 8th June 1864,—that is to say, of the substitution of *gazonnement* for *boisement*, of grass for trees, and a liberal distribution of indemnities; and at the same time it considered it duty, while doing this, *to reduce the subvention from 3000 francs to 500*.

In the month of November the imperial decree enforcing the law of 8th June was issued.

In 1865 the success of the first works began to bear fruit,—the reaction had begun. In the General Conseil of the High Alps the prefect again brought up the question of *reboisement*, which for two years he had, from prudential considerations, avoided or touched on slightly. The Commission of the Conseil made it the subject of a long report, in which they extolled the benefits of "*that law which people had cursed, when they ought to have blessed it.*" The Conseil called for the prompt execution of the works, cast blame on the conduct of the Conservator of the Forests, who had taken up his residence at Gap, while by a decree of the Emperor he was required to reside at Valence. From this report it appears that two communes—but two only—had demanded the indemnities to which they were entitled, and reflected on the tardiness of others to profit by a law so advantageous.

In the Session of 1866 it was reported that a newly-appointed Conservator, M. Sequinard, had fixed his residence at Gap, that his enlightened experience and benevolent firmness had gained for him the complete sympathy of the Conseil, that the service of *reboisement* discharged its functions in a satisfactory manner, that the works carried out in 1865 gave the following results :—

Extent *Reboisée*, - - -	664 Hectares.
„ *Regazonnée*, - - -	3400 „
Number of *barrages* constructed, -	2797 „
Length of *clayonnages*, or barriers of hurdles, - - - -	26,500 „
Length of *barrages vivants*, - -	22,267 „
„ Drains cut, - -	727 „
„ Roads, - -	12,695 „

On which had been expended 103,196 francs.

But the Conseil passed a vote that the indemnities to the communes should be *obligatoires et non facultatifs*, imperative and not simply permissible, and that proprietors who may have re-planted their grounds with trees should be released from having to reimburse to the State the advances which shall have been made to them; and *ultimately the Conseil voted a credit of 500 francs*.

The opposition may be supposed to have been factious and the demands unreasonable, but they were not altogether so. M. Sequinard, the Conservator of Forests in the district says, in one of the reports made by him about this time, that they were not. After stating what were the requirements of the enterprise, he goes on to say : " The basis of material operations having been determined, it is proper to enquire who in equity should bear the expense of the work.

" The laws of 1860 and 1864 have laid down on this subject rules which do not satisfy the population of the mountains, and which have made them in many communes very hostile. They consider that the extinction of the torrents, being demanded by the general interests, ought to be executed entirely at the expense of the State, as are all other works pertaining to the public interest, such as canals, roads, &c. They find it unjust that nothing should be required of the wealthy populations of the valleys, to protect whom is the prime object of the regeneration of the mountains, and that, discounting the profit of which they have by degrees been deprived, they should withdraw from this useful operation. Further still, they believe that they have a right to indemnity for the trouble which the extinction of torrents will necessarily occasion them in their habits of life and in their means of livelihood."

It is impossible, says M. Sequinard, to withhold an acknowledgment that these allegations are not without foundation. Then, after having made some reservations suggested by the immoderate use of the pasturage, an abuse which, tending to the ruin of the mountains, is the primary cause of the inundations which come, with brief intervals, to desolate the valleys, and after having established with true sagacity that the legislation of 1860 and 1864 was not responsible for this,—having given to the communes, under the form of subventions, the amounts necessary for the consolidation of the ground in ravines,—he concludes thus: "In view of the preceding considerations, the Conservator asks to be authorised on his next official circuit to promise to the communes interested that in future the State will undertake the charge of all the works which the extinction of torrents imperatively demand, will indemnify in a just degree the communes which shall have suffered thereby, and will aid those which shall labour seriously at the *reboisement* of their mountains, *on condition that the communal lands shall be placed under the régime of the Forest Administration.*"

Of M. Sequinard's qualifications for his office as Conservator of Forests in the High Alps the highest testimony is borne by those who know him. M. Cézanne writes: "It is in the reports of M. Sequinard that we must look for what may be called the philosophy of these operations. The extinction of torrents is now a science, and its principles are deduced by M. Sequinard in explicit and substantial theorems."

"If," says M. Sequinard, "*boisement* be the only means of extinguishing torrents, it is not indispensable everywhere to fix and consolidate the soil. It is nature that teaches us that, except in the cases of landslips and rents and of some few spots, the ground will consolidate itself if it be protected against the abuse of pasturage. So, the extinction of torrents requires only (1) that the ravines be replanted with trees; (2) that the pasturage be brought under regulation; and (3) that the most exposed spots, principally at the head of ravines, should be strengthened by partial *boisement*, trees of high growth being selected, so as to interfere as little as possible with the use of the pasturage.

"In ravines copsewood is of more effect than forest trees, from the difficulty of getting out the produce; high trees cover the soil only imperfectly, and they protect it less manifestly than young close coppice in a state of brush.

"Trees of small dimensions do tempt to trespass, but planted in horizontal strips the wider apart the steeper the declivity, the young trees promote, by the freshness and moisture which they maintain, the growth between the strips of herbaceous plants, which after a few years may be depastured, which is essential to the prosperity of a pastoral country. Moreover, in almost all the localities the people feed the cattle in winter with the leaves and young shoots cut green in the beginning of September.

"Thus all interests are secured, (1) by replanting the ravines in alternate strips with broad-leaved trees, which shoot readily again from the stump, and which may be exploited in a brief rotation—which conditions are fulfilled by acacias, elms, ashes, maples, and *les bois blancs*, [a designation applied conventionally to woods of inferior quality and of a soft contexture, irrespective of colour, as alders, elders, poplars, and willows]. (2) By *gazonnement* of the ground between the strips; but it is not enough to fill up actual gaps, it is necessary to prevent the formation of others.

"The abuse or immoderate use of pasturage being the main cause of disintegration of the soil, it becomes of essential importance to regulate this. It is a weighty and imperative duty for the communal proprietors to do this. In point of fact, if the communes be proprietors, the successive generations have only the usufruct of the ground, and they should act the part of a good father of a family—that is to say, be improving and not destroying the heritage; and it pertains to the prefectoral authority—their tutor—to regulate their enjoyment of the usufruct; the abuse of this, then, is not permitted, and ought to be withstood energetically."

In the Session of 1867 the Commission of the General Conseil of the High Alps reported with grateful expressions the efficacious and productive impulse which had been given to the service by M. Sequinard, and reported thus of the work: "Experience has spoken, and if we decree to-day, or have already decreed, the regeneration of the mountains, the success of this great work is henceforward a matter of certainty. Results almost beyond what were expected obtained on many spots permit us to reckon absolutely on the final result. They permit us to foretell from the present the day when, with means for adequate action being given, our grand slopes shall be regenerated, the ground on steep declivities shall be consolidated, and the main torrents which desolate our Alps shall be extinguished, or at least repressed.

"The ruling principle from this time forward in the operations of the *Commission des Reboisements* is the substitution, wherever it may be practicable, of *regazonnement* for *reboisement*. *Gazonnement* cannot fail to ameliorate in a not distant future the lot of the pastoral populations, while *boisement* would deprive them for an age of the enjoyment of lands on which it might be adopted with the most marked success." *The Conseil voted a subvention of 500 francs.*

In the Session of 1868 the report of the prefect established that "the work of *reboisement* and *gazonnement* was making year by year great progress in the department, and that wheresoever works of this kind had been undertaken the population of the localities, being enlightened in regard to the design and object of the operations, and in regard to the means by which this was being accomplished, had shown themselves satisfied." And as for the Commission of the Conseil, their report rose to the height of a lyric in praise of the undertaking. The Chief of the *Service de Reboisement*, M. Costa de Bastelica, had said in a report: "Come see, and you will be satisfied and pleased." The Commissioners were taken to the périmètre of the Sapet, and reporting to the Conseil of this visit they expressed themselves thus:—

"As for your Commission, the same who had seen the works in this périmètre in their embryotic state, they know not whether most to praise the admirable harmonious bearing which is characteristic of the works as a whole, or the marvellous results already obtained. . . . Thus are we able to say with the same faith which animates our noble Conservator, '*After such a baptism, the success obtained gives us the hope that henceforward the work of the consolidation and of the regeneration of the mountains is assured. We are certain that we shall save our country, if you second our endeavours, and if we receive sufficient credits.*'" *The Conseil raised the subvention from 500 to 1000 francs.*

In this same Session of 1868 there were presented to the Conseil a good many projects of extinction to be executed in the environs of Briançon. These were the first projects submitted for consideration in this part of the department; so the local opposition was very keen.

"In accordance with the apprehensions with which the périmètres of Chagnes, Vachères, and Saint-Marthe, &c., at present in full course of being restored, had been decried with all the violence of a passion by the populations of the localities affected by these, the population and the municipal councils of the communes of Cervières, of Briançon, and of Monétier, together with the council of the arrondissement of Briançon, protested against the approval of these new périmètres."

The Commission considered that the resistance was not justified, and the Conseil without hesitation approved the projected schemes.

In addition to the expression of such views and sentiments by the *Conseil General des Hautes Alpes*, many other General Councils have given expression to views and sentiments of a corresponding character, and in the Session of 1871 many of them gave expression to their sympathy with the enterprise.

In that year the General Conseil of the High Alps renewed the expression of its sympathy in the work of the regeneration of the mountains.

In the General Conseil of the Department of the Loire, at their sederunt of the 27th October 1871, the Commission reported :—

"The *reboisement* of our mountains, so necessary to prevent great inundations, and still more necessary to secure in the future valuable resources to our great coal basin, is going on slowly but steadily. Already 2724 hectares have been replanted. The total expense of these *reboisements* has risen to 390,000 francs.

"The department has contributed of this sum 29,000 francs ; proprietors, 114,000 francs ; the State, 245,000 francs. The sub-inspector of the forests expects that numerous works of *reboisement* will be executed in 1872 ; he earnestly prays the General Conseil to continue in the budget the usual allocation. Your Commission shares the desires of that official. They propose that you should pass a vote in the following terms :—The General Conseil, recognising the fact that it is in the highest degree for the public interest to encourage the *reboisement* of the mountain, vote the credit of 3800 francs, as asked by the sub-inspector of the forests, for the service of *reboisement* in the department of the Loire."

In the General Conseil of the department of the Lower Alps, at their sederunt of the 3rd November 1871, it was reported :—

"The *reboisement* and the *gazonnement* of the mountains have received, within the last few years, a lively impulse in our parts. With a view to giving to this service, the importance of which cannot have escaped you, a new impulse, and to obtain for our district that share to which her position gives her a title of the credit opened in the budget of the State, which the Assembly maintains at 1,500,000 francs, the Commission desires to express to you their satisfaction with the zeal and intelligence of the agents of all grades with a view to their encouragement, and ask of you to adopt a resolution in the following terms :—

"Considering (1) that the regeneration of the mountains by *reboisement* and *gazonnement* is for the valleys of the department a vital question, affecting and determining their very being ; and that this measure affects beyond this the general interest, as seen from a point of view higher and vaster still than that, which is already so important, of the inundations in the lower regions of the rivers ; (2) that the works executed in the first decade, provided for by the law of 28th July 1860, present at this moment results which one could scarcely have dared to expect in regard to the protection of the roads, as well as in regard to the extinction of torrents ; that these works, concentered very properly during the first period of trial on a surface which, relatively to the state of ruin in which the greater part of the high mountains find themselves, was somewhat restricted, ought from this time forward to receive great extension, in order to the protection of almost all the valleys menaced ; (3) That the department of the Lower Alps is, beyond contradiction, one of those most devastated by torrents, the ravages of which are increasing every year, in consequence of the denudation of the unstable soil on the enormous slopes of the high mountains ; (4) that, despite its public resources, the department has never ceased to take part in the work undertaken, and has constantly supplied an important subvention."

The General Conseil, wishing to give anew a testimony of sympathetic concurrence with the *service de reboisement*, granted for the operations of 1870 a sum of 2500 francs, and passed the following resolutions :—

" 1. That the credits to be opened, in continuation or renewal of those granted in accordance with Art. 14 of the law of 28th July 1860, and 6th June 1864, be voted for a new period of ten years. That these credits be raised to a greater amount, and one more conformed to the importance now established of the results to be obtained, and already obtained, from those useful works.

" 2. That indemnities to be granted for temporary deprivation of pasturage be largely allowed to the communes which shall demand them, in such a way as to give entire satisfaction to the interests injured.

" 3. That the State purchase the wooded lands which may come to be alienated by communes, wherever the acquisition of these may be made to come in aid of the work of *reboisement*.

" 4. That to the *service de reboisement et gazonnement* be granted credits sufficient to ensure the prompt execution and large development of the works pertaining to this useful service.

" The Conseil, unanimously adopting the flattering words of the spokesman for the agents of *reboisement*, and the resolution which he submits to the assembly, carries those resolutions by acclamation."

These recent testimonies, says M. Cézanne, writing in 1872, are of high import; they demonstrate the true and deep interest attaching to this work, since at a very recent day, after the terrible events which have sown our soil with such ruins, the population calls instantly for the immediate resumption of the fruitful works of the Forest Administration, and enable us to form an opinion in regard to what were originally and what are now the feelings of the population in the localities interested in these operations. This signal conversion may be attributed, without doubt, to the character of the work itself; but the prudent and judicious conduct of the Forest Administration, its increasing regard to local interests, and the good wages which it has caused to be paid in the villages, have contributed largely to this success. And he remarks, that the Forest Administration has always had the honour and the happiness to be directed by eminent men devoted to the public good.

After MM. Forcade de la Roquette and Vicaire, M. Faré, struck with the importance of this question of *reboisement*, and the magnitude of the benefit it might effect, devoted himself with zeal to the work ; and in the most remote gorges of the Alps the peasants have seen him, as Director-General, studying the sores of the mountain, and prescribing the remedy.

M. Faré, in closing his report of operations in 1867 and 1868, remarks,— " If the experience acquired by a practice of nearly ten years' continuance has made us acquainted with difficulties of which no suspicion was entertained at the outset, it has also caused many fears to disappear.

"The populations, formerly hostile, now examine and discuss our works, the progress of which they have followed with marked interest. The intervention of the Administration is in general accepted without resistance, even by those whose habits of life have to submit to a temporary violent restraint.

"The special commissions called to give advice on the projects almost always approve them, which testifies to the good spirit by which are animated the forest agents charged with the preparation of them.

"But the task so difficult which the Forest Admistration then undertook, when everything had to be created in connection with *reboisement*—agents, science, and means of accomplishing it—would have been impracticable but for the enlightened and devoted support of the General Conseils.

"Not only have these assemblies examined with lively interest, and supported by their advice the projects which have been submitted to them, but what is more, a great many of them have given their co-operation in the work of *reboisement*. The money voted by them for this purpose amounted in 1868 to 62,600 francs, and in 1867 to 67,144 francs 77 cents. These sums may be small in comparison with the whole expense incurred, but they acquire no small importance from the terms in which they have been voted."

And a similar appreciation of the importance of the work has been manifested in the National Assembly. The late wars interrupted operations, and entailed a reduction in the pecuniary provision for the prosecution of the works, as the resumption of operations. But confidence in the work was undiminished.

In the National Assembly, at the sederunt of 27th June 1871, M. Eugène Tallon, deputy of Puy-de-Dômè, laid on the table, in name of a sub-commission of agricultural works, a report treating specially of the *Regeneration des Montagnes* undertaken in execution of the laws of 1860 and 1864. It related more particularly to the regions of the central plateau of France. The following is the conclusion of the report —

"FIRST PROPOSITION, *Révision of the Law of 28th July 1860*.

"The Commission, determined by the considerations which have been stated, submit the recommendation that in the renewal of the law of 28th July 1860 there should be introduced into it the following modifications :—

"1. To suppress compulsory *reboisement*, or *le reboisement obligatoire*, on communal lands or lands belonging to private parties, and only to admit it exceptionally on communal lands in virtue of a decision of the General Conseil, given after investigation, and on advice of the municipal conseils of the communes interested.

"2. To maintain the grants of subvention, in money or in kind, and of indemnities for the benefit of communes and of private proprietors who shall consent to submit their lands to the Forest régime.

"3. To give priority to works of *gazonnement*, and to carry out these everywhere where it shall be possible to do so.

"4. To modify the composition of the commissions charged with determining the périmètres of the lands for which subventions are granted ; to leave to General Conseils the choice of members of these commissions, which shall be principally composed of agriculturists ; and in the case of communal lands, to submit, after previous formal inquiry, to General Conseils for approval the fixing of périmètres.

" 5. To raise the amount of subventions, and to transform them in part into reductions on the entire amount of imposts in favour of proprietors of wooded lands.

" 6. To insure the direct payment of indemnities for pasturage to the dispossessed parties using these, by payments made to each of these according to the return of personal estate prepared by the Forest Administration and approved by the General Conseils.

" SECOND PROPOSITION, *Modification of the Budget submitted.*

" To maintain in the budget a credit of 3,500,000 francs, allotted for works of forest roads, *boisement,* and *gazonnement.*

" THIRD PROPOSITION, *Modification of the Ministerial Organisation.*

" To transfer the general direction of the forests from the Ministry of Finance to the Ministry of Agriculture and Commerce.

" To transfer in consequence to the Ministry of Agriculture the credits connected with the General Direction of Forests relating to them.

" Such are the reforms which a careful study of the legislation, combined with an equal solicitude for the general interest of the country and respect for private property, have determined the Commission on Public Works to submit to the National Assembly.

" The first proposition will be presented as a parliamentary initiation. We express the desire that ere long it will take the form of a law which will give legitimate satisfaction to the protests of the populations interested.

" With regard to the two other propositions, we ask at present that they be transmitted to the special commissions on the organisation of the public service and on budgets." And it was ordered accordingly.

Such proceedings in General Conseils and in the National Assembly may be considered a fair indication of public opinion in regard to the enterprise, and indicative of its being in accordance with the conclusions to which those more immediately affected by these operations, which were being carried on or had been completed, had been brought by what they had seen and experienced of the results.

CHAP. VII.—PRESENT POSITION AND PROSPECTS OF THE ENTERPRISE.

It has been intimated above that a chapter of the history of this enterprise was closed with the commencement of the war in 1870. By the present position and prospects of the enterprise I understand the state of the enterprise in which operations have been resumed after the interruption thereby occasioned.

The monographs in regard to different périmètres in different departments of each of the three regions in which the more important of the operations have been carried on, which have been given, may suffice to give a definite idea of the state of the works. The proceedings of General Conseils and of the National Assembly, and the statement by M. Cézanne which have been cited in the latter part of the preceding chapter, show the spirit in which the work has been resumed.

The enterprise is great, and it has been resumed with crippled means, but not with less sanguine expectations of success. Of the magnitude of the evil against which they are contending some idea may be formed from accounts which have appeared in English journals of inundations which have in the course of the current year occasioned great loss of life and property in France.

As the men engaged are mainly engaged in attacking the evil at its source, their reports relate chiefly to the evil as it presents itself in the mountains and in the underlying valleys, or on the verge of the plain upon which they debouche, undermining fields and covering fertile fields with the detritus, undermining houses and covering the sites of villages at a lower elevation with the *debris*; but the evil stops not there, and in such inundations as have been referred to the evil is seen in another form. These inundations, it is reported, surpass any which have occurred since the operations of *reboisement* and *gazonnement* were commenced, and they have been spoken of at the meeting of the British Association for the advancement of science, held this year at Bristol, as supplying evidence that these operations have been proved to have been in vain. It is not thus that they are looked upon by those who are conversant with what has been effected. It has been stated that from the first, 140 years was the time reckoned necessary for the accomplishment of the work, and of these only fourteen, or a tenth of the whole, have yet passed; and though the most urgent cases were attended to first, it may be assumed that not much more than a tithe of the work has been executed, leaving all in confident expectation of " a good time coming."

Such has been the expectation of those engaged in the work from the first, and such it still is. With M. Surell the future was a *tabula rasa*. Of what would be he had while prosecuting his study no indication; but he saw what would be, if things were left to themselves, and he saw what might be, if his suggestions were followed up by others carrying out in practice what with him, situated as he was, could only be words, and counsels, and warnings, and admonitions, and entreaties,—" Leaves, nothing but leaves !"

Of what might be he then wrote thus: " It would be easy to draw a fascinating sketch by combining in one picture the numberless benefits which would flow from the execution of these works. We should have the Department of the Alps brought back as from the grave, her features entirely renovated, and prosperity succeeding everywhere to desolations and ruins, these fearful beds of dejection concealed under waving harvests, and majestic woods hanging on these *revérs* which are to-day crumbling and emaciated. We should have the mountains in three zones, rising one above another to different heights, the various products of which would be for the country a triple source of wealth : the lower zone, comprising the valleys and the brows of the lower mountains, would be reserved exclusively for cultivation; higher, where the slopes begin to be steeper, the ungrateful soil and the cold air would display a girdle of thick mountains, which would follow the undulations of the chain rising upward towards the crests ; and there, in fine, would commence the pastoral meadows, undulating plateaux carpeted with green sward, where numberless flocks and herds had now become innocuous. The forests grown thus on the most mobile portions of the mountains, between the cultivated ground of the base and the impending rocks of the summit, would serve as boulevards to the valleys, and would protect them against the fall of the upper portions. The inhabitants would enjoy at one and the same time the advantage of cultivated fields, of forests, and of flocks. Each of these products, wisely confined to the region which suits it, would leave a free field to the adjoining product. The flocks would no longer trespass on the cultivated fields, nor the cultivated fields encroach upon the forests; and the territory, thus utilised in its various parts, would yield all it can yield.

"Without speaking of the happy change which these new forests might introduce into the climate, might we not reckon, on good grounds, on the reappearance of a great number of springs which the felling of the woods have caused to dry up, and which the restoration of these would most probably bring to light again. These springs would spread around them fertility and freshness; whilst the waters of the torrents, become tranquil, would furnish to agriculture fertilising slime and moisture in abundance, to industrial works force of inexhaustible power, which doubtless would then excite astonishment that it was allowed so long to run to waste without benefit to man.

"The destruction of torrents and of ravines, and the general stability of the ground, would allow of good district roads being opened at little expense. These roads, at present expensive and constantly torn up, rendered more solid and increased in number, would carry life into the deepest recesses of the mountains. They would even facilitate in many quarters the working of lands which the inconvenience of communication often renders difficult, and sometimes impracticable.

"Then, also, there would be nothing to hinder the multiplication, at little expense, of works of irrigation. At present one cannot resort to these important works without trembling on account of the difficulties, sometimes insurmountable, presented by the courses of torrents; and when at last these obstacles are overcome, one sees rise given to new difficulties through the extreme want of cohesion in the soil. The storms, in carrying away the ground, cut up the channels; the friable *revérs* across which it flows allow the water to filter away until they dry up; and the crumbling down of the grounds fills up the canals. These difficulties experienced in the construction, in the cleaning, and in the maintenance of them, are such that they have often occasioned a recoil from the execution of canals likely to be most useful. From the day that these drawbacks shall be taken away these works will no longer present any difficulty, nor will they be of costly execution, and they might be easily spread over all parts of the territory.

"Easy communications, combined with the presence of forests, of watercourses, and of mineral riches which are shut up in the bowels of the mountains, will attract thither industrial operations which hitherto have never found there a home. This will give employment during the winter, and will retain there the population which generally deserts the country at that time of year. On the other hand, the increase of the products of the soil, in diffusing here more case, will relieve the inhabitants from the necessity of seeking a livelihood elsewhere. Thus will come to an end the wretched custom of emigration, which disperses families from their domestic hearths, and condemns them to an unpleasant wandering and solitary life.

"The State, in this transformation, will have seen her roads improved, the maintenance of them become more easy and more perfect, and their creation more economical; there will be gained a very extensive area of taxable lands, and of fine forests in proximity to its harbours. In fine, the Treasury will reap that increase of revenue which always follows the prosperity and numerical increase of population."

And with loving fervour he pleads with eloquence the cause of the Highlander as the cause of the community, showing that if one member suffers the whole body suffers with it. The results have justified his anticipations.

> "Lo! former scenes, predicted once,
> Conspicuous rise to view;
> And future scenes, predicted now,
> Shall be accomplish'd too."

Everywhere, so far as it is known to me, the results obtained have been such as to warrant the victors to prepare even now to raise the shout of triumph. Meanwhile they are labouring to complete and to perfect their work; but, looking on the future in the light of the present, they foresee what is coming, and rejoice in all the confidence of hope.

M. Cézanne, in concluding his sequel to the treatise of Surell, writes:— "In fine, an epoch of reparation begins, and, thanks to the labours of the eminent authors whose names have been presented to us in the preceding pages, the coming generation may hope to see the definitive decline of the torrential era."

And writing of the present, after describing the formation of lines of plantations on the terrace-like banks of *berges vives*, or *berges vivants*, and speaking of the astonishment of visitors on seeing, pointing heavenward, the verdant shoots of the elm, the maple, and the acacia, growing on the dry schists, and of the walnut and the oak on the dry and solid buttresses, while the alder, the poplar, the ash, the osier, and the white willow of the Alps grow on the more moist depths of the ravines, he goes on to say: "These works, so ingenious in their very simplicity, form a network of horizontal lines, like to the alleys of a garden. The green edgings and linings develope themselves amongs the innumerable sinuosities of the Combes, embracing, from the rocky bed of the torrents to the very summit of the mountain crests, those ravines which were but lately inaccessible, and presented an aspect full of horror.

"One on seeing what has been done understands immediately how such a combination should be effectual; every liquid molecule, so to speak, is seized individually, the thin sheet of water flowing downwards is retarded in its course by a thousand thirsty little plants, by the lines of cultivated herbage, and by the hedges of shrubs and trees; it is compelled to tarry for a little on each terrace to slake the thirst of the ground, and when it reaches the lower end of a furrow it spreads itself out on the flattened bed there prepared for it, stopped at every *barrage*, it loses its vital force on every hand, and finally, from resting-place to resting-place, and from descent to descent, it arrives, after a thousand retardations, and still limpid, in the channel which conveys it on to the river."

"'The violence of torrents is occasioned by the combination of an infinitude of elements infinitely minute, and the system of extinction consists in extinguishing each of these elements without neglecting one; it is an accumulation of infinitesimal littles. The secondary ravines are blocked, the minute ramifications are intercepted, the lesser flanks are filled up; and, finally, there are spread over the surface of the soil, in order completely to diffuse them, the innumerable threadlets, divided and subdivided, like the fibres of a root, which are manifestly the root of the evil.'

"These are the statements of Surell. But there is one of the precepts of the master which it is right it should be known has not been carried out, and the visitor who sees so many precautions taken against drought, who hears the workmen and the foremen crying out for rain as their most efficient helper, may be astonished to see in the bed of the torrent a stream of water going to lose itself in the river without any one attempting to use it for irrigation. And yet amongst these terraces, some formed with a gentle slope might have been formed, at little expense, into irrigating runnels. It seems that the difficulties experienced in fixing the moist parts of hills have inspired exaggerated dread of the accidents which water might produce, and the local inconvenience has caused to be forgotten the more general and more important advantages referred to. It is, however, necessary to guard against prejudging in a question of such delicacy: experience alone can decide whether the irrigation of hills be advantageous or hurtful, and therefore it is matter of regret that that experience has not been catechised on this point of primary importance in any of the périmètres."

At the end of 1869 there had been spent upon the High Alps alone 1,074,136 francs 57 centimes, more than a tenth of the 10,500,000 francs allotted for the whole work. So large share of the funds at command having been allocated by the Administration for works in this department shows how serious the evil had become in that district. The total area of the surface given up to the Administration in the High Alps amounts to 85,962 hectares, spread over 13 périmètres; 16,903 hectares have given occasion for works of restoration, *boisement*, or *gazonnement*; 13,460 hectares have been only interdicted to flocks and left to a natural *gazonnement*.

Now, over against the expense incurred must be set the benefits which have thence resulted. But with the feeling that these are considerable, it is difficult, says M. Cézanne, to represent them in figures.

M. Gentil, *ingenieur en chêf des ponts et chaussées* in the High Alps, writes: —" The aspect of the mountain has been suddenly changed; the soil has acquired such stability that the violent storms of rain in 1868, which have brought such disasters on the High Alps, have been innocuous on the regenerated périmètres.

" The mountain has in a short time become productive; there, where a sheep could scarcely live destroying all, are now to be seen an abundant herbage fit for the scythe.

" This mode of *mise en valeur*—one may say, of culture—is remarkable in this, that it furnishes to the population the very thing which they need, and furnishes this to them with little delay. The population of the High Alps are essentially a pastoral people; what is required by them is provision for the support of flocks; this they find in these périmètres—be it in the herbage which shall be mown—be it in the leaves of the ash trees and the elm trees planted on the levelled terraces; and further, the acacia will yield soon wood which will be employed in the culture of the vine.

" M. Sequinard has sought above all, in the creation of vegetation, to satisfy the actual and immediate wants of the inhabitants; in acting so, he has secured the concurrence and support of the populations, who can very well comprehend that, in a not distant future, they will find, thanks to the Forest Administration, important and more abundant resources for the feeding of cattle. These populations were, on the contrary, openly hostile when they feared they might have to submit indefinitely to the prohibition of depasturing, in a remote hope of forests which might be enjoyed by some future generation, and these forests the growth of which was considered very uncertain, and very problematical—and not unreasonably so, for the attempts to rear them made previously had all proved abortive.

"In running over the périmètres in course of restoration, it may be seen how much the soil has been changed for the better, and consolidated; one may, without being accused of entertaining utopian dreams, foresee that soon some portions may be brought under cultivation, and brought by attention and irrigation into the condition of natural meadows.

"In consequence of the consolidation of the soil, and of vegetation, the torrential characters so well described by M. Surell have disappeared. The waters, even in time of rain, are less terrible, and are better fitted for use in irrigation. This has been testified by all the proprietors who make use of the waters of torrents in the irrigation of their lands.

"There are no longer sudden and violent floods; and the waters on reaching the cones of dejection are no longer charged with solid materials, and they naturally dig out for themselves a passage through these deposits. In taking up and carrying further the materials brought there, they uncover stones of great size, and these form a fixed and solid bed. The shiftings of the bed of the current are less formidable, and less dangerous; and at little expense the proprietors of lands on the banks can protect their property from injurious effects of these.

"But it is befitting that I should give cases and figures. I shall, therefore, cite definite facts which relate to our roads, or to our undertakings.

"At Sainte-Marthe [as has been stated by others] there was discussed, in 1861–1862, a proposed erection of a dike on the cone of dejection on the left bank of the torrent. This dike, estimated to cost about 40,000 francs, had as its design to protect the imperial road No. 94 and the properties on the river bank against invasion by the torrent. These works would have been in reality but a temporary remedy; the dike would have been, after some years, buried under the dejections of the torrent. To-day, the torrent of Sainte-Marthe is completely extinguished; nothing now comes down from the mountain. *The proprietors and the engineers no longer think about dikes; simple fencing walls suffice to protect the lands on the river bank.*

"The torrent of Pals, in the commune of Rizoul, traversed the departmental road No. 4, and the imperial road No. 94. In 1865, I brought under discussion a project of works to enclose this torrent with dikes, to fix the bed of it thus, and to conduct it in a straight line to Guil, thus avoiding the imperial road No. 94—it would have cost at least 25,000 francs. Since that time the basin of reception has been restored and consolidated, the torrent is extinct, the change of bed has become unnecessary, it is only requisite now to construct on road No. 94 an aqueduct for the passage of the waters of the Pals; a work costing 1000 francs has sufficed here when there was being anticipated an expense of more than 25,000 francs.

"The torrent of Rioubourdoux, near Savines, was one of excessive violence; it carried away a great deal of material, and the establishment of a bridge for the passage of the imperial road No. 94 was considered as a difficult and dubious undertaking; the passage of the cone of Rioubourdoux was also an uncovered one, interrupted at every rain and every storm. The Forest Administration has *mis en defends* the basin of reception, and has commenced works of consolidation. The régime of the torrent has been modified; it has been possible, without very great expense, to fix definitively the bed upon the cone, and to construct a bridge. The waters no longer bring down material from the mountain.

"The expense of the construction of the road, of the fixation of the bed by dikes, and of the creation of a bridge, has been about 40,000 francs; the enquiries made formerly led to the estimate of the expense of the works to be 60,000 francs. *The change in the régime of the torrent has rendered practicable the execution of definitive works, and, moreover, has allowed these works to be executed at less expense.*

"These three cases specially concern our roads; they are the only ones I can specify, giving details and figures. The other périmètres operated on by the Forest Administration are distant from roads, and do not relate to these so particularly, consequently I cannot give figures, but I have not the shadow of a doubt that they have had analogous results.

"Be that as it may, the cases I have cited are in my opinion very striking, and supply one means of measuring the advantages gained. As for the advantages by which lands situated near to the cones profit, these are immense. Not only are the proprietors relieved from the construction of expensive, and, at best, precarious dikes and embankments, but their property, having no longer the risk of being suddenly buried under gravel, takes a definite value, and cultivation is carried on with the assured hope of being followed by a harvest. This security is itself a very great benefit; the proprietor, counting on the future, will not dream of expatriating himself, as numbers have done.

"The successive extinction of the great torrents which threw themselves upon the principal valleys will lead, inevitably, to a marked amelioration in the *régime* of those water-courses, and this amelioration will extend to a lower level of the river's course.

"It appears to be established, or at least it is the opinion of the inhabitants, that the waters coming from the restored périmètres are less subject to sudden variations in their delivery, and channels of irrigation connected with them are fed in a more regular manner. It appears to me to be both natural and probable that it should be so; but I cannot in regard to this give any definite numbers. I cannot adduce any experiment, or conclusive and testing observation. I confine myself to repeating what has been told to me, and what I believe to be true.

"At the close of 1867, for 81,012 hectares which had been restored and consolidated, M. Sequinard estimated the expense to have been 4,113,000 francs. These figures are taken from the *Annuaire des Hautes-Alpes* for 1869. The mean expense, then, would be about 50 francs per hectare; but I always make this reservation, that the expense varies very much with the périmètres—in some it rises as high as 100 francs, in others it falls so low as 30 francs per hectare; this depends on the nature of the soil, on the state of desintegration, &c., &c.

"In the High Alps the total area to be restored is about 200,000 hect. (that is I believe the number given in 1840 by M. Surell); the expense for the whole department, then, will be 10 millions at the most, and 8 millions at the least—taking the above stated mean, and taking into account that the works will cost in the future, by reason of the experience acquired, less than was provided at the commencement of the operation.

"In 1840 M. Surell estimated the whole expense at 6 millions (100,000 francs per annum for sixty years). The difference is accounted for if it be considered that, within these thirty years, labour has become more scarce and higher priced.

"I ought to mention, in fine, that the system adopted by M. Sequinard gives, after some years, appreciable products which it would be right to take into account. These successive benefits will be considerable; and they would have no existence if it were arranged to create only forests, and plantations of forest trees.

"The expenditure of 8 or 10 millions may appear to be enormous; but I do not believe it to be out of proportion with the results. M. Surell has demonstrated this most explicitly; and I have nothing to add to what he has said. But one may go a little further; I am certain that in five years the périmètres, by their herbage and their wood, will yield revenues which, added to the savings, and the benefits resulting from the transformation of the *régime* of the torrents and of the water-courses, will cover by far the greater part of the expense of restoration."

Thus far, M. Gentil. M. Cézanne adds,—"After such testimony it is not left open to us to doubt that the operation is good, for it seems to satisfy everybody—the Administration, the professional men employed, and the populations." In another connection he remarks that—"The object and design of the enterprise was not, what has been called for by some students of forest science, to carry out *reboisements* everywhere, and to re-establish the ancient forest domains of France; but the enterprise was confined to a measure to curb and master torrential rivers, and thus protect at once the mountains which these were attacking, and the plains which they were devastating by inundations.

"The mountains in general, and in particular those of the High Alps, are essentially pastoral lands; flocks are the sole source of wealth for the inhabitants—they are the life of the country; the pasturage required to be extended, not curtailed. Such were the views of M. Surell. And on the same mountains, *gazonnement* has been carried out in some situations, and *reboisement* or *boisement* in others.

"The planting of woods promises a return at a period too remote to allow of its being required of the existing generation, excepting in places in which it is indispensable as a means of retaining the soil, and of allowing the ulterior development of the turf; and the arrangements made by the law of 1860 spread the work of *reboisement* over a period of somewhere about 140 years."

A reduced expenditure may, if not counteracted by an enlarged expenditure, in more favourable circumstances, necessitate a prolongation of time for the completion of the enterprise. In every other respect the prospects of the future are as bright as ever.

CONCLUSION.

It has been my desire and my endeavour, in preparing the statements contained in the preceding pages, to supply such information as might enable any student of hydraulic engineering or of forest science to form an intelligent idea of what has been done in France in carrying out works of *reboisement* and *gazonnement*, with a view to arresting and preventing the destructive consequences and effects of torrents, and of the consequences and results which have followed, and, in conclusion, I submit for the consideration of any who may inhabit lands exposed to ravages of torrential floods, whether like measures in like circumstances be likely to produce like effects. It might be as imprudent to apply to some one case exactly the same measures which have proved successful in some one or other of the cases which have been brought under consideration as it would be for a nurse to follow in every case in which her services might be required precisely the same course of treatment which she saw followed with success in some one case for which prescriptions and directions were given by a physician of renown. Better do so, perhaps, than disregard altogether what she may then have seen; but it holds good in medicine as well as in law that the case being altered that alters the case, and it holds good in the treatment of torrential floods by *reboisement*, as well as in medicine, that each particular case demands particular consideration and a particular application of the general principles to be followed in its treatment. We have found it laid down in the treatises which we have laid under contribution that not only each torrent but each affluent requires to be specially studied and to be specially treated.

To distinguish things which differ is as necessary as to perceive the analogies in which things agree. A rash generalisation, extending the process of induction—or what seems to be such—beyond the facts ascertained, is to be deprecated; but it is otherwise with deduction from what is actually ascertained, and I speak advisedly when I say, that a prompt and judicious application of measures which may be suggested by what has been accomplished by *reboisement* and *gazonnement*, in combination with *barrages*, in France, might prevent much destruction of property and of life.

While the preceding pages have been passing through the press there has occurred in France one of these periodical inundations which it is sought to prevent, and which present a form of the evil with which the inhabitants of newly-settled lands—for whom more especially I write—are more familiar than they are with the form of the evil for which *reboisement* and *gazonnement* were previously employed as remedial measures; and it may be that to some of them, as to others, it may appear that this is an indication that the measures have failed to produce the effect that was anticipated, and for which they have been employed; and I feel as if my work would be incomplete if I passed over this view of the case in silence.

There have been many inundations in France—as there have been elsewhere—in the course of this year; but the reference is to the inundation in the valley of the Garonne, which proved destructive to life and property in many a town and village, and that to such an extent in and around Toulouse as to awaken a thrill of sympathy throughout the civilised world. This inundation was attributed to a late fall of snow on the Pyrenees and on the Cevennes having been followed by a long-continued heavy fall of rain, and a warm westerly wind, which melted both the newly-fallen snow and much which was lying there, and thus added to the rainfall what by itself might have produced a torrential flood.

Great damage was done not only along the lower-lying lands through which the water flowed off to the sea, but also in the high-lying lands on which the flood originated.

In the Pyrenees the flood may be traced from Bagnères de Bigorre downwards to the sea by the ravages of its waters. This town is in the region of Baréges and the Brèche de Roland, and the *Pic-du-Midi*. It has been famous as a watering-place since the days of the Cæsars. Twenty-seven public springs bubble up within view of the town, one of which is so copious as to turn a mill-wheel. From all of these the hot water rushes with force, and there are lesser springs which supply baths in many of the houses. As a place which is well known it supplies a good starting-point for our study.

Much of the water resulting from the fall of rain and the melting of the snow in that region was carried off by the Adour, which empties itself into the sea at Bayonne. At Tarbes a large artillery foundry suffered much from the inundation occasioned by the overflow of this river, and valuable machinery for boring guns is said to have been hopelessly destroyed. There and at other places along the channel of this river houses were undermined or overthrown, and all growing crops were destroyed. Three villages close to the railway from Tarbes to Toulouse—viz., Roques, Anterive, and Pinsanguel—were reported to be literally annihilated. At Sarginnet all the houses were inundated, several fields were washed away, and the bridge was demolished.

But more of the water produced by the rainfall and the melting of the snow seems to have flowed away by the Garonne, which rises a little further to the east, and following a much more lengthened course, flowing past Bordeaux, is lost in the Gironde ere it reaches the sea; and still more, perhaps, was carried away by the Arriége, an affluent of the Garonne, rising still further to the east: the Oriége (aurifera) as it was anciently called, on account of the gold found in the detritus chiefly between Campiegnac and Foix, the chief town of the department to which the same name has been given.

At Muret, on the Garonne, situated a little above the confluence of that river and the Arriége, the destruction of property was great. The suspension-bridge was destroyed, and many houses wrecked. At Moulis, on the Tarn and Garonne, the church fell. At Golfech only four houses and the church were left standing. At Lamagistère many houses fell, and the bridge was carried away.

At Verdun, on the Arriége, fifty houses were destroyed, and eighty lives were lost, besides 500 head of cattle. Numerous are the notices of whole villages having been washed away—or of one, or five, or twelve, or some similar number of houses alone standing, and showing what was the site of the village, while numerous bodies have been found.

At St Lizier three-fourths of the town were under water.

At Foix, with its picturesque castle, perched upon a rock, and characterised by a lofty donjon tower, the destruction was considerable.

But at Toulouse, situated below the confluence of these two rivers, the destruction of life and property was such as to cause that to be scarcely thought of, and it is with the name of this city that the inundation has become associated throughout the world, wherever it has been heard of.

Toulouse, like many other important cities situated on the banks of a large river, may be said to be composed of two towns. It was in one of these, the Faubourg St Cyprien, that the flood created the greatest destruction of life and property.

In one of the first telegrams received in England it was stated :—" The St Cyprien quarter of Toulouse is a perfect sepulchre. 215 bodies have already been found. The waters exceeded the usual height by nine mètres, or 20 feet, and the flooded quarters were for a long time inaccessible on account of the extreme violence of the torrent. Several persons who endeavoured to save life, including the Marquis d'Hautpoul, perished in the attempt. There are upwards of 20,000 persons in Toulouse deprived of all means of subsistence. The railway traffic around the town is interrupted. The disasters in other parts of the south are equally great." And subsequent communications confirmed all that was thus reported.

The Paris correspondent of the *Times* gave the following particulars of the commencement and progress of the flood at Toulouse :—" Up to Wednesday nobody had any idea that, owing to the heavy rains of the previous week, the Garonne would overflow its banks with such rapidity. On Tuesday the river indeed was unusually high, and some slight damage was expected, such as usually happens in winter when the snow melts, but had anybody suggested precautions against disaster these would have been deemed quite unnecessary. On Wednesday, however, the prospect changed; the river became a torrent, and by 10 A.M. it reached the level of the flood of 1855. It continued to rise, and measures were taken with desperate ardour to hem in the waters, but the attempt was too late. They invaded all the low-lying quarters of the town, and at 2 P.M. two of the arches of one of the bridges and twenty houses were swept away, and the swimming baths and lavatories moored to the banks were hurried down the stream, dashing against the houses alongside in their course. At 5 P.M. the water rose over the parapets protecting the populous quarter of St Cyprien. In an hour later it was ten feet deep. Boats were hastily got out to rescue the inmates of the falling houses, but several of them were carried away by the current and dashed to pieces or swamped. Two boats, manned by eight soldiers, were dashed against each other, and sank in a pool formed by the gardens of the Civil Hospital. Out of thirty persons on board only one woman was saved. Several persons who tried to escape from the quarter on the left bank on horseback were carried away by the flood and perished. In the evening the whole quarter of St Cyprien was cut off from the rest of the town, the three bridges being carried away. All Wednesday night were to be heard the crash of houses and the cries of the victims. On Thursday morning the whole population was in the streets, all classes vying with one another in their efforts to rescue the victims. The town presented a heartrending spectacle; 5000 to 6000 poor creatures, half-naked, bruised, and benumbed, were conducted on foot or carried in vehicles or on litters to the military hospital; many of them were **women leading their children by the hand.** By the afternoon the waters **had fallen six feet** in St Cyprien, but both above and below Toulouse they extended like an immense lake, dotted with the bodies of human beings and animals, and with articles of furniture, while roofs of houses and trees here and there appeared above the flood. At 4 P.M. the clergy of St Nicholas went in procession to the Church de la Daurade to supplicate the Virgin for mercy, and services were going on all day in all the churches. About 100 corpses were discovered on Thursday and Friday, and were buried, after being photographed, by their friends. One of the victims was M. Wohlfart, a retired major, who had entered a house to save two children, when the walls fell in and buried him. Bodies were discovered in many instances in alleys and gardens. Five victims composing one family were found in a first-floor room locked in an agonizing embrace. Those who had escaped were to be seen stationed at the entrance of their ruined tenements, and giving information as to their fellow occupiers. In the belfry of St Nicholas Church 60 persons took refuge. The flood reached the altar table, and not far off a clerical student was searching all day among the ruins of a large house for his parents and sister, listening intently for any sound of life. All the granaries on the banks of the river are destroyed, and the entire quarter may be said to have disappeared, for all the houses still standing are damaged or tottering. The new Carmelite Church was flooded, a lady who was confessing being drowned, while the confessor effected his escape. The girls at a convent school were rescued on Thursday. They had passed a dreadful night, going up from storey to storey as the flood rose, and passing the time in prayer, expecting every moment to be their last. The bodies of four women were found, each with an infant in her arms. Another woman was found by her dog. In one room ten victims were found, two of whom were still breathing, and received the last sacraments before expiring. The older houses, on account of the greater solidity of their walls, offered the greatest resistance to the flood, and those supported by the old fortifications were not carried away. Large numbers of persons have applied at the Mairie for food and shelter. The troops rendered great service, displaying the greatest courage and daring, and they have been warmly thanked by the Municipality. The villages round Toulouse have suffered considerably, and several persons perished.

"It is announced that at Toulouse alone, 900 persons have perished, and the outbreak of an epidemic is feared. The loss of life will never be precisely known, for the Garonne has carried away many bodies, and some have been recovered in the costume of districts 20 leagues distant from Toulouse. It is believed that 600 houses have been swept away in the town, and 2000 in the environs. 1200 soldiers are still engaged in clearing away the ruins, and only a fifth of the inundated buildings will be saved. Water for drinking was distributed yesterday in the town, which had had none since Wednesday."

"Nothing else," says a letter from Paris in the *Daily News*, "is talked about or thought of here save the floods. Politics are completely suspended. The lowest estimate of deaths is now 2000, and the rapid swollen Garonne carries away, unsuspected, many dead bodies to distant parts. At Toulouse the river rose 15 feet higher than during the great flood of 1855. Typhus fever is feared. A terrible military measure is proposed to prevent accidents in the crumbling suburbs of St Cyprien, —viz., the complete destruction of the district by bombardment."

A special telgram from Toulouse in the *Standard*, dated Sunday, contained the following:—"The Garonne has now subsided almost as rapidly as it rose, and on gazing on the scene of ruin and destruction the waters disclosed as they receded, it is difficult to realise that such an amount of destruction could have been the work of a few hours. Readers will, however, be able to form some notion of the nature of the calamity if they will understand that St Cyprien, the ravaged suburb, stands with regard to Toulouse very much as Southwark does to London. The Garonne runs between Toulouse and St Cyprien, and on three sides hems in St Cyprien in its bend, every part of the suburb being considerably below the normal level of the river, from which it is protected by embankments. On Thursday last, after several days' heavy rain, and receiving an influx of water beyond all precedent from the mountain streams that feed it, it rose upwards of eight mètres, swept away the two suspension bridges, and bursting over the embankments on the south side, gradually laid the whole of the place under water. Many of the houses, being lath and plaster, speedily collapsed under the rush of the waters. For nearly ten hours it was impossible to afford assistance to the sufferers, and 35,000 men, women, and children were clambering out of the way of the waters. The greater number were eventually saved through the pluck of the garrison and the heroism of a few civilians, who, like the Marquis d'Hautpoul, fell victims to their zeal. It is not known, and cannot be for some time to come, how many bodies may be lying under the *debris* of the shattered houses. One of the local papers talks of 15,000 victims. It is probable that that number, and more, of artizans have lost their little all, but the dead bodies as yet recovered may be stated at 310. The site of the catastrophe just now is a scene of desolation, but it is not picturesque. The soil being clay you have to wade in the streets knee-deep in water, while now and then a wall totters and falls to the ground. Soldiers in fatigue dress are busy clearing the ruins, but it will be a long operation, and there is a rumour that dynamite will be resorted to. A strong detachment is stationed at the only bridge left standing, and no one is allowed to go over but male inhabitants of the suburb, who are working in seas of mud to try and save some of their goods and chattels from the wreck, and these are provided with a special permission. The body of the Marquis d'Hautpool was recovered this morning, a few miles down the river. His funeral at six o'clock this afternoon was attended by all the local notabilities. Toulouse is just now crammed to overflowing. From all parts of the neighbourhood people are flocking in to inquire for missing relatives and friends."

The correspondent of the *Gaulois*, writing on the 26th, said,—"The Garonne has this morning again taken its normal course, but has unfortunately committed more ravages in re-entering its bed than in quitting it. The ground has so sunk under the influence of the waters that the few houses which had resisted the inundation crumbled to pieces like houses of cards. The ground is everywhere overlaid with a coat of slime. One sees only ruins. Where there was a house, there is now a pit; where there was a street, one finds a shapeless mass, composed of pieces of walls, chimney pots, bales of merchandise, broken marbles, and rubbish of every description, in the midst of which are discovered every other instant human remains—pieces of flesh and crushed limbs." The correspondent had seen between two stones a man's head horribly disfigured, and a little farther on an arm separated from the trunk, and half buried in the mud.

The *Debats* published a description of the aspect of Toulouse when the inundation was at its height. The inhabitants assembled on the quays and on the stone bridge, looking helplessly on the scene of desolation, and following with terrified eyes the work of destruction. Property of all sorts was being swept away by the angry waves—piles of timber, carts, heavy planks, &c., were hurled against the piers of the bridge. Half-an-hour after the precaution had been taken to close the bridge of Saint Pierre it gave way with a crash, and it was followed by the baths and large public washhouses of Tournay and St Pierre. At last, as if all the elements were combining against the unfortunate town, the large rolling mills of Bezacle were discovered to be in flames ; while another fire broke out at Port Garandin in a house that was entirely cut off from any chance of succour. The manufactories at Bezacle and in the Rue des Amidonnius were abondoned on the water rushing into them, and were of course greatly damaged. On the Quai de Tonnes the rise of the water was so rapid that many families could only be rescued by means of the windows. The tug-boat stationed at St Pierre was carried away by the force of the current, and was capsized. Four or five persons were on board at the time, and as the vessel was borne away they uttered cries of despair.

The special correspondent of the *Times*, writing from Toulouse on the Monday night, thus described the devastation in Toulouse :—" Nothing can present a greater contrast than the north and south sides of the river at Toulouse—the one with its usual aspect undisturbed, and its inhabitants pursuing their ordinary avocations ; the other, like a place which has been bombarded. The Faubourg St Cyprien is, in fact, a town of itself, and is to the upper part of Toulouse what the Surrey side of the Thames is to the Middlesex shores. The quarter is densely peopled, or rather had 25,000 inhabitants, most of them of the working classes, though outside the town walls are villas and cottages belonging to wealthier people. Through the quarter ran avenues of trees, and around it were fields of corn and market gardens. To describe what St Cyprien is now is almost impossible. It is a town of ruins. The Garonne is now running in its natural bed, but all over the inundated quarter are pools of water and rank river mud ; trees are uprooted, gardens are mere swamps, and streets half-dried water-courses, with here and there great cracks and yawning gaps. In some places the houses are heaps of rubbish, in others the walls are left standing, with pictures or articles of attire hanging on them. In one street through which I passed only three houses were left standing, and this is probably the case in a dozen other thoroughfares. Everywhere gangs of soldiers were at work demolishing walls, collecting furniture, and making the roads passable, while the owners of the ruins were, some helping with a resolute fury, others sitting on beds and mattresses gazing vaguely at the rubbish which represents their homes. Some of the more energetic families were picnicing and cooking their midday meals in the desolate gardens, the women bright if not cheerful, the children playing about as if nothing had happened. The despondent were just the contrary, the women sitting with their faces buried in their hands, and the boys and girls lying huddled together among the broken beams and heaps of plaster. The roadsides were lined with all kinds of articles, from four-post bedsteads down to blankets and bonnets ; and in several places the small shopkeepers were drying, or rather attempting to dry, in the sun their stock in trade. Outside the barriers the scene is the same, with the addition of large swamps, which were once vineyards and cornfields, and of little piles of gaily-painted boards, once forming portions of summer-houses. The town wall itself, a solid line of concrete, has in one place been thrown down quite flat, and a little further on two great beams, at least 30 ft. long, have completely barred one of the gateways. Nothing, however, shows the force of the flood so much as the ruins of the suspension bridge. Nearly half the bridge itself has been firmly and securely laid by the water on the bank, where, indeed, I at first mistook it for a landing stage. Only one bridge is now standing— viz., the old stone one at St Pierre. Had it been carried away, the losses on the St Cyprien side would have been even more terrible than they were."

Below Toulouse, between that city and Agen, the Garonne receives the waters of the Tarn, coming from the department of Lozére and the mountains to the south, and along the course of this river there were similar devastations. At Mount-Auban the water rose 40 inches above what it did in 1835, which was the highest flood of the century, and all the farms around were destroyed. At Moissac, near Mount-Auban, the destruction was fearful, and the river was found to have definitively forsaken its former bed, and to flow four kilomètres away in an ancient channel. Castel-Sarrasin, between the Tarn and the Garonne, was gutted entirely, and the number of victims was supposed to be about a hundred; and similar were the accounts which flowed in from all parts of the devastated region. Everywhere were dead bodies being found, or seen drifting down the stream.

Below Agen the Garonne receives the waters of the Lot, coming from the northern part of Lozére, and the district between the Tarn and the Lot had the same tale to tell.

At Bordeaux the river was not overflowed, but it brought down trees, hay, animals, and several dead bodies. An incident more touching than terrible occurred. An infant in a cradle, supposed to have come some distance, and floating down towards the sea, was saved.

Such was the inundation of the valley of the Garonne. I have spoken of it as a form of the evil which it is sought to remedy with which the inhabitants of newly-settled lands are more familiar than they are with the form of the evil in the mountains for which *reboisement* and *gazonnement* were primarily employed as remedial measures. Illustrations crowd upon me. I confine myself to a selection from those supplied by the history of the Colony of the Cape of Good Hope, being with them well acquainted.

Captain Hall, in his *Manual of South African Geography*, says (p. 95),— "In Great and Little Namaqualand, the Kalihari Desert, and the whole of the region situated on the southern slope of the Nieuweveld and Roggeveld Mountains, whole years may elapse without the phenomenon of a running stream, and yet the magnitude of the dry water-courses of the Buffalo, Hartebeest, and Oup or Borradaile River, all tributaries of the Orange, show how immense must be the torrents that sometimes sweep along them. The writer of this has seen the bed of the Great Fish River perfectly dry, and within twenty-four hours a torrent thirty feet deep and several hundred feet wide was roaring through it. In February 1848 the Kat River suddenly rose upwards of fifty feet in the course of a few hours, sweeping seventeen feet above the roadway of a stone bridge at Fort Beaufort, supposed to have been built high enough to leave a clear waterway to the highest flood ever before remembered. The Gamtoos, Gauritz, and all the other rivers draining the Karroo, are also subject to very sudden rises, although generally but dry water channels. The periodical rains falling in the mountains near its sources, between September and March, also swell the Orange River to a great extent, and large portions of land along its banks are then inundated."

In an official report which I made to the Government of the Colony in 1864, it is stated :—" I have seen the Tarka, the Fish River, the Keiskamma, and the Buffalo in their might. I have crossed the bed of the first-mentioned in a box suspended from a rope stretching from trees on the opposite banks, while the river torrent was tearing along below, twenty-two feet deep, as ascertained by measurement, where forty-eight hours before the depth was only eighteen inches. I have been told by a gentleman who had given special attention to the subject, of the mean rise of a number of rivers in the same district being twenty-eight feet; I have been told by the same gentleman of a maximum rise of sixty feet; and I have gone over the scene of devastation occasioned by the sudden rise of a river to a height of seventy feet above its usual level."

The case referred to occurred in the vicinity of Hankey, on the Gamptoos river and its affluent, the Klein Riviere, on the 1st of October, 1867; and by the inundation thereby occasioned the promising village of that name, one of the most promising of the institutions for the conversion and civilization of the natives of South Africa, was reduced to a heap of ruins. It was the residence of a people remarkable for their misfortunes and for their enterprise. In 1830 they formed a water-course several miles in length over a very difficult country, for the purpose of leading out a small river upon their garden ground. The work was twice completed and twice destroyed by floods. A few years before the occurrence of this inundation a still bolder scheme was projected and carried out with complete success by the missionary, William Philip. This was the excavation of a tunnel through a hard sandstone ridge which separated a reach of the Gamptoos river itself from a considerable extent of excellent ground near them, half surrounded by one of its bends, through which a copious stream, with a fall sufficient to work machinery, as well as to irrigate the soil, had just begun to flow when their friend and guide was snatched from them in the prime of his life and usefulness.

This tunnel I have visited—I know of no work of the kind in the colony except itself; and I have learned that it subsequently received a corresponding extension by the brother of the projector, the missionary now in charge of the station, who, with the men on the station as workmen, has carried the surplus supply from the tunnel under the river at a lower point, and brought it up on the further side to irrigate lands lying below the level whence the waters are obtained, but considerably above the level of the river there. In accordance with the energy manifested by this people in the execution of such works, and with like perseverance, they had continued to devote all the time and labour they could spare from the occupations by which they support their families to public improvments, to buildings, and to the extension of cultivation in the shape of gardens and cornfields, when this terrible flood again devastated their lands, destroyed their dwellings, and I may say decimated their numbers.

The valley inundated, more than three-fourths of which are the grounds of the Institution, is somewhat tortuous, bearing some resemblance in shape to a letter S with the body disproportionately expanded, between four and five miles in length, with an average breadth of above a mile. At its lower extremity is a kind of gorge through which the river passes. Here the waters rose, from inability to escape as rapidly as they came, partly apparently in consequence of some stoppage occasioned by the retention and accumulation of detritus and debris carried thither by the waters; and the rise of the waters inundated all the land above to the same height though not to the same depth. It was during the night that it reached and overwhelmed the village. The house of the missionary and the chapel stand on an elevated ground overlooking the plain, where dwelt the people. The missionary, Mr Durant Philip, brother of the enterprising projector and executor of the tunnel, and a man of like energy,—both of them sons of Dr Philip, who, aided and sustained by the co-operation of his wife, had spent health and wealth in befriending the Hottentots in a time of need,—the missionary, roused at dead of night by cries of alarm, was for a moment paralysed, nonplussed, brought to a stand. Nothing could save but a boat, but the mooring of the boat was by the river's bank, a mile away, now covered fathoms deep by the rising flood, and all was dark around. But he was brought to a stand only for a moment. Something must be done, and done at once. Was there nothing to be had? No, nothing but a soap box! The soap box was emptied in a trice. It would make at least the skeleton of a boat. Deals were found or were wrenched from the floor. But nails! Nails there were none of the size required or none to be found; but there was a gross of screws. Well, these must suffice. But it is weary work and slow to screw these home, and they must be driven home with blows. With all that willing hands and heaving hearts, and learning and intelligence and skill could do, it was nearly night again before the boat so built could be launched, and all the while one and another and another were perishing.

 The editor of the *South African Commercial Advertiser*, John Fairbairn, a man still held in reputation amongst the benefactors of South Africa, in writing of the sufferers who, not many months before—at that time rising by industry and economy above their previous trials—had contributed £25 for the relief of the destitute Scotch and Irish suffering from the failure of the potato crop through disease, a sum considered, with regard to their numbers and wealth, or rather poverty, a grateful sacrifice equal to the most liberal efforts of the rich, says,—" The descriptions of this great calamity set in a most affecting light, not only the sufferings, but the *character of the people*, many of the incidents being most honourable to human nature, while they show the force and power of religion, which can neither be extinguished by the tortures of life, nor overcome by the terrors of death. Immersed in a raging flood that was rapidly thinning their numbers, these poor people, for the space of nearly twenty hours, exhibited the most touching proofs of filial piety, conjugal affection, and faithful friendship, with hope triumphing in the very moment of dissolution; nor does there appear to have been a single instance of courage failing, or of despair undermining virtue."

 All that is thus alleged is borne out by a touching narrative by one of the Hottentots, which not only tells of what he and others suffered, but brings the whole scene vividly before one acquainted with the locality as I am. The following is a translation of the narrative as given to Mr Philip :

 " On Friday night William Landman and others came to warn me that I should come away; and I would have done so, but my mother being a heavy woman I felt unable to carry her so far, and that too in a pouring rain. **The water had never been known to rise so high as the spot on which I** was, though it had been surrounded, and so I remained. William Landman persuaded the wives of William Smit and Philip Marais, and their sisters and youngest brothers, to accompany him, but the brothers laughed at his warnings, and even opposed the departure of their wives and sisters. I passed the night between asleep and awake, till I heard the water pouring over the dam sluit, then I began to fear danger. I ran immediately to the road, and I saw it was still possible to escape if I could only get assistance in carrying over my wife and mother. J. Jacobs, whose house stood on the other side, was just then leaving with his family. So I roused the Smits, but they only got up and sat by the fire talking. Returning to my house, I roused Lucas and urged him to come and attempt to cross; but he answered,—'Where should we go in the dark? Let us wait till daylight. Even in the great flood, in Mr Kitchingman's time, this knoll was never covered with water.' When I went the second time the road was impassable, and the water was coming on towards Smit's house. I roused the young men with this information, and we soon found we were surrounded on all sides. We dragged chests and other things to the knoll, and carried some of the fire with us. There we stood (six women and ten men), gazing speechless at it and at one another. Lucas never spoke another word; but I never allowed my heart to fail me. It soon reached us and rose above our waists. I then bound up a mat for my wife, and told her to keep it across under her chest, and she would be able to keep her head above water till help came. I then took up my mother and held her in my arms till I could hold her no longer. She was the first carried away. Then Lucas drifted away from us and sank at once.

"About this time the three Smit's swam off, each pushing a chest before him to keep himself up. The boy Karl Baan went from one to another, now holding his mother up as he saw her sinking, then his little sister, until they all sank. I now saw a roof floating towards us and resolved to reach it. My wife had drifted away while I was holding my mother up, and she had got the mat under one arm, instead of across her chest, so that she just turned over and over with it, till she was carried out of my sight among the thorn trees. I now tried to reach the roof, but my strength was quite spent ere I was half way. I turned over then upon my back, resolved to keep myself afloat, paddling with my hands and feet, to let it reach me. When I felt my chest recovered and my arms rested, I turned again upon my face and found that the roof was within two strokes of me. I reached it, but felt my legs so benumbed that I could not bend them to climb on to it, and drawing myself up with my arms I writhed up my lower extremities like a snake on to it. When I was on it I found that Karl and Sarah Baan were following me, and not far off. Sarah was the nearest and she called out, 'Help me, dear uncle!' I said, 'I have not power to help you; but don't strike so wildly; be calm; don't tire yourself, and pray God to help you.' When she was only a little way off she sank, but came up again some way lower down; and Karl came now near the roof, and cried as she had done before him, 'Help me, dear uncle!' I said the same thing, but drawing a lath from the roof, just as he rose from sinking once, I pushed the end of it into his hand, and when he had grasped it I drew him up on to the roof beside me. The roof had floated towards Sarah, and she was just sinking the second time when I placed the lath so that she just seized it with her finger and thumb, and I drew her towards me and put her on the roof.

2R

"At this time Lydia, old Lucas' wife, was floating on a mat about 30 yards from me, the only one that was left. She now commenced singing the hymn, '*Jezus neemt de zondaars aan!*' Jesus receiveth sinners; and when she had sung it through, exclaiming "O Great God!" she laid her head down upon the mat as upon a pillow—and sunk. After I had reached the roof, I saw that Smit had just reached the thorns (about 600 yards), and Hendrik and William, his brothers, were following about 200 yards behind him. Smit called out the names of his brothers, and urged them to come on. They replied that they were coming. I then lost sight of Smit in the thorns, and Hendrik and William soon after. I never heard them call throughout the day. I had heard the hammering in the morning, and I felt that they were doing something for us, but saw nothing of the rafts. We were very numb from cold; and, being resolved to abide by the roof, even though it should go out to sea, I set about pulling out the thatch and erecting a shelter against the rain and cold, under which the children might creep, while I covered myself with a calfskin. We found some oranges and meal. We ate the oranges but kept the meal. Our roof had now drifted fast against the thorn trees. I continued calling for help throughout the whole day, and told the children to do so when I was tired. I felt it must be a boat you were making, as it took so long a time to make, for I heard the hammering. The time passed by very heavily, but in the evening I heard, in answer to one of our cries—'Yes, help is now on the way to you.' I then fell asleep, and was awoke by the call of Philip Bonnan and Daniel Lucas, coming to our help. They came alongside after a little while, and I wept with excitement at my deliverance. I thought as I sat upon that roof of Noah saved in the ark; but felt it was not because I was a righteous man like Noah that God had saved me. I wondered why I had been saved and others better than myself allowed to perish. I felt that God was sending me like a letter to announce the circumstances in which the rest had died, and I wonder still that the strong swimmers should have been taken, and us, who were no swimmers, left."

The following account is given by Mr Philip of the next day's sequel to these saddening events:—

"Oct. 3. When I rose this morning the first thing I observed was the garden ground almost clear of water, and people walking in the valley which had yesterday been one flood of water. In the distance the Gamtoos river was still high; but subsiding rapidly. The bodies of the three Smits had been found not far from one another, just where the people fancied they had seen them—quite dead.

In the course of the morning nine other bodies were found. Five or six were lying close together on the knoll, and bringing our boat down upon a waggon, we fetched them through. In the afternoon they were carried to a grave on the side of a hill behind our house, and, wrapped in shrouds,—twelve corpses at one burial, to be laid side by side in one grave. A number of people from the other side of the river took advantage of the boat and came to the funeral, so that when the bell had tolled and we commenced the service a good number were present. It was God that was speaking, and man's words were to be few, that he might the rather hear. Two had been members of our church. Beginning with prayer, I then read the 39th Psalm, and attempted to impress upon the hearers the utter vanity of all mortal things, and the repose of the soul upon God as the only imperishable possession. There was much stifled sobbing and weeping;

but, the rain coming on, we were obliged to desist, and having again prayed and sung, the bodies were laid in the grave, mats were laid over them, and the earth closed over their heads. Dr Philip, who was present, then concluded with a few solemn admonitions, and we dispersed."

In a paper cited in the Preface it is stated : " I have before me details of destructive effects of torrents which have occurred since I left the colony in the beginning of 1867. Towards the close of that year there occurred one, the damage occasioned by which to roads and to house property at Port Elizabeth alone was estimated at from £25,000 to £30,000. Within a year thereafter a similar destructive torrent occurred at Natal, in regard to which it was stated that the damage done to public works alone was estimated at £50,000, while the loss to private persons was estimated variously at from £50,000 to £100,000. In the following year, 1869, a torrent in the Western Provinces occasioned the fall of a railway bridge, which issued in loss of life and loss of property, and personal injuries, for one case alone of which the railway proprietors were prosecuted for damages amounting to £5000. In Beaufort West a deluge of rain washed down the dam, and the next year the town was flooded by the waters of the Gamka ; and in 1871, Victoria West was visited with a similar disaster." The loss of property here was great, and fifty-three dead bodies were recovered after the waters had subsided. On the same day similar destructive effects were produced by the fall of torrents of rain near Oudtshorn. In the following year, 1872, heavy rains fell in all parts of the colony, and in Capetown they caused a flood which for a time turned several of the streets into rivers, while many of the houses and stores had water in them to the depth of four feet ; and last year, 1874, still more disastrous effects were produced by torrential floods. According to the report given by one of the colonial newspapers, the damages done could not be estimated at much less than £300,000. According to the report given by another the damage done to public works alone was estimated at £350,000. From a statement made in Parliament by the late Commissioner of Crown Lands and Public Works, I am led to conclude that there must be some error in this latter statement ; but in any case, evidence is given by the facts adduced that in South Africa floods and inundations similar in character to that of this year in Lanquedoc are not unknown. In a volume on the Hydrology of South Africa I have given details of the effects of some of these floods, and of others, which, like that here given of the inundation of Hankey, present considerable similarity to those given of the inundations in France—only less destructive of life and property than was this, because there was less of life and property there to destroy.

In addition to what is there given, I may cite in illustration of what I now affirm the following account, given in the *Eastern Province Herald*, of the inundation of Port Elizabeth, in Nov. 1867 :—

" On Tuesday last there was a stiff south-easter on in the bay, and in the course of the day and during the night a good deal of rain fell. Next day the wind and sea showed no signs of abatement, and the steam-tug St Croix, which had parted from her anchor, but was happily brought up by another within one hundred yards of the rocks near the Bethel, seemed to be in imminent danger. There was but one man on board, and he, poor fellow, hung out an old handkerchief as a flag of distress. The Sailor's Friend noticed this signal, and bore down to him. This was not done without some trouble, though, for the waves—swollen into huge masses, and tipped with frothy foam—careered hither and thither in the most uncertain way possible, so that it was difficult to know how to take them. Then the sailors were almost blinded by the spray, which was driven before the wind like drift-sand. But the Sailor's Friend—the pluckiest thing of its kind afloat—served the St Croix with a warp, though she was floundering near the breakers; and then flew away like a bird before the wind. The one man on board hauled with what strength he had until he secured the warp, and the vessel, though tossing and heaving fearfully, was made secure. As, however, it was not deemed prudent at this juncture to let the man remain on board, the life-boat bore down to her and took him off. This little feat was not unattended with danger either, for a cargo-boat, which had just before broken from her moorings, was tossed from one wave to another until it came within fifty yards of the life-boat, which was at this time near the bows of the St Croix. For a moment both were lost in a trough of a sea; and when they were next seen the cargo-boat was within ten feet of the bows of the life-boat; suddenly, however, a huge wave took her by the keel and hove her, as it seemed to us, right over the lifeboat—for, as they rose on the crest of the next wave, the cargo-boat was seen just astern of the lifeboat, which now sped as fast as stout hearts could ply their oars to the Breakwater, which she reached in safety. The cargo-boat, which belonged to the Union Company, now dashed on till her progress was stopped by the sea-wall. Here some three or four boatmen jumped aboard of her and made her fast. One other cargo-boat, belonging to the Algoa Bay Company, we believe, drifted ashore, but whether either of them sustained much damage we have not heard. The events which we have been chronicling took place on Wednesday morning; and we may as well mention that the roads generally, and the stores on the east side of Main Street particularly, suffered to some considerable extent from the rains of the previous night.

"But what had happened up to this time was as nothing to that which was to follow. The leaden aspect of the sky all around betokened a storm of a yet more violent character. But from what point of the compass it would come no one could tell. But that it would come seemed inevitable, and what could be done in the way of clearing drains and making all secure was done, not only by the municipality people, but by all who had property to protect. Towards evening the wind veered more to the south-west, and moaned in fitful gusts, as if impatient of restraint. The rain, which came gently at first, increased by degrees until it fell down in sheets. This was about nine o'clock, and all around was dark as Erebus. Presently the lightning began to flash in vivid sheets, and the distant rumbling of thunder, mingling with the roar of breakers on the shore, was enough to make the stoutest heart shudder. There was not a house, probably, in the town that was proof against the inroads of the storm. How long would it last? seemed an anxious question. Would the wind, which had now increased to a gale, beat it down? There was a fierce conflict for the mastery, but both held on with unabated violence. As midnight approached the thunder and lightning ceased, but still that dismal hollow roar upon the beach was heard mingling with the rushing, shrieking, howling storm, in fearful chorus. At this time anxiety for the safety of the shipping was intense. The Port-office people were on duty, but we fear had their services been called into requisition, which happily they were not, they would have been but of little avail owing to the fury of the tempest.

"As the night wore on the storm seemed to increase in violence rather than to abate its fury, and few people, we venture to say, had an hour's repose throughout the night. Every one had his own particular grief thrust suddenly upon his hands, and some had more than they could attend to. It was a busy, melancholy time with all, and the worst of it was that the storm raged with such fury that no one knew what misery it would bring or how it would all end.

"Out at the South End, and more particularly in Rudolph Street, the storm was playing fearful havoc. The rushing waters came on as a mighty torrent from the rising ground beyond, and carried away pailings and stoeps; and then gathering force, ripped up the road, and drove the sand of which it was composed before it in its onward course to the sea. It was a fearful time for the poor residents here. Now that a gully had been opened the waters from all the surrounding places seemed to find an egress down this doomed street until the gully assumed the dimensions of a river. First one and then another house was undermined, and down it came with a fearful crash into the seething bubbling stream, now some twenty feet wide by eight and twelve feet deep. It might be supposed that the *debris* from these disasters would dam up the stream. Not so, however. The rush of waters, too powerful to be stayed, swept all impediments before it, or if diverted for a moment only made the breach wider. The alarm, as we have said, was intense, and this was increased by the darkness of the night. No one could ensure an hour's security. The waters of this newly-made river, which had been strong before, now became stronger, and took a wider range in their course, and, as a consequence, house after house, to the number of twenty, fell victims to its inroads. One poor child, named Harvey, ignorant of the danger which threatened her, left her mother's house in search of help, and, falling into the rushing torrent before the door, was carried away to the Harbour Board Bridge, where she was found some six hours afterwards buried in the sand. It is difficult to depict the horror of the scene. Here were poor houseless women and children—almost clotheless—running about in the dark for shelter; while strong hardy men were running hither and thither with lanterns to render all the help they could; and it is to the credit of the occupants of all the houses left standing that they were thrown open for the shelter of those who had been so suddenly and so fearfully deprived of their homes.

"The street—as our readers will easily conceive from the above description—is a complete wreck, and the loss to the poor people, mostly Malays, who had expended their means in erecting these houses for their own occupation, will be something that will take them years to replace.

"The tramway bridge erected by the Harbour Board at the foot of this street was also carried away by the fury of the storm, while the bridge which spans Baaken's River lost the southern buttress, and may be said to be in a shaky condition.

"Before leaving the south side of the town, we must mention the fact that Mrs Hayes and her child, a boy about seven years of age, had a narrow escape. They were found buried beneath a wall in a most exhausted state, and removed to the hospital. A man named Martin Devitt also escaped, as it were by miracle. He had just left his stable, where he had gone to see how his horse fared, when it fell in and killed the animal.

"The road in South Union Street, opposite Mr Webster's house, is sadly cut up. The gas pipes are laid bare, the holes in some places being from six to eight feet deep. Then higher up, near Neslin Castle, there is another immense hole, through which the water seems to have passed with great force in front of a number of small cottages situated in a valley at the back of the castle. But for the energy of one Gover, and some willing hands he got around him to divert the current, the whole of these cottages, in all human probability, would have been swept into the sea. There was great distress here in the small hours of the night, and Gover took in thirteen children and three women who were running about in the utmost alarm in their night dresses. . . . We may say of Southend that wreck and disaster was apparent on every side—more, indeed, than we have had time to inspect or have space to chronicle.

"Military Road, Castle Hill, White's Road, Donkin Street, Constitution Hill, Russell Road, and, indeed, all the approaches to the hill, showed in an especial degree the devastating effects of the storm. All the drains are washed away, and the huge stones of which they were composed are scattered about like pebbles, or thrown down into deep gullies, through which the water rushed in its headlong course to the streets beneath. White's Road, Constitution Hill, and Russell Road, look as if they had been riven by an earthquake. This is no exaggeration. At the top of White's Road, near Mr Buchanan's new house, the drain is washed away, and just beyond there is a gully six feet deep, cut out by the fury of the rushing waters. Then, below this, there are zig-zag cuts from four to five feet deep, through which the water forced its course to the barrel drain below. So great was the rush of the torrent here that the drain, becoming surcharged, burst with a loud explosion, ripping up the road in all directions, and nearly bringing down the wall near the theatre. The water thus set free now careered on in its mad course, and burst through the window into Messrs Anderson & Co.'s store, which was inundated to the extent of nearly two feet. From top to bottom, White's Road is a complete wreck, intersected as it is with holes and gullies varying from twelve to four feet deep.

"Donkin Street, near the Mechanics' Institute, presents the appearance of a huge ditch. The volume of water here must have been immense. The Mechanics' Institute is cracked asunder, and but that Mr T. Griffiths lent some planks, and Mr Macgregor shored it up, it would have fallen into the ditch below. Next to the Mechanics' Institute is a shed in the occupation of Mr Sherman, in which some iron is stored, and the shock of the riven earth beside it brought down a portion of the wall. Then the water from this quarter drove down such piles of earth into the passage near Mr S. White's shop, and around the front door, that admission to the premises was obtained with the utmost difficulty. The vaults of the Port Elizabeth Bank were flooded, but beyond the inconvenience occasioned, the papers or premises sustained no damage. On the whole, however, though Donkin Street has suffered much, it is not so badly damaged as White's Road. Constitution Hill escaped comparatively scatheless.

"Not so, however, Russel Road. What has been said of White's Road obtains here, save that the damage is even greater. It is a complete wreck —so much so, indeed, that a man cannot pass through it on horseback, and it is almost impracticable to pedestrians. The rush of waters here carried away the walls enclosing Mr Powell's cow-house, and hurled a stone 500 lbs.

weight some considerable distance. . . . . The houses below Mr Powell's were shaken to their foundations, and those in which Mr Archibald and Mr Atkinson lived have been vacated because, in their present state, they are untenantable. The road, from one end to the other, is a series of winding rivers, varying from four to eight feet deep.

"As may be imagined, the rush of water from all these approaches to the Hill caused considerable damage to the houses and stores on either side of the streets stretching from Market Square to the North End. The premises right and left of Main Street, extending from Messrs Geard & Co.'s to J. O. Smith & Co.'s, suffered more or less severely, and great is the destruction of property in some of the stores—especially in the cellars—in which some had fine goods packed. Mr H. B. Christian was engaged with seven men for hours damming up his lower store, or the loss there would have been immense.

"From J. O. Smith & Co.'s on the sea side, and Sherman & Co.'s on the hill side, down to Mr Crago's place, Mason's Hotel, and Janion's, the houses were mostly literally swamped; and the heaps of rubble, sand, rags, bones, turned out and piled up in the streets, showed to what an extent the inhabitants must have suffered. Then, lower down, Bishop's (late Stolle's) butcher's shop fell in with a crash. Between John Tee's place and Mr Pearcey's the water came down like a cataract, driving before it huge stones and rubble, and depositing it to the height of a good-sized mountain near Mr Cunningham's. Rising to the height of some ten feet in Mr Farrell's yard, the water burst through the window at the back, and this being fortunately in a direct line with the passage leading to the street, it found easy egress in a roaring torrent. Then, considerable damage was done to the stoep near Mr John Geard's store, at the corner of Korsten Street. At the upper end of Frederick Street, the back wall of Mr Walker's stable was driven in, when a perfect mill-stream rushed beneath, making its way across South Street, considerably damaging Mr Pearson's shop in its course. Mr Pearson, with his wife and family, left the building during the night, thinking that if they remained they would be washed out. From this point the stream rushed on, hissing and roaring, down towards Solomon's Row, entirely destroying the first and second houses on the right. Then, this impediment removed, it rushed down the row, with the roar of a waterfall, to the beach.

"Out towards Mr Dent's place the country had all the appearance of a huge lake.

"Returning towards the Town-hall, we may state that the Market Square is literally cut to pieces, and that a house belonging to Mr Inngs, situated in Military Road, was knocked over with a fearful crash, the occupants narrowly escaping with their lives.

"Yesterday morning, we regret to say that a poor fellow, named Simon Maddan, lost his life while endeavouring to recover some drift-wood near Baaken's Bridge. He rushed into the stream, which did not appear more than two or three feet deep, and he must have been sucked up by the sand, for he instantly disappeared, and his remains have not since been recovered. He was a well-behaved hard-working man, and had been in the service of the Harbour Board some eight years. A man called out to warn him of his danger, but it was too late—he was already in the stream. He has left a wife and five or six children totally unprovided for.

"The beach speaks with terrible eloquence of the roughness of the storm.

As we have said, the large craft rode it bravely out, thanks to Providence, their cables, and the stout holding ground of the bay. But a number of small craft sunk at their moorings within the breakwater, having been actually swamped by the force of the rain. Here is a list of them . . .

"Several persons had very narrow escapes, more perhaps than we know of. The Rev. E. Pickering fell into the ravine formed by the current at the top of White's Road, and after floundering about up to his neck in water, managed to scramble out comparatively unhurt. The water broke his fall, or he might have been killed where he fell. Then Messrs Amyot and Thompson fell into a gully, and escaped with difficulty. Lieutenant Shaw, 86th, while passing from a friend's house to his cottage was suddenly brought up by a flash of lightning, which revealed to him the awful fact that, had he moved a step further, he would have fallen over a fearful precipice, and been dashed to pieces. Then Mr Suter, who was passing up Russell Road to his home, just before the storm was at its height, was knocked down by a huge stone, which was impelled from the Hill, and nearly carried away by the volume of water which followed it; and we hear that, shortly after, Mr French, who lives in Russel Road, had to make his escape with his family by the back window, such was the rush of the torrent into his house. The escapes from almost certain death were truly miraculous. Those who happened to be out in the storm had to make their way home waist deep in water, which poured down from the hills in all directions.

"Out at the Fishery the flood was immense. The Shark's River dam was partly broken, and Lippert's dam was carried away, while the house near it was stove in. . . .

"The damage to the roads and house property here is estimated at from £25,000 to £30,000, while the injury to the stock and crops in the district, it is feared, will be something distressing."

Such are forms of the evils with which the inhabitants of not a few newly settled lands are familiar. It may be difficult, as is stated by a writer in the *Spectator* on the inundations in the valley of the Garonne, for Englishmen who have never quitted their own country to comprehend the destruction such can inflict, and above all to realise the special horror—a horror like that caused by an earthquake—which water can inspire in those who suffer from its ravages. England has been visited by terrible calamities, like the floods in Morayshire, in August 1829, and the bursting of the Sheffield reservoir, but even such calamities are of rare occurrence. Elsewhere it is different. At the very time of the inundations in Lanquedoc it was reported from Bohemia, Carinthia, South Tyrol, and Banot, that similar inundations had occurred in these districts, with corresponding calamities—of railways being injured; bridges, horses, and herds of cattle carried away; houses totally destroyed; and men, women, and children drowned. According to one account,—" A thunderstorm, with hailstones, caused terrible destruction in Buda-Pesth. The hills and the roofs of the houses were covered with ice two feet thick; the torrents rushed into the streets of Ofen. Five hundred persons are missing, and at least one hundred have been drowned, or killed by the falling houses."

Next day it was reported,—"The disaster in Ofen is greater than was feared. 120 corpses have been found, but many dead bodies have been carried away by the Danube. Baron Bela Lipthay, a distinguished member of the Conservative party, recently nominated deputy of Ofen, is missed."

This was in the end of June, and the autumn brought to the newspapers numerous accounts of inundations in other parts of France, in various parts of England, in America, in India, and elsewhere. It seemed as if the torrential era were giving place to a new era of inundation, and by some it was rashly alleged that evidence was thus supplied of the failure of *reboisement* to accomplish the prevention of torrential floods. I say *rashly* alleged, for all the official reports, issued with the sanction of the Government, of which translations have been given in the preceding pages, with their testimony in regard to facts accomplished, remain unaffected by any thing which may have subsequently occurred; and time ought to have been allowed for reports to have reached us in regard to the results of the surveys, and inspections, and reasonings by the professional men acting under the Forest Administration of France. It is not for me, a foreigner, led incidentally in the prosecution of other studies to give attention to the subject, to rush into the field with my opinion, ere those who are professionally engaged in the work, and those who are officially responsible for the work, have reported, as I doubt not they will in due time, upon these inundations in relation to the enterprise in which they are engaged. But in the interests of those for whom I write I may again call attention to the facts that 140 years were from the first reckoned necessary for the accomplishment of the work—and of those only 14, or one-tenth of the period, have elapsed—and, though the most urgent cases were attended to first, it may be assumed that not much more than a tithe of the work has been completed, and much of that—by far the most of it—in regions in which it could not affect in any way the river-courses by which the inundations in question were produced. And in connection with this I would call attention to the facts reported by M. Laydecker, when Director-General of the Forest Administration, in regard to what was observed in connection with devastating floods which occurred in 1866, and cited at length in a previous part of this compilation (ante p.p. 224–229) from which it appears that few, if any, of the works then executed had been carried away; that the effects which had been produced by these works, in preventing disastrous consequences from the floods, had been all that had been expected, and had been more than was hoped or feared during the prevalence of the floods, or would have been thought likely to have been the case had the coming of such floods been foreseen, before they came, after the execution of the works was completed.

And a corresponding report I expect to follow the survey of the works after these disastrous inundations. I anticipate that that report will bear that wherever works of *reboisement* and *gazonnement* have been executed properly, they have accomplished the object for which they were undertaken; that wherever nothing had been done, there the evil has been seen in full force. Knowing what I do of the work, I consider it probable that, had the works contemplated been executed in all the basins of reception drained by the affluents of the inundating rivers, the inhabitants of Toulouse and other places would have had timely warning to prepare for the coming flood; and this, instead of rushing headlong, tearing up and carrying away all before it, might have taken fourteen days instead of four to pass a given point in its course, prolonging the flood, but to a corresponding extent reducing its depth or height and force.

It may be felt as an objection that these are but conjectures; let us then leave all conjectures whatever, take up the facts of the case and look at these in the light of what is known.

The accounts given of inundations in South Africa may enable sufferers from inundations elsewhere to identify the form of the evil with which they are familiar, with the form of the evil seen in the late inundations in France, these presenting them with an intermediate form having much in common with both; and the identification of this with the torrents of the Alps and other mountain regions of France is not more difficult. Surell spoke from the first of torrents and torrential rivers as essentially identical; and the evils calling for remedial measures, though varying in the degree of importance attached to them, have been the destruction of the mountains, the covering up of fertile lands in the valleys with sterile detritus, and the inundation of the plains beyond by the superabundant waters. In these inundations we have one of the correlated effects. And it is in view of this that I have proposed to look at the phenomena of this inundation in the light of what has been ascertained in France in regard to torrential floods and the means of extinguishing them.

The Alpine torrents are traced by Surell to two sources—the melting of snow about the beginning of June, and storms of rain occurring about the end of summer. The inundations in question have been occasioned by similar causes, but by these operating simultaneously, and this in the Cevennes and in the Pyrenees at the same time. In accordance with what has been stated, when a basin drained by a river is covered with vegetation the flow of the water is retarded, diffused, and protracted; but when mountains upon which the rain falls are devoid of vegetation, the rain rushes off as does water on the roof of a house,—and thus was it here.

The *Journal des Debats* thus explains the phenomena of these inundations:—

"It is the chain of the Cevennes which causes these immense disorders. Between the sources of the Loire and the Hérault the Cevennes are 3,700 feet high. All this surface is composed of granite impermeable to the rains. The river waters rush over this ground with immense rapidity, but do not enter it. The chief streams rising there are the Dour, the Ervieux, the Ardèche, and the Gardon, affluents of the Rhône; on the west, the Lot and the Tarn, affluents of the Garonne; on the north, the Loire and its tributary the Allier; on the south, the Hérault. The Ardéche, whose basin is only 2429 kilomètres, has enormous rises. At the bridge d'Arc the stream rises to nineteen mètres above the lowest level, and pours down at a rate af 7000 cubic mètres per second, almost as much as the Loire at Tours. An equal violence is registered in the Dour, the Ervieux, the Gardon, the Isère, the Drôme, and the Durance. Since everything depends on the rainfall, it is obviously impossible to calculate with certainty beforehand. Every year the Cevennes cause vast 'spates' in the largest rivers in France—the Rhône, the Loire, and the Garonne. All the streams of the region are torrents. The southern part of the Cevennes, the Black Mountain, and the Corbières exercise a great influence on the small Mediterranean streams between the Rhône and the Pyrenees. A rain of 200 millimètres, which has no perceptible effect elsewhere, causes in these parts a sudden flood."

In general the rains fall there in May, and being then comparatively cool, **they melt but little of the snow, and flow away as they fall. But when they** fall in June, as this year they did, they are somewhat tepid, and, falling upon the snow, melt it rapidly, and the watery produce is added to the rainfall; thus two sources of flood are combined, and disastrous consequences not unfrequently follow. And thus, as has been stated, was the late flood produced. Persistent rains from the north-west fell upon the Cevennes and the northern slope of the Pyrenees. This was preceded there in some cases by a heavy fall of snow; and there was over all the higher-lying lands the snow which had fallen in the course of the winter. This snow was dissolved; all the tributaries of the Garonne were flooded simultaneously; and we see the result.

In such a case time is everything. It may make all the difference between the loss of life and property and perfect safety to both, if a body of water such as was here precipitated from the mountains shall rush past a given point in four days or take fourteen for its flow—flowing in flood, but never rising above the height of the containing banks. And it may make a very great difference, though not so great, if a flood and inundation come suddenly in the night, without notice or warning, and if it come after twelve or twenty four hours' notice of its coming.

Thus it is with floods in the Seine and in the Loire. Warning is given by telegraph all along the course of these rivers that a flood is on its way, and the inhabitants on their banks are prepared when it comes.

But this could not be done in the case of the late inundations. There is an observatory at Pic-du-Midi, a spur of the Pyrenees, and it seems that General de Nansouty, who commands there, would have been able to give timely warning of the coming inundation had the observatory been in telegraphic communication with the threatened towns and villages,—at all events along the course of the Adour. He did warn the people in the neighbouring valley of Campan what was to be expected from a heavy fall of snow in the mountains, which snow had suddenly commenced to melt under the influence of the rain and westerly wind; and on the first appearance of danger, on the night of the 22nd June, M. Beylac descended the mountain during the most fearful weather to spread the alarm; but the floods in all the tributaries of the Garonne were so sudden that to give warning was impossible. Had the *bassins de reception* of all these been wooded it would have been otherwise, but they were to a great extent devoid of vegetation.

Very different had been the case had warning been given along the course of the Garonne of the coming flood from one to twelve or twenty-four hours before it reached the different towns and villages destroyed; and very different had been the case had the waters which swept along in a torrential wave taken fourteen days to flow past any and every point on its course! It may be, that never would it have risen so high as to imperil a single house, and that in consequence of the timely warning given not one life would have been lost! It is said by a writer I have quoted,—" If this observatory (that on the Pic-du-Midi), now isolated on the peak, were bound to the plain by telegraph, the General might transmit to the officials of the Ponts et Chaussées previsions of the last importance. In the same manner a station should be made on the Corbiéres. As soon as the quantity of rain falling on these cliffs became dangerous the authorities would be warned." Yes; but this, if combined with a complete *reboisement* and *gazonnement* of the mountains, would give the longer time to prepare for what was coming. And it may be asked, Why has this not been done? An answer is forthcoming, and that not the answer which might be expected, that, as stated, "Between the sources of the Loire and the Hérault the Cevennes are 3,700 feet high. All this surface is composed of granite, impermeable to the rain, and to plant such either with herbage or with trees, is impossible;" but the answer, that the work is being done as fast as money and men and material can be found, and that already, previous to this inundation, all that could be done up to that time had been accomplished. It is often easy to tell, after an event has occurred, how it might have been prevented; and it may be that had these inundations been foreseen, operations which would have to some extent modified or prevented them would have been prosecuted with the vigour called forth by a race against time, in preference to some others which have not been ineffective, but the execution of which might without series consequences have been postponed; or, at all hazards, grants on a scale of magnitude equalling or exceeding those made previous to the war would have been made, and the difference between these and the amounts actually granted spent exclusively on the valleys and basins of reception drained by the upper waters and affluents of the rivers by which such devastation has been wrought. The legitimate use now to be made of such reasonings is, to prepare for the future in accordance with the suggestions suggested by the past. And this, I have no doubt, will be done.

The flood of 1875 has proved the most destructive and the most sudden flood of the century; but though floods of such a magnitude are unfrequent in the valley of the Garonne, scarcely do twenty years pass without the occurrence of a flood of serious importance; and during this century so frequently and regularly have they occurred at such an interval as to suggest as probable the existence of some unknown meteorological law. To some it appears that the periodicity involves a cycle of ten years rather than twenty.

The *Journal des Débats*, writing of the late inundation, says,—" We lately spoke of a probable law regulating the recurrence of rains and floods. What matter whether it be quite correct or no? If experience prove it true in the majority of cases, why should we not take it into consideration? All the engineers of the South know that very dry years correspond with sudden deluges of rain, and consequent floods. The unusual dryness of this year ought to have given warning. And the cycle of ten or twenty years now come round should have set people on their guard. Such empirical observations are too lightly disregarded. No hint should be neglected when events of this gravity are at issue. There are memorable dates of which the recollection should never be suffered to die. The cycle appears to differ somewhat according to the region. Thus, for the Loire and the Rhône, it seems to be a year later—'46, '56, '66, '76. Are the banks of the Rhône not in danger next year?"

Much interest attaches to the meteorological question involved. Here we have to do mainly with the practical question raised, and the justification of a considerable expenditure of money, of labour, and of thought, on averting the disastrous consequences following a phenomenon which appears to be, not the result of accidental coincidences, but of frequently recurring coincidences, the recurrence of which is apparently not only regulated but insured with all the certainty attaching to phenomena occurring in connection with the operation of physical laws.

Designs of certain measures to prevent such occurrences were proposed after the last great flood in 1855, but they were pronounced too costly; and now it is considered doubtful whether, what many considered the most obvious expedient—the excavation of an overspill canal, specially designed to carry off surplus waters—would be either sufficient or possible.

"The arrival of the flood," says a writer in the *Spectator*, " is so rapid, the mass of water so vast, the formation of a lake in the low-lands between the slopes and the bed of the river so instantaneous, that any canal it might be possible to cut might, on the recurrence of the fatality, so to speak, be *drowned* under the advancing wave, as the Garronne was itself. It is thought that by greatly deepening the channel of the Garonne beyond the confluence of the mountain streams aid might be afforded ; but that decree, though most beneficial against an ordinary rise in the waters, would be worthless against a flood of this kind ; while a dike, even if it could be constructed, would not be a safe reliance. A dike against ever-present water may be a perfect defence; but a dike against a flood which comes in its highest fury only once a century, and in a dangerous form only once in twenty years, is pretty certain to be neglected. If the boats of a ship were always required, they would always be ready ; but being wanted only in extremity, even the fear of death, of ruin, and of lost reputation, does not suffice to compel ship-captains to keep them in order. Planting the slopes makes the channel deeper, and the rains more regular ; but the expedient is a slow one, and requires determined attention, which even governments in the end become unwilling to pay."

It may seem to have been so in France, but I believe it has been so more in appearance than in reality.

The law of 1860 was enacted for 10 years. This period expired in 1870. It was impossible to review it then, and the works of *reboisement* and *gazonnement*, previously maintained by extraordinary budgets, then fell upon the ordinary budget ; and for 1871 there was granted a credit for 3,500,000 francs, of which 1,500,000 francs were for *boisement* and *gazonnement* ; and the draft budget for 1872 reduced the 3,500,000 given under this head to 1,563,000 and the 1,500,000 allotted for *gazonnement* and *reboisement* to 763,000.

Such, writes M. Cézanne, is the sad consequence of war ! France is reduced to augment the military expenses, which are ruinous, and to diminish the outlay on public works which are productive !

The war is now, it is true, a thing of the past ; but the effects of the war remain. There is still a war expenditure deemed necessary, and so necessary that it is treated as a first claim upon the country, to which all improvements must, excepting in so far as they are imperatively called for, be deferred. And, as a consequence, we find not a million a year spent, as before the war, on the work of *reboisement*.

But I am not aware of any one connected with the Administration, the Government, or the Legislature, having lost faith in *reboisement* and *gazonnement* as a means of counteracting the evil, though they have had to limit operations in consequence of the demand made on them for money to meet what were considered more urgent claims. And as the inundations of 1855 led to practical effort being given to the suggestions of Fabre and Surell, I anticipate that the inundation of 1875 will, in the light of the results obtained, lead directly or indirectly to the operation being resumed and prosecuted as it was during the first decade of the work.

"In ordinary times," says the writer in the *Spectator*, " the snow on the heights of the Pyrenees melts gradually, and trickles down in hundreds of runlets over a granite soil, which absorbs nothing, to the larger streams which fill the two rivers which unite a short distance from Toulouse into the Garonne, and make the prosperity of the rich surrounding plain. When, however, from any cause, the snow melts too rapidly, as is believed to have occurred this year, the heat and rainfall having been both unusually great, and lasting for three weeks on end, the channels cannot convey the water, which rushes in broad torrents to the streams, which again, owing to some configuration of the soil, cannot convey away the unwonted mass of fluid. The water collects into a lake, sometimes miles in length and breadth, and forty feet deep, a veritable reservoir, and then bursts through the open mouths left by the rivers into the valley of the Garonne, with as resistless a force as the great Sheffield reservoir burst into the little vale below it. The Garonne fills and fills until it overtops its lower bank, and then, as the supply increases hourly, its sweep over the lower ground becomes as resistless as that of a slow storm wave. The effect is not quite so severe because of its gradual approach; but the Garonne must have rushed over St Cyprien, bringing a mass of water equal to that embraced in a reservoir twenty miles long, by ten miles wide and thirty-eight feet deep. This year heat, rainfall, and wind seem to have united, and on June 23rd the Garonne was filled in an hour, and in six hours the upper valley had been turned into a bursting lake, and a flood which, like an earthquake, makes its victims think the laws of nature overturned, and that there is no help even in heaven, came rushing towards the city. Within six hours of the first alarm of an unusual rise in the water, the Garonne had swept away every bridge of Toulouse except one, the old stone bridge of St John, and flowing on in an unbroken rush into St Cyprien, rose above the streets so rapidly that the terrified inhabitants were compelled to take refuge in the upper stories. Scores of persons appear to have been strangled by the flood, all the slaughterers in the great abattoir, for example, being killed at once; but the great loss of life arose from another cause, which recalled the idea of earthquake to the wretched people. The rushing water felled the weaker houses as giant shells would have done, and undermining the foundations of the stronger, till through one entire night houses were toppling as in an earthquake, and the awful scenes at and around Cucuta, in New Granada, on May 18th, when 16,000 persons perished at once by earthquake, were repeated in Languedoc. Escape of the house, once shook, was of course hopeless. There were the walls above and the waters below, and the stream outside in which a boat could scarcely live. Nearly 1000 persons are known to have been killed in St Cyprien alone by the falling houses, trees, and monuments, or to have been drowned in escaping from the upper stories, or capsized in boats which put out into the streets to rescue the sufferers, sometimes—to the credit of human nature be it spoken, if not of human reason—with a priest on board to grant absolution to the dying as they swept past. The villages beyond Toulouse, and presumable on lower ground, were in some instances swept away bodily, the church in one instance being the only building left standing, and in another a mill so injured that it must be blown up. The ravages extended over 100 miles, and at one time fears were entertained for Bourdeaux itself. The destruction of property is, of course, greater than that of life. Neither vineyards nor houses can run away. The quarter of St Cpyrien, with its 30,000 people, had, in the words of the official report, "*ceased to exist*," and its whole population is houseless, without furniture, clothes, or food. In St Cyprien and the villages 100,000 persons are supposed to be destitute. The crops over hundreds of square miles are destroyed, and in many places the very ground has been swept away. It is calculated that the actual loss in cash reaches *four millions* sterling, and that years must elapse before the suffering districts can again resume their old appearance. According to another estimate, the loss caused by the floods in the south of France will amount to 300,000,000 francs, and it is estimated that 3000 persons have lost their lives." It was reported that by the Government the loss was estimated at three millions sterling. And subsequent reports have shown that it was somewhat in excess of this amount.

Before this volume can come into the hands of many exposed to the devastating effect of torrential floods, it may be known what the Forest Administration of France will do in such circumstances. And, in anticipation of this, I would submit for the consideration of all interested in the matter the expediency of giving careful consideration to the report expected, and to the practicability and expediency of adopting similar measures.

It may seem to be impracticable to do so; but what is impracticable for one man to do may be quite practicable for another to accomplish; and if the thing be possible the impracticability may be only in appearance. *Buissonement* may be practicable where *boisement* is not, and *gazonnement* where *buissonement* is not. I have been told of mountain crests of granite upon which neither herbage, nor bush, nor tree can be grown, as if that were a condition of things in which the measures adopted in France must be inapplicable; but the details given show that it is not the mountain crest, but a lower-lying zone to which *reboisement* is applied; and that a zone of forest extinguishes torrents formed above. I only know of one form of impracticability before which I am silenced, the impracticability of finding the money requisite for the execution of operations so extensive as might be requisite to meet and counteract the evil in some given country or locality. Where this is alleged I am silenced, but I am not convinced. The only impracticability I know is that of convincing communities that the outlay would be remunerative.

It does not, however, comport with my purpose to argue out this point. My purpose was and is simply to report what has been done, what have been the results. But I may, without departure from this, state that if the expense of carrying out such operations be great, so also are the losses occasioned by torrential floods such as they are employed to remedy. And the conservation and extension of forests, coppice woods, shrubbery, and herbage may bring accessory benefits, increasing the comforts and amenities of life, having a pecuniary value which might be reckoned an important offset against the expense of such operations.

THE END.

www.ingramcontent.com/pod-product-compliance
Lightning Source LLC
Chambersburg PA
CBHW030308240426
43673CB00040B/1102